Molecular Sieves

SYNTHESIS OF MICROPOROUS MATERIALS

Volume I

Molecular Sieves

Edited by

Mario L. Occelli   *Georgia Tech Research Institute*
Harry E. Robson   *Louisiana State University*

VNR VAN NOSTRAND REINHOLD
————— New York

Library of Congress Catalog Card Number 91-45163
ISBN 0-442-01116-4 (set)
ISBN 0-442-00661-6 (v. 1)
ISBN 0-442-00662-4 (v. 2)

Manufactured in the United States of America

Published by Van Nostrand Reinhold
115 Fifth Avenue
New York, New York 10003

Chapman and Hall
2-6 Boundary Row
London, SE1 8HN, England

Thomas Nelson Australia
102 Dodds Street
South Melbourne 3205
Victoria, Australia

Nelson Canada
1120 Birchmont Road
Scarborough, Ontario M1K 5G4, Canada

16 15 14 13 12 11 10 9 8 7 6 5 4 3 2 1

Library of Congress Cataloging-in-Publication Data
Synthesis of microporous materials/edited by Mario L. Occelli, Harry
   Robson.
        p.   cm.
     Includes bibliographical references and indexes.
     Contents: v. 1. Molecular sieves—v. 2. Expanded clays and other
   microporous solids.
     ISBN 0-442-01116-4 (set).—ISBN 0-442-00661-6 (v. 1).—ISBN
   0-442-00662-4 (v. 2)
     1. Molecular sieves. 2. Zeolites. 3. Layer structure (Solids.
   4. Clay minerals. I. Occelli, Mario L., 1942-   . II. Robson,
   Harry E., 1927-   .
   TP159.M6M59   1992
   660' .2842—dc20   91-45163
                                                           CIP

Contributors

R. Aiello, Department of Chemistry, University of Calabria, Rende, Italy

F.M. Allen, Engelhard Corporation

S.J. Andrews, ICI Chemicals and Polymers Ltd., England

Didier Anglerot, Groupe ELF-Aquitaine, GRL, Lacq, 64170 Artrix, France

Michael J. Annen, Department of Chemical Engineering, Virginia Polytechnic Institute, Blacksburg, Va 24061

A. Arafat, Faculty of Science, Helwan University, Cairo, Egypt

Reza Asiaie, Department of Chemistry, The Ohio State University, Columbus, OH

S.V. Awate, National Chemical Laboratory, Pune 411 008 India

Sean A. Axon, Department of Chemistry, University of Cambridge, Cambridge CB2 1EW U.K.

A.K. Barakat, Faculty of Science, Helwan University, Cairo, Egypt

Alexis T. Bell, Center for Advanced Materials, Lawrence Berkeley Laboratory, University of California, Berkeley 94720

Meena Bhalla-Chawla, Exxon Research and Engineering Company, Annandale, NJ 08801

Stuart W. Carr, Unilever Research, Port Sunlight Laboratory, Merseyside, L63 3JW, U.K.

J.L. Casci, ICI Chemicals and Polymers Ltd., England

K.J. Chao, Tsinghua University

S.H. Chen, Tsinghua University

Abraham Clearfield, Department of Chemistry, Texas A&M University

Edward W. Corcoran, Jr., Exxon Research and Engineering Company, Annandale, NJ 08801

P.A. Cox, ICI Chemicals and Polymers Ltd., England

Mark E. Davis, Department of Chemical Engineering, Virginia Polytechnic Institute, Blacksburg, VA 24061

Thierry Des Courieres, Centre de Recherche ELF-France Solaize, St. Symphorien d'Ozon, France

E.G. Derouane, Department of Chemistry, University of Namur, B-5000 Namur, Belgium

R. De Ruiter, Delft University of Technology, Delft, The Netherlands

Francesco Di Renzo, Ecole Nationale Superieure de Chimie, 34053 Montpellier, France

B. Duncan, Zeolite Research Program, Georgia Tech Research Institute, Atlanta, GA 30332

Prabir K. Dutta, Department of Chemistry, The Ohio State University, Columbus, OH 43210-1173

M.J. Eapen, National Chemical Laboratory, Pune 411 008, India

Francois Fajula, Ecole Nationale Superieure de Chimie, 34053 Montpellier, France

Francois Figueras, Ecole Nationale Superieure de Chimie, 34053 Montpellier, France

Katherine K. Fox, Unilever Research, Port Sunlight Laboratory, Merseyside, L63 3JW, U.K.

Z. Gabelica, Facultes Universitaires N.D. de la Paix Namur, B-5000 Namur, Belgium

R. Galiasso, Intevep S.A., Caracas 1070A, Venezuela

G. Giannetto, Universidad Central, Fac. Ingen., Caracas, Venezuela

Thurman E. Gier, Department of Chemistry, University of California, Santa Barbara, CA 93106-9510

D.M. Ginter, Center for Advanced Materials, Lawrence Berkeley Laboratory, University of California, Berkeley 94720

Yoshiaki Goto, Department of Materials Chemistry, Ryukoku University, Otsushi 520-21 Japan

M.M. Hamil, Engelhard Corporation

William T.A. Harrison, Department of Chemistry, University of California, Santa Barbara, CA 93106-9510

David T. Hayhurst, College of Engineering, University of South Alabama, Mobile, AL

Heyong He, Department of Chemistry, University of Cambridge, Cambridge CB2 1EW, U.K.

P. Donald Hopkins, Amoco Oil Company, Naperville, IL 60566-7011

Elke Jahn, Ruhr-Universitaet Bochum, Institut fuer Mineralogie, Bochum 1, Germany

J.C. Jansen, Delft University of Technology, Delft, The Netherlands

P.N. Joshi, National Chemical Laboratory, Pune 411 008, India

H. Kacirek, Institute of Physical Chemistry, University of Hamburg, Hamburg 13, Germany

Henri Kessler, Ecole Nationale Superieure de Chimie, 68093 Mulhouse Cedex, France

Jacek Klinowski, Department of Chemistry, University of Cambridge, Cambridge CB2 1EW U.K.

Mitsue Koizumi, Department of Materials Chemistry, Ryukoku University, Otsushi 520-21, Japan

Waclaw Kolodziejski, Department of Chemistry, University of Cambridge, Cambridge CB2 1EW U.K.

A.N. Kotasthane, National Chemical Laboratory, Pune 411 008, India

S.M. Kuznicki, Engelhard Corporation

Hans Lechert, Institut of Physical Chemistry, University of Hamburg, Hamburg 13 Germany

S.M. Levine, Engelhard Corporation

J. Lievens, Tsinghua University

K. Lillirud, Department of Chemistry, University of Oslo, Oslo, Norway

J.C. Lin, Tsinghua University

L. Maistriau, Department of Chemistry, University of Namur, B-5000, Namur, Belgium

Mahmoud Mansour, Department of Chemical Engineering, Cleveland State University, Cleveland, OH

Pascale Massiani, Universite Pierre et Marie Curie, Paris, Cedex 05, France

C. Mayenez, Facultes Universitaires N.D. de la Paix Namur, B-5000 Namur, Belgium

Nancy McGuire, Union Carbide, Tarrytown Technical Center

M.T. Melchior, Exxon Research and Engineering Company, Annandale, NJ 08801

Abdallah Merrouche, Ecole Nationale Superieure de Chimie, 68093 Mulhouse Cedex, France

R. Monque, Intevep S. A., Caracas 1070A, Venezuela

D. Mueller, Zentralinstitut fuer Anorganische Chemie, Berlin, 1199, Berlin, Germany

Yumi Nakagawa, Chevron Research and Technology Company

A. Nastro, Department of Chemistry, University of Calabria, Rende, Italy

Tina N. Nenoff, Department of Chemistry, University of California, Santa Barbara, CA 93106-9510

J.M. Newsam, Biosym Technologies Inc., San Diego, CA 92121

M. Agnes Nicolle, Ecole Nationale Superieure de Chimie, 34053 Montpellier, France

Joel Patarin, Ecole Nationale Superieure de Chimie, 68093 Mulhouse Cedex, France

Jaime Perez, Department of Chemistry, Texas A&M University, College Station, TX 77843

C.J. Radke, Center for Advanced Materials, Lawrence Berkeley Laboratory, University of California, Berkeley 94720

G.N. Rao, Catalysis Division, National Chemical Laboratory, Pune 411 008 India

P. Ratnasamy, Catalysis Division, National Chemical Laboratory, Pune 411 008 India

Francois Remoue, Centre de Recherches de Pont a Mousson, Pont a Mousson, France

Kathleen Reuter, Union Carbide, Tarrytown Technical Center, Tarrytown, NY 10591

J. Richter-Mendau, Zentralinstitut fuer physikalische Chemie, Berlin, 1199 Berlin Germany

Joao Rocha, Department of Chemistry, University of Cambridge, Cambridge CB2 1EW, U.K.

Hiroshi Saegusa, Department of Materials Chemistry, Ryukoku University, Otsushi 520-21 Japan

Donald S. Santilli, Chevron Research and Technology Company, Richmond, CA 94802-0627

Johanna B. Savader, Exxon Research and Engineering Company, Annandale, NJ 08801

M.D. Shannon, ICI Chemicals and Polymers Ltd. England

S.P. Sheu, Tsinghua University

V.P. Shiralkar, National Chemical Laboratory, Pune 411 008, India

Gary Skeels, UOP, Tarrytown Technical Center, Tarrytown, New York 10591

Michel Soulard, Ecole Nationale Superieure de Chimie, 68093 Mulhouse Cedex, France

K.G. Strohmaier, Exxon Research Engineering Company, Annandale, NJ 08801

Galen D. Stucky, Department of Chemistry, University of California, Santa Barbara, CA 93106-9510

R. Szostak, Zeolite Research Program, Georgia Tech Research Institute

K.A. Thrush, Engelhard Corporation

M.M.J. Treacy, NEC Research Institute, Inc., Princeton, NJ 08540
M.M.J. Treacy, NEC Research Institute, Inc., Princeton, NJ 08540
H. van Bekkum, Delft University of Technology, Delft, The Netherlands
Robert A. Van Nordstrand, Chevron Research and Technology Company, Richmond, CA 94802
D.E.W. Vaughan, Exxon Research Engineering Company, Annandale, NJ 08801
K. Vinje, Department of Chemistry, University of Oslo, Oslo, Norway
H. Weyda, Institute of Physical Chemistry, University of Hamburg, Hamburg 13 Germany
Stacey I. Zones, Chevron Research and Technology Company, Richmond, CA 94802

Contents

Preface I

On September 25–30, 1988 in Los Angeles, California, the first ACS Symposium on zeolite synthesis emphasized the importance that gel chemistry, zeolite nucleation, crystal growth, crystallization kinetics, and structure-directing phenomena have in understanding zeolite (and molecular sieve) synthesis. The objectives of a similar ACS Symposium held in New York on August 25–30, 1990 where expanded to include papers on pillared clay synthesis and on the synthesis of other microporous materials. Today these microporous solids have become an essential part of the catalysts used in petroleum refining, an industry that in 1990 had sale of $140 billions (U.S. Department of Commerce; U.S. Industrial Outlook 1991).

It was Baron A. Cronstedt, a Swedish Mineralogist, who first reported in 1756 the occurrence of natural zeolites. Thus, the first chapter is a reproduction of Cronsted's article together with a translation by Ir. G. Sumelius. The rest of the book contains chapters illustrating the effects that the various synthesis parameter have on the crystallization of new phases (such as titanium silicates) and on phase composition (such as gallophosphates). Progress toward the synthesis of large pore molecular sieves is represented by several papers on VPI-5 (the first molecular sieve containing 18-member oxygen rings). Chapters dealing with zeolite characterization and testing have also been included to illustrate the importance of modern characterization techniques and of test reactions in understanding crystal properties.

The generous financial contribution received from industrial sponsors, is greatfully acknowledged. We thank the many colleagues who acted as referees and the authors for the effort they gave in presenting their research at the symposium and in preparing the camera ready manuscripts. We would also like to thank Professors A. Clearfield, T. J. Pinnavaia, Z. Gabelica, H. Kessler, S. Yamanaka, and to Dr. G. W. Skeel and Dr. D. E. W. Vaughan for helping to chair this symposium. Special thanks are due to Professor H. van Bekkum for providing to us Baron Cronstedt's paper and figure.

Mario L. Occelli   Harry E. Robson
Yorba Linda, CA   *Baton Rouge, LA*

Molecular Sieves

1

Attempted Translation of the Original Old-Swedish Paper by Cronstedt

Translation: Ir. Göran Sumelius, Finland

> *Source:* "An Attempt to Mineralogy or Arrangement of the Realm of Minerals," Wilska Press: Stockholm, 1758
>
> 1756. April–May–June
> *Observation and description of an unknown species of rock, called ZEOLITES, by Axel Fr. Cronstedt.*

A. F. Cronstedt. 1722–1765. Courtesy of the Swedish Academy of Science.

Among the rocks which I have collected and of which I have tried to learn the properties, the hereafter reported species, shows in the fire a strange and characteristic behaviour, which makes it impossible to classify it under any known brand, except for a certain class, depending on the elements found in rocks: lime-, silicon-, clay and talc soil.

I have obtained samples from two sources, from two sources, from Mr. Adlerheim of the Svappavari copper-mine in Tornea, Lapland, and from Mr. Schindel in Iceland, but not enough to melt it together with other rocks in the melting pot. Yet the following can be reported with certainty:

1. The colour of the rock from Svappavari is light yellow, that from Iceland is white, partly half transparent, partly opaque.
2. The texture and the morphology of the particles is different in both samples. The species from Svappavari can be found as spheres and cylinders, made of threadlike pyramids, with their tops pointing to the centre. The Icelandic material is partly composed of particles like chalk, it is opaque too, partly appearing in wedges. The beamlike species seems to be a crystal-cavitation or a beginning of crystallization in a composite form, like the spar of limestone, rock-crystal of quartz, the special rocks of garnet and (black) turmalin which all show an irregular configuration when they had no room to develop freely.
3. The degree if hardness is generally similar to spar or compact limestone, and under these circumstances it does not spark with steel: it releases bubbles of gas with fire-(96%*)Spiritus.
4. In the flame of the blow-pipe the material ferments and swells like borax, a property which the rock from Svappavari shows more distinctly; the mentioned pyramids get apart and divide in many threads. These threads keep together and turn into a white spongelike substance, and melt with a phosphorescent light to a white glass. If the heat is increased, the glass becomes clear and coloured, when the air bubbles have disappeared, which seems to influence the opaqueness.
5. Fire makes it soluble in Borax and sale fusibili microcosmico makes it soluble, but slowly, without fermentation.
6. Soda Salt attracts it and dissolves it violently: and with carbon it is possible to drive out the Svappavari rock, but not the Icelandic one, to pure glass. Because the first appears even in copper glass, traces of copper can be seen when the glass first becomes heliotrope and opaque. Copper is also seen in the green fire. However, it can not be used, only accidentally.

After this deduction, it is not possible to place it among the familiar spar species, where eye-measurement and hardness seems to put it, especially as it does not ferment with sale fusibili, and melts easily with Soda Salt. This is in contrast to characteristics of those rocks who have lime as element. *That is my humble opinion.*

Asbestos-type materials do not have these properties. And beam turmalin, which is often confused with it, melts easily, like the whole turmalin family, but not under these circumstances. Still it reminds most of turmalin, for it melts equally easily in mixed metals or in the Earth which is their element, which then is different from the silicon-earth, and rather would be called "vitrifying," if it is possible for rocks or Earth to have this name.

The quantity of pure qualities, which is not abundant yet, gives only the opportunity to try and use it in some handaflôgd."

To prevent extensive epithets, which involve all kinds of troubles and names for a particular property which many species have in common, I take the liberty, in a brave mood, to call it ZEOLITES.

ZEOLITES

Zeolites, i.e., petrified, are described in the Treatises of the Swedish Scientific Academy of the year 1756. In the particular article the following properties are discussed:

1. It is slightly harder than "Flusser" and limestone, but can be pierced with steel without sparking.
2. It melts easily under release of gas, like Borax, turning into a white frothy glass, which is very difficult to make compact and transparent.
3. It dissolves better in Alcali minerale or Soda Salt than in Borax or Sale fusibili microcosmico.
4. It does not release bubbles of gas with the latter salt like Lime nor with Borax like Gypsum.
5. It does not effervesce in acid liquids such as Oleo Vitroli (=Sulfuric Acid*) and Nitric Acid, but it slowly dissolves. If concentrated Oleo Vitroli is poured out on zeolite powder, the powder heats up and sinters.
6. At the moment of melting it gives off a phosphorescent glow.

OCCURRENCE IN PETRIFIED FORM

I. Invisible Components. Z. particulis impalpabilibus
   A. Pure. Z. Purus.
      a. White. Iceland.
   B. Mixed with Silver and Iron.
      a. Blue. Lapis Lazuli. Buchareiska Calmuckiet.
         From experiments the following results were obtained:

1. During calcination it keeps its colour for a long time, but it turns brown upon prolonged heating.
2. It is easily melted to a white foamy glass, which can be swelled up with a blowpipe.
3. It does not froth with acids, but
4. in boiling Oleo Vitroli it slowly dissolves and looses its blue colour. Precipitation with Alkali fixa gives a white soil, which leaves Silver particles (bigger or smaller, depending on the sample) when slagged with Borax.
5. From slagging experiments with Lead 4 Lod Silver per Centner mineral were obtained.
6. Nitric Acid does not allow determination of the Silver content as unambiguously as Oleo Vitroli.
7. Spiritus Salis Ammoniaci does not colour a solution of crude well-calcined Lazurite blue, implying that Copper cannot be accused of causing the blue colour. This is confirmed by the stability of lazurite in fire and the colour of the glass produced from it.
8. It is slightly harder than other Zeolites, but not as hard as Quartz and other Silicon species: the purest blue kind can be crushed with steel to a white powder, although it can be polished like Marble.
9. The Magnet has little attraction on thoroughly calcined Lazurite and, unlike Copper, it gives lead glass a greenish gleam, like Iron mixed with Lime.

Observations: Lapis Lazuli is rarely pure; mostly it is mixed with Lime, threads of Quartz and Iron Pyrites. Nevertheless, the samples used for these experiments are pure, as a trained eye can see, and it is hoped that those who are skilled will use their expertise and discover the cause of the persistent blue colour, for it does not precipitate on Copper or Iron which sometimes are blue as well; but it is of a kind that disappears in fire and in Alkali. Although literature reports the use of Silver to prepare Ultramarine, this never can lead to the observed properties when a Silver alloy and things that contain Alkali Volantile are used similarly to the process that makes Copper blue. From these observations it can be concluded that the Stones cannot be classified under any other Earth than this one.

2. Spar-Zeolite. Zeolites Spatosus. Looks like lime-spats, but shows more complicated patterns and is much brittler.
3. Crystalline Zeolite. Zeolites Crystallisatus

Are more common than Spar-Zeolites and can be found as:

A. Joint crystals shere-shaped, pointing inwards. Zeolites pyramidales concreti, ad centrum tendentes.

    a. Yellow. Swappawari in Tornea Lappland

    b. White. The Gustafsbersmine in Jemteland

B. Separate truncated prismatic Crystals. Crystalli Zeolitis distincti figura prismatic truncata.

    a. White. The Gustafsbergsmine in Jemteland.

C. Hair-like crystals. Crystalli Zeolitis capillares. Are partly joint and partly separated, in the latter case they are like "Peridotite" and are sometimes, at some finding-places, called Lifenblᶍte. They can be found:

    a. White. The Gustafsbergsmine in Jemteland

Observations on Zeolites. This species behaves in fire like "Stone-marrow" (paragraph 85) and after many experiments on both, they both belong to the same class and probably can be classified under the better known Earth.

Tetra Porcellanca Luneburgica, mentioned by Bruchman and classified as a Gypsum species by H. Wallerius, belongs to this class as well; but it has not been found in sufficient quantities to be compared with the previously mentioned ones, and we only have it available in narrow Cylinders and Crystal-cavities; therefore, it has not been possible to melt it with any other Rocks than Fluorinespar and this is not easily achieved. The latter sample gives an opaque Glass with the same colour as Alkali Vitri, threadlike cracks and rough surface. The property to ferment in fire, like Borax, is also exhibited by these Crystals (see paragraph 111); yet it changes further forming a white film of Glass around it, before it becomes hard-melted.

2

The Chemistry of NaY Crystallization from Sodium Silicate Solutions

D. M. Ginter, A. T. Bell, and C. J. Radke *Center for Advanced Materials, Lawrence Berkeley Laboratory and Department of Chemical Engineering, University of California, Berkeley, CA 94720*

Zeolites were crystallized at 100°C from aluminosilicate gels prepared from sodium silicate solutions. The composition and structure of intermediate and final products were observed by elemental analysis, NMR spectroscopy, FTIR spectroscopy, and powder X-ray diffraction. Aging of Na-lean gels had no effect on crystallization, with NaP zeolite forming preferentially. Addition of a small amount of a Na-rich seed gel, however, enables crystallization of NaY zeolite from synthesis gels that remained Na-lean overall. Precursors to NaY nuclei are observed in the seed gel, and must be distributed uniformly in the synthesis gel in order to crystallize pure NaY. NMR and FTIR spectroscopies reveal that the solid phase of the seed gel contains NaY fragments with a Si/Al ratio of 2. These fragments contain an abundance of hydrated Na^+ ions which stabilize the NaY structure. NaY of varying Si/Al ratios were crystallized directly from seed gels that were partially neutralized with varying amounts of dilute sulfuric acid. The Si/Al ratio and growth rate of NaY are set by Si/Al ratios, solubilities, and reaction rates associated with dissolved aluminosilicate species. These factors are determined by the alkalinity of the synthesis gel.

INTRODUCTION

A number of recipes for the synthesis of NaY zeolite have been reported in the patent literature [1-9]. In the earliest of these [1], colloidal silica sols or reactive solid silicas were required to produce NaY zeolite with Si/Al ratios in excess of 2.3. Room-temperature aging of the synthesis gels produced from such silica sources was found to increase the rate of formation of high-silica NaY upon heating and reduce the amounts of excess silica and soda needed in the batch [1,10]. Subsequent patents demonstrated that high-silica NaY could be prepared using less expensive silica sources, such as sodium silicate solutions, by seeding the synthesis gel with an aged seed gel [3-9]. In the absence of seeding, these synthesis gels crystallize primarily NaP [11], but seeding enables routine crystallization of NaY with Si/Al ratios up to about 3.

Irrespective of the materials from which the synthesis gel is prepared, the crystallization of NaY can be divided into two important stages: nucleation and crystal growth. Sharp differentiation between these two processes is difficult to achieve because they occur concurrently in a structurally complex environment. Nevertheless, recent studies have shed some light on the physical and chemical processes occurring during nucleation [12-16] and growth [17].

Nucleation in colloidal silica gels is attributable to the formation of an amorphous Na- and Al-rich aluminosilicate solid during aging at room temperature that reacts to form NaY nuclei upon heating [14,18]. In seeded syntheses, aluminosilicate precursors to NaY zeolite are presumed to form in the seed gel [3-5]. Little is known, however, about the composition and structure of these precursors, or the mechanisms by which they form and are converted to NaY nuclei. NMR studies of sodium aluminosilicate solutions with compositions near those of seed gels [19,20] have shown that Al atoms incorporate into Al(nSi) environments ($n \geq 2$) in dissolved aluminosilicate species at the expense of $Al(OH)_4^-$ anions. Studies of seed compositions have shown that the seed loses its effectiveness for nucleating NaY when it contains too little Al or Na [15], or when it has been aged for too long or too short a period [15,16].

A further requirement for successful crystallization of NaY from seeded feedstocks is dispersal of the nucleation centers throughout the synthesis gel, something that becomes increasingly difficult as the viscosity of the gel increases [8]. The growth of NaY and the Si/Al ratio of the final product are strongly dependent on gel alkalinity [17]. With decreasing pH, the rate of crystallization decreases and the Si/Al ratio in the crystals formed increases.

The objective of the present investigation is to study the formation of nuclei precursors in seed gels and the effects of seeding on the crystallization of NaY zeolite. The compositions of the liquid and solid components of the seed and synthesis gels were followed by elemental analysis, and the solid phase was also characterized by ^{29}Si MAS-NMR spectroscopy and FTIR spectroscopy. The crystallinity of the solid phase was determined by X-ray diffraction.

EXPERIMENTAL

Seed and Feedstock Gels

The following procedure was used to prepare a seed gel with an overall stoichiometry of $10.67 Na_2O, 1 Al_2O_3, 10 SiO_2, 180 H_2O$ [8,9]. An aqueous sodium silicate solution (30.27 wt% SiO_2, 9.40 wt% Na_2O, $\rho = 1400$ kg/m^3, $\mu = 250$ mPa·s, pH = 11.3) was slowly added to a stirred sodium aluminate solution which contained the remaining reactants. The aluminosilicate solution thus formed was clear initially, but slowly formed a gel.

The feedstock for NaY synthesis was prepared by vigorously mixing the sodium silicate solution described above with a sodium aluminate solution to make a viscous aluminosilicate gel having the overall stoichiometry $4.28 Na_2O$, $1 Al_2O_3, 10 SiO_2, 180 H_2O$. For most of the studies undertaken, the feedstock gel was aged for one day prior to use.

Seeding Procedure

Seeding of the Na-lean feedstock was achieved by slowly blending in the Na-rich seed gel in an amount corresponding to 5% of the overall aluminum content to produce a synthesis gel with the overall stoichiometry $4.6 Na_2O, 1 Al_2O_3, 10 SiO_2, 180 H_2O$. The freshly prepared synthesis gel was homogenized using a turbine mixer rotating at 1600 rpm and then poured into 250-cm^3 polypropylene bottles.

Seeded synthesis gels were also prepared directly from seed gels. After aging for 1 d, the seed gel was partially neutralized with dilute H_2SO_4 to produce a gel with the overall stoichiometry $x Na_2O, 1 Al_2O_3, 10 SiO_2, 180 H_2O$, (10.67 - x) $(Na_2SO_4 \cdot 10 H_2O)$, where x = 4.6, 3.3, or 2.5. This gel was homogenized and then aged at room temperature for 1 d before heating to 100°C to induce crystallization.

Sample Characterization

Gel samples were taken periodically during crystallization. Each sample was separated by centrifugation at 2500 rpm for about 10 min into a supernatant liquid and a wet solid. The solid and liquid phases were analyzed by a variety of techniques. The Si and Al contents of the liquid phase were obtained by inductively coupled plasma emission spectroscopy (ICP), and the Na content was obtained by atomic absorption spectroscopy. The wet solids were filtered and rinsed with demineralized water. Part of the moist filter cake was characterized by ^{29}Si MAS-NMR spectroscopy using a home-built 180 MHz spectrometer equipped with a Doty MAS probe operating at 35.77 MHz. Samples were spun at up to 2.6 kHz in zirconia rotors sealed with Ultem end caps containing Viton

O-rings [21]. Each spectrum was acquired using 1024-90° pulses and a 10 s recycle delay. Lorentzian line broadening of 25 Hz was applied to each spectrum. The spectra of the washed gels were referenced to the peak for monomeric silicic acid (0 ppm) in a sodium silicate solution containing 3 mol% SiO_2 and 2 mol% Na_2O.

The remainder of the filter cake was dried at 110°C for ~ 16 h. The elemental composition of the dry solid was obtained by dissolving it in KOH and then performing inductively coupled plasma emission spectroscopy to determine the Si and Al contents. Samples for infrared spectroscopy were prepared by pressing a disk of a mull containing 1% of the sample and 99% KBr. Transmission FTIR spectra were acquired using a Digilab FTS-80 FTIR spectrometer. The zeolite phases present and the degree of crystallinity of a given phase in the dried solid were determined by powder X-ray diffraction.

RESULTS

Sodium Silicate Solution

The ^{29}Si MAS-NMR spectrum of the sodium silicate solution is presented in Fig. 2-1. The broad resonances centered at -8.5, -17.0, and -25.5 ppm are characteristic of Q^1, Q^2, and Q^3 silicon environments in dissolved silicate anions [22], and the especially broad resonance centered at -36 ppm is characteristic of Q^4 environments in bulk silica [22]. The average connectivity of the Si atoms is 2.85, as calculated from the distribution of area among these peaks.

Seed Gel

Addition of the silicate solution to the aluminate solution produced a highly alkaline (pH ~ 14) aluminosilicate solution, the viscosity of which was lower than the original silicate solution. While initially clear, this solution became cloudy after 3 h of aging at room temperature, and formed a gel after 18 h. After 1 d of aging, a sample of this gel (sample A) was separated by centrifugation with marginal success, yielding a small amount of a hazy supernatant over a large quantity of a gelatinous solid. A sample of the gel that was aged for 1 d and heated to 100°C for 1 h (sample B) was slightly easier to separate by centrifugation, yielding similar liquid and solid products.

Listed in Table 2-1 are the Si/Al ratios of the solid phases and the structure of any X-ray detectable crystalline material in the solids obtained from samples A and B. The compositions of the corresponding liquid phases are also listed. The liquid phase concentrations of Si and Al, and the Si/Al ratios of the solid phase, are similar for the two samples, but the liquid phase concentration of Na is higher for sample B than for sample A.

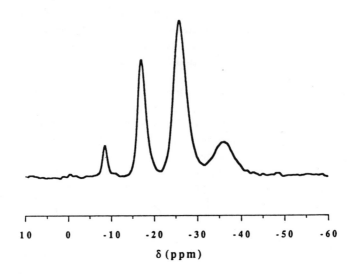

Figure 2-1. ^{29}Si MAS-NMR spectrum of the sodium silicate solution (30.47 wt % SiO_2, 9.40 wt % Na_2O.) Add -72 ppm to reference to TMS.

Table 2-1 - Seed Gel

| Sample | Solid | | | Phase | Liquid | | |
| | Si/Al ratio | | | | Concentration, M | | |
	ICP	NMR	FTIR		[Na]	[Si]	[Al]
A. Aged 1 d.	2.06	1.96	2.0	NaY	5.23	2.49	0.070
B. Aged 1 d, Heated 1 h.	1.95	---	---	Possible NaY	5.79	2.38	0.074

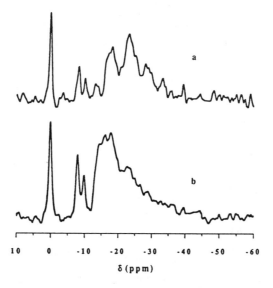

Figure 2-2. ^{29}Si MAS-NMR spectra of the rinsed solids obtained from the seed gel mixture (10.67 Na$_2$O, 1 Al$_2$O$_3$, 10 SiO$_2$, 180 H$_2$O) after aging at room temperature for 24 h and (a) before heating and (b) after 1 h at 100°C. Add -72 ppm to reference to TMS.

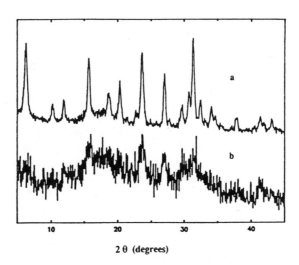

Figure 2-3. Powder X-ray diffraction patterns of the rinsed and dried solids obtained from the seed gel mixture (10.67 Na$_2$O, 1 Al$_2$O$_3$, 10 SiO$_2$, 180 H$_2$O) after aging at room temperature for 24 h and (a) before heating and (b) after 1 h at 100°C.

The ^{29}Si MAS-NMR spectra of the wet solid phases are presented in Fig. 2-2. For sample A, sharp peaks are observed at 0, -8.3, and -10.2 ppm (and possibly a minor peak ca. -3.7 ppm). These features are characteristic of Q^0, Q^1, and Q_A^2 [22] (and Q^1(1Al) [16,19,20,23]) Si atoms in silicate (and aluminosilicate) anions present in the interstitial fluid trapped in the gelatinous solid. Broad peaks are also seen at -13.4, -17.8, -23.4, -28.8, and -33.3 ppm. These features are characteristic of Q^4(4Al), Q^4(3Al), Q^4(2Al), Q^4(1Al), and Q^4(0Al) Si atoms in faujasite [24]. The distribution of area among these peaks is consistent with that of NaY zeolite having a Si/Al ratio of 1.96 [24]. The corresponding spectrum of sample B (spectrum b) also shows peaks due to silicate anions. However, in contrast to the spectrum for sample A, the distribution of peak area in the region between -12 and -35 ppm is shifted downfield, and the peaks are less clearly resolved. Peaks for Q^4(4Al), Q^4(3Al), and Q^4(2Al) Si atoms in faujasite can be discerned together with a peak at -16 ppm, due to Q^3(3Al) Si atoms [25]. This latter feature overlaps the peaks for Q^4(4Al) and Q^4(3Al) Si atoms.

Figure 2-3 shows the powder XRD pattern for the solid obtained from samples A and B after rinsing and drying. The pattern for sample A shows sharp peaks characteristic of crystalline NaY zeolite over a broad feature due to amorphous solid material. While peaks characteristic of NaY are present in the XRD pattern of sample B, their intensities are lower, and their widths are greater than those in the pattern of sample A.

The FTIR spectra corresponding to the XRD patterns in Fig. 2-3 are shown in Fig. 2-4. The spectrum of the rinsed and dried solid from sample A exhibits distinct absorbances at 460, 570, 770, and 1000 cm^{-1}, characteristic of NaY with a Si/Al ratio of about 2 [26,27]. The spectrum of the solid obtained from sample B is less defined than that of the solid obtained from sample A.

Crystallization

In the absence of seeding, heating the feedstock gel resulted in the formation of NaP, but no NaY. The solid phase was completely amorphous until NaP was first detected after ~24 h of heating at 100°C. Complete conversion of the solid phase to NaP was achieved after 30 h of heating. Aging of the gel at room temperature for up to 5 weeks prior to heating had no effect on the crystallization kinetics. Table 2-2 lists the Si/Al ratio of the solid as well as the composition of the liquid phase for a gel aged for 5 weeks and then heated at 100°C for 30 h (sample C). Identification of the solid phase as NaP is supported by ^{29}Si MAS-NMR and FTIR spectroscopy. Figure 2-5 shows that the ^{29}Si MAS-NMR spectrum of sample C exhibits well defined peaks at -22.1, -27.4, -32.5 and -37.7 ppm, characteristic of Q^4(3Al), Q^4(2Al), Q^4(1Al), and Q^4(0Al) Si atoms in NaP [28]. The Si/Al ratio determined from the distribution in peak areas is 2.87, which is roughly the same as the Si/Al ratio of the solid phase as determined by

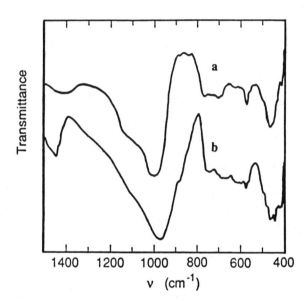

Figure 2-4. Transmittance FTIR spectra of the rinsed and dried solids obtained from the seed gel mixture (10.67 Na_2O, 1 Al_2O_3, 10 SiO_2, 180 H_2O) after aging at room temperature for 24 h and (a) before heating and (b) after 1 h at 100°C.

Table 2-2 - Crystallization/Seed Gel Addition

Sample	Solid			Phase	Liquid		
		Si/Al ratio			Concentration, M		
	ICP	NMR	FTIR		[Na]	[Si]	[Al]
C. Unseeded, Aged 5 w, Heated 30 h.	3.11	2.87	---	NaP	1.96	2.09	0.005
D. Seeded, No aging, Heated 8 h.	2.55	---	---	NaY, with significant NaP impurity	2.45	2.12	0.005
E. Seeded.	3.48	---	---	Amorphous	2.52	1.95	0.004
F. Seeded, Aged 1 d, Heated 2 h.	2.98	---	---	Amorphous + NaY (10%)	2.14	2.09	0.025
G. Seeded, Aged 1 d, Heated 6 h.	2.56	2.30	2.4	NaY	2.81	2.17	0.008

elemental analysis, 3.11. The FTIR spectrum of the solid is shown in Fig 2-6. The bands at 1035, 710, 615 and 435 cm^{-1} are characteristic of NaP [26].

Crystallization of NaY from the feedstock gel was accomplished by mixing in a portion of the seed gel in an amount corresponding to 5% of the total Al content. The aged feedstock gel became fluid with vigorous shearing, and addition of the seed gel caused a visible thinning of the gel. When heating immediatedly followed seeding, both NaY and NaP were detected in the primarily amorphous solid phase after 4 h at 100°C. After 8 h of heating, the amorphous solid was completely converted to crystalline products. NaY is the major crystalline phase, but a significant amount of NaP is present as an impurity. The Si/Al ratio of the solid and the composition of the liquid phase are listed in Table 2-2 (sample D). The ^{29}Si MAS-NMR spectrum of the rinsed solid is shown in Fig. 2-7. The peaks at -17.8 and -23.1 ppm are characteristic of Q^4(3Al) and Q^4(2Al) Si atoms in NaY [24], whereas the peaks at -27.2, -32.2, and -38.0 ppm are characteristic of Q^4(2Al), Q^4(1Al), and Q^4(0Al) Si atoms in NaP [28]. The FTIR spectrum of the dried solid, presented in Fig. 2-8, exhibits peaks at 1125, 1020, 785, 720, 580, 500, and 455 cm^{-1} characteristic of NaY [26], as well as a peak at 615 cm^{-1} characteristic of NaP [26].

The last three entries, E-G, in Table 2-2 track the changes in gel composition immediately after addition of seeds, after aging the seeded mixture for 1 day, and after heating at 100°C. The solid phase of the gel produced immediately after seeding (sample E) has a Si/Al ratio of 3.48 and is X-ray amorphous. The liquid phase of sample E had a dissolved Al concentration of 0.004 M and concentrations of dissolved Na and Si of 2.52 and 1.95 M, respectively. Aging 1 d and heating for 2 h (sample F) result in the appearance of 10 wt% NaY in the solid phase which now has a Si/Al ratio of 2.98. During this time, the Al concentration in the liquid phase increased to 0.025 M and the Si concentration increased to 2.09 M, while the Na concentration decreased to 2.14 M. Conversion of the amorphous solid to NaY is nearly complete after 4 h of heating, and is complete after 6 h of heating (sample G). The final NaY product has a Si/Al ratio of 2.56. The Al concentration in the supernatant fluid has now dropped to 0.008 M at the end of crystallization, while the Na and Si concentrations have increased to 2.81 and 2.17 M, respectively.

Figure 2-9 shows the ^{29}Si MAS-NMR spectra of the wet solids obtained from samples E through G. The spectrum of the wet solid obtained after blending in the seed gel (sample E), shows the peaks at 0, -8.3, and -10.2 ppm. The intensities of these peaks are lower than those in Fig. 2-2, indicating that the solution trapped in the interstices of the gel contains a smaller concentration of silicate anions than that found in the interstitial fluid in the seed gel. The main feature, due to the amorphous aluminosilicate solid, extends from -12 to -40 ppm, indicative of a broad range of Si connectivities. Aging for 1 d and

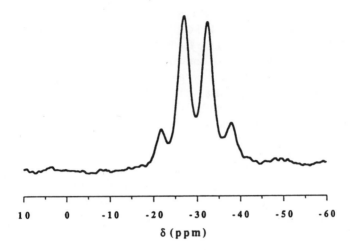

Figure 2-5. ^{29}Si MAS-NMR spectrum of the solid product obtained from the feedstock gel (4.28 Na$_2$O, 1 Al$_2$O$_3$, 10 SiO$_2$, 180 H$_2$O) after 5 weeks of aging at room temperature and 30 h of heating at 100°C. Add -72 ppm to reference to TMS.

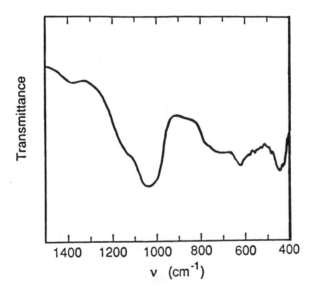

Figure 2-6. Transmittance FTIR spectrum of the solid product obtained from the feedstock gel (4.28 Na$_2$O, 1 Al$_2$O$_3$, 10 SiO$_2$, 180 H$_2$O) after 5 weeks of aging at room temperature and 30 h of heating at 100°C.

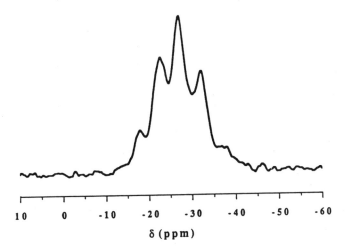

Figure 2-7. ^{29}Si MAS-NMR spectrum of the solid product obtained from the feedstock gel after the addition of seeds (overall 4.6 Na$_2$O, 1 Al$_2$O$_3$, 10 SiO$_2$, 180 H$_2$O) and 8 h of heating at 100°C. Add -72 ppm to reference to TMS.

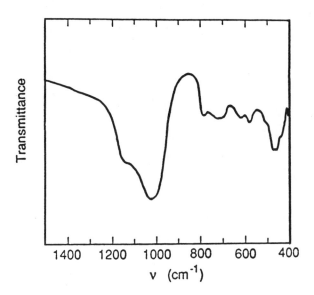

Figure 2-8. Transmittance FTIR spectrum of the solid product obtained from the feedstock gel after the addition of seeds (overall 4.6 Na$_2$O, 1 Al$_2$O$_3$, 10 SiO$_2$, 180 H$_2$O) and 8 h of heating at 100°C.

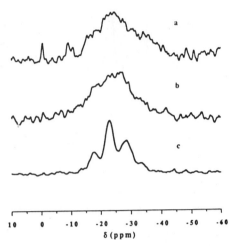

Figure 2-9. ^{29}Si MAS-NMR spectra of the solid products obtained from the feedstock gel (a) immediately after the addition of seeds (overall 4.6 Na$_2$O, 1 Al$_2$O$_3$, 10 SiO$_2$, 180 H$_2$O), and after 1 d of aging at room temperature followed by (b) 2 h of heating at 100°C, and (c) 6 h of heating at 100°C. Add -72 ppm to reference to TMS.

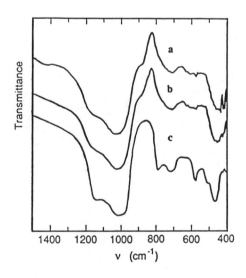

Figure 2-10. Transmittance FTIR spectra of the solid product obtained from the feedstock gel (a) immediately after the addition of seeds (overall 4.6 Na$_2$O, 1 Al$_2$O$_3$, 10 SiO$_2$, 180 H$_2$O), and after 1 d of aging at room temperature followed by (b) 2 h of heating at 100°C, and (c) 6 h of heating at 100°C.

heating for 2 h (sample F) leads to the disappearance of the peaks at 0, -8.3, and -10.2 ppm. Although this solid contains 10 wt% NaY, no peaks characteristic of crystalline material are evident above the broad feature due to amorphous aluminosilicate. The slight downfield shift of the center of gravity of the spectrum for sample F is consistent with the decrease in the Si/Al ratio of the solid. After the completion of crystallization (sample G), distinct peaks are seen at -13.7, -17.8, -22.9, -28.5, and -33.6 ppm, characteristic of $Q^4(4Al)$, $Q^4(3Al)$, $Q^4(2Al)$, $Q^4(1Al)$, and $Q^4(0Al)$ Si atoms in NaY. The distribution of area among these peaks is consistent with that of NaY having a Si/Al ratio of 2.30 [24].

Figure 2-10 shows the FTIR spectra of the rinsed and dried solids obtained from samples E, F, and G. The FTIR spectra of samples E and F (spectra a and b) exhibit broad features centered at 1030, 715, and 450 cm^{-1}, and are similar in appearance to the spectrum of albite glass with a Si/Al ratio of 3 [29-31]. The small, sharp feature at 570 cm^{-1} observed in spectrum a can be ascribed to vibrations in double-six membered ring structures in NaY [26]. The spectrum of the product from this batch (spectrum c) exhibits features near 1140, 1020, 785, 715, 575, 500 and 465 cm^{-1} characteristic of NaY [26]. The positions of the peaks seen in this spectrum are consistent with NaY having a Si/Al ratio around 2.5 [26,27].

Addition of sulfuric acid to the aged seed gels results in partial neutralization of the base, precipitation of silica, and a significant increase in the gel viscosity. The base in excess of that neutralized (x = total Na_2O - H_2SO_4 added, 1 mole Al_2O_3 basis) was varied. Table 2-3 lists the gel compositions after acid addition and 1 d of aging, followed by heating to 100°C, for gels in which x = 4.6 and 3.3. The structure of the solid phase deduced from XRD is also indicated.

When x is reduced to 4.6, heating for 2 h (sample H) produces a solid with an overall Si/Al ratio of 2.95, which contains 13 wt% NaY in amorphous aluminosilicate. Heating for 4 h results in nearly total conversion of the amorphous material to NaY, with complete conversion after 6 h (sample I), to produce a NaY product having a Si/Al ratio of 2.43. The concentration of Al in the liquid phase decreased from 0.021 M to 0.006 M as the solid phase of the gel was converted to NaY. The concentration of Na in the liquid phase increased from 4.31 M to 4.38 M, and the concentration of Si increased from 1.50 M to 1.59 M during the same period of time. The ^{29}Si MAS-NMR spectrum and the FTIR spectrum of the solid are shown in Figs. 2-11 and 2-12, respectively. The peak positions and appearance of both spectra identify the solid as NaY. The Si/Al ratio of this product as deduced from the distribution of peak area in the NMR spectrum and the positions of the FTIR peaks is 2.25 and 2.4, respectively [24,26,27].

Table 2-3 Crystallization/Seed Gel Neutralization

Sample	Solid			Phase	Liquid		
	Si/Al ratio				Concentration, M		
	ICP	NMR	FTIR		[Na]	[Si]	[Al]
H. x = 4.6, Aged 1 d, Heated 2 h.	2.95	---	---	Amorphous + NaY (13 wt%)	4.31	1.50	0.021
I. x = 4.6, Aged 1 d, Heated 6 h.	2.43	2.25	2.4	NaY	4.38	1.59	0.006
J. x = 3.3, Aged 1 d, Heated 2 h.	4.80	---	---	Amorphous	4.23	1.05	0.010
K. x = 3.3, Aged 1 d, Heated 96 h.	3.16	2.87	3.0	NaY	4.60	1.42	0.005

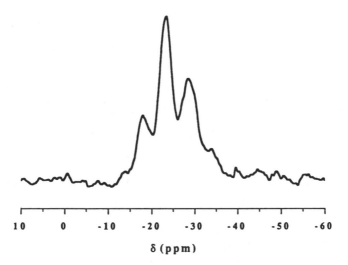

Figure 2-11. ^{29}Si MAS-NMR spectrum of the solid product obtained from the partially neutralized seed gel (4.6 Na$_2$O, 1 Al$_2$O$_3$, 10 SiO$_2$, 180 H$_2$O, 6.07 Na$_2$SO$_4$·10H$_2$O) which was aged at room temperature for 1 d after neutralization and then heated for 6 h at 100°C. Add -72 ppm to reference to TMS.

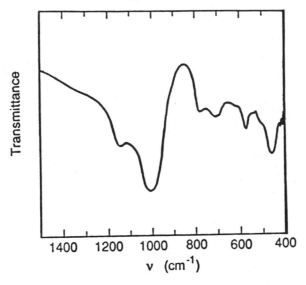

Figure 2-12. Transmittance FTIR spectrum of the solid product obtained from the partially neutralized seed gel (4.6 Na$_2$O, 1 Al$_2$O$_3$, 10 SiO$_2$, 180 H$_2$O, 6.07 Na$_2$SO$_4$·10H$_2$O) which was aged at room temperature for 1 d after neutralization and then heated for 6 h at 100°C.

Reduction of x to 3.3 results in significantly slower zeolite formation and an increase in the Si/Al ratio of the final product compared to crystallization with a free base content of 4.6. As seen in Table 2-3, no zeolite is evident after heating for 2 h (sample J), and the maximum yield of NaY is achieved only after 96 h at 100°C (sample K). During this time, the amorphous solid material which has an initial Si/Al ratio of 4.80 is converted to NaY with a Si/Al ratio of 3.16, the Al concentration in the liquid phase decreases from 0.010 M to 0.005 M, and the Na and Si concentrations increase, from 4.23 M to 4.60 M, and from 1.05 M to 1.42 M, respectively. The ^{29}Si MAS-NMR spectrum of the final product and the corresponding FTIR spectrum are shown in Figs. 2-13 and 2-14, respectively. Here again, the product can clearly be identified as NaY zeolite. The Si/Al ratio determined from the distribution of peak area in the NMR spectrum is 2.87, while the value determined from the positions of the IR peaks is 3.0 [24,26,27].

Neutralization of the seed gel to achieve x = 2.5 resulted in a material from which it was impossible to obtain any zeolite product, even after 18 days of heating. The ^{29}Si MAS-NMR spectra of the solids obtained from this gel after 2 h and 432 h of heating show only a single broad feature characteristic of an amorphous aluminosilicate. The absence of any long range order is also evident in the FTIR spectra, which only contain broad features, and in the XRD patterns, which contain no peaks due to crystalline material.

DISCUSSION

The results of the present investigation show that, independent of the extent of aging, NaY zeolite cannot nucleate in unseeded feedstock gels. Consistent with the literature [11] the only crystalline phase formed from the unseeded feedstock gel is NaP. NaY can be produced, though, by dispersing a small amount of the seed gel into the feedstock. Therefore, NaY nuclei must have formed from precursors in the seed gel. When heating to 100°C immediately follows seeding, the amorphous solid phase of the gel is completely converted to crystalline products within 8 h, most of which is NaY zeolite, but a significant amount of NaP also appears as an impurity. When the seeded gel is aged for 1 d prior to heating, complete conversion to NaY is accomplished in 6 h, without the formation of NaP. This suggests that perfect mixing (i.e., uniform distribution of precursors) is not accomplished during homogenization. Heterogeneities that exist between the feedstock and the dispersed seed after mixing are reduced during aging such that the precursors are more uniformly distributed, and pure NaY crystallizes.

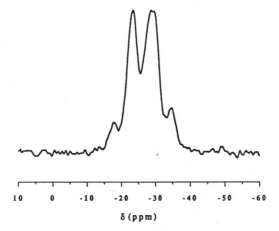

δ (ppm)

Figure 2-13. ^{29}Si MAS-NMR spectrum of the solid product obtained from the partially neutralized seed gel (3.3 Na$_2$O, 1 Al$_2$O$_3$, 10 SiO$_2$, 180 H$_2$O, 7.37 Na$_2$SO$_4$·10H$_2$O) which was aged at room temperature for 1 d after neutralization and then heated for 6 h at 100°C. Add -72 ppm to reference to TMS.

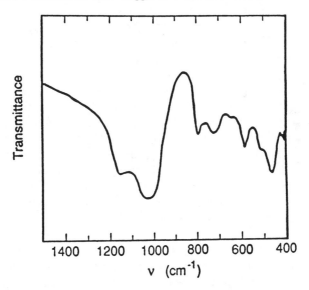

ν (cm^{-1})

Figure 2-14. Transmittance FTIR spectrum of the solid product obtained from the partially neutralized seed gel (3.3 Na$_2$O, 1 Al$_2$O$_3$, 10 SiO$_2$, 180 H$_2$O, 7.37 Na$_2$SO$_4$·10H$_2$O) which was aged at room temperature for 1 d after neutralization and then heated for 6 h at 100°C.

The reasons why precursors to NaY zeolite form in the seed gel can be deduced from a consideration of the physical and chemical changes which occur during the preparation of the seed gel. Because of its relatively low alkalinity (Na/Si = 0.6), most (~ 96%) of the Si atoms in the starting sodium silicate solution are contained in species with Q^2, Q^3, and Q^4 connectivities, and only 4% of the Si atoms are contained in species with Q^0 or Q^1 connectivities. Upon mixing of the sodium silicate solution with the Na-rich aluminate solution to form the seed mixture, aluminate anions react preferentially with the low connectivity (i.e., Q^0 or Q^1) silicate anions [23,32]. The increase in the Na/Si ratio from 0.6 to 2.13 upon mixing of the two solutions results in an increase in pH (from 11.3 to 14), causing the oligomeric silicate anions to undergo depolymerization. The low molecular weight silicate anions released continuously during this process react with aluminate anions in solution to increase the concentration of aluminosilicate anions until, ultimately, the solubility limit for amorphous aluminosilicate is reached [33,34]. The low initial concentration of low molecular weight silicate anions and the slow production of such anions by depolymerization of higher molecular weight silicate species accounts for the observation that the solution is clear upon mixing and slowly turns cloudy during aging. Studies of dissolution of amorphous silica at high pH (>13) have shown that Q^3 Si atoms depolymerize at room temperature on a time scale of about 40 min [18,35,36]. It is not surprising, therefore, that it takes 3 h for the solution to turn cloudy. As the solution is aged further, the oligomeric silicates continue to depolymerize, creating more amorphous aluminosilicate particles, until eventually the cloudy solution becomes a gel after 18 h of aging.

As indicated in Table 2-1, the solid isolated from the aged seed gel has a Si/Al ratio of 2.06. The Na content of the solid can be deduced from the fact that the Na/Si ratio in the liquid phase (2.10) is roughly that of the overall mixture (2.13). Therefore, the solid is estimated to have a Na/Si ratio of about 2.1. The high Na content of the solid indicates that Na+ ions balance not only the charge associated with Al in the solid, but also the charge associated with deprotonated silanol groups. The water content of the solid is estimated to be half the total solid weight, based on a component balance and the measured liquid phase density. Most, if not all, of this water is contained in the hydration spheres of the Na+ ions.

Rinsing of the amorphous solid separated from the seed gel removes much of the surrounding liquid, thereby decreasing the pH and the ionic strength. Rinsing also results in a partial removal of Na+ ions. These changes lead to reaction in the solid phase and liberation of Na+ and OH- ions as shown below:

$$\equiv\text{Si-O}^- \text{ Na}^+ \ + \ \text{HO-T}\equiv \ \rightarrow \ \equiv\text{Si-O-T}\equiv \ + \ \text{Na}^+ \ + \ \text{OH}^- \quad (\text{T = Si or Al})$$

The positions of the $Q^4(nAl)$ peaks in the ^{29}Si MAS-NMR spectrum of the rinsed solid (spectrum a in Fig. 2-2) suggests that this reaction creates a NaY framework having a Si/Al ratio of about 2. Drying of the rinsed solid at 110°C results in the appearance of highly crystalline NaY zeolite (spectrum a in Fig. 2-3), having a Si/Al of 2. It may, therefore, be concluded that aluminosilicate species produced in solution react to form an amorphous, highly hydrated solid, which upon rinsing and drying crystallizes to form NaY zeolite. The formation of NaY zeolite, rather than some denser phase, is a consequence of the stabilization of the open zeolite pore structure by the water hydrating the Na^+ ions [13,37]. The failure to nucleate NaY zeolite from the unseeded feedstock is most likely due to the fact that the solid formed in this case has a low Na^+ content, and hence is less hydrated. This water-deficient solid more easily nucleates the denser NaP.

The NaY nuclei precursors in the seed gel must be small to redistribute in the seeded feedstock upon aging. Characterization of the solid obtained from the aged seed gel after heating for 1 h (sample B) reveals evidence for this assertion The broadness of features in the XRD pattern (spectrum b in Fig. 2-3) and FTIR spectrum (spectrum b in Fig. 2-4) and the appearance of a $Q^3(3Al)$ peak in the ^{29}Si MAS-NMR spectrum (spectrum b in Fig. 2-2) suggest that the faujasitic domains formed by the polymerization of precursors (note the increase in Na concentration upon heating, see Table 2-1) are very small (< 50 Å).

<u>Crystal Growth Chemistry</u>

Whereas the nucleation of NaY is dependent on precursors formed in the seed gel, the species responsible for NaY crystal growth must be derived from the feedstock gel (or the gel produced upon partial neutralization of the seed gel). This conclusion is based on the fact that the synthesis gel obtained by seeding the feedstock contains very little seed gel (5% of the total Al and Si). It follows, therefore, that the rate of NaY formation and the Si/Al ratio in the final product should depend on the composition of the synthesis gel. The data presented in Table 2-3 clearly demonstrate this to be the case. In agreement with previous investigations [12,17], it is found that as the alkalinity of the gel decreases, so does the rate of NaY formation, and the Si/Al ratio of the crystals formed increases. An understanding of the physical and chemical transformations occurring during crystal growth can be deduced from the results of the present study.

We begin by noting that the rates of NaY formation and the Si/Al ratio of the final zeolites are virtually the same for synthesis from an aged seeded feedstock and from a partially neutralized seed gel (compare samples G and I in Tables 2-22 and 2-3, respectively) provided the value of x (i.e., the alkalinity) is the same. What this means is that the rate of NaY growth and the Si/Al ratio of the zeolite are governed by the gel alkalinity, more than anything else.

MAS-NMR and FTIR spectroscopies and XRD show that the freshly seeded synthesis gel is completely amorphous, and elemental analysis shows that the Si/Al ratio of this solid is significantly higher than that of the final zeolite (i.e., Si/Al = 3.48 immediately after seeding versus Si/Al = 2.56 in the zeolite). As heating proceeds, evidence for the formation of crystalline NaY is observed and the Si/Al ratio of the solid decreases monotonically. Concurrently, the concentration of dissolved Si increases monotonically, the concentration of dissolved Al passes through a maximum, and the concentration of dissolved Na passes through a minimum. These trends can be attributed to dissolution of the amorphous aluminosilicate, and the incorporation of dissolved species into growing zeolite crystallites, the nuclei having been formed upon heating from the nucleation precursors produced in the seed gel. This picture of NaY growth is similar to that proposed by Kacirek and Lechert [17], who were able to observe growth, independent of nucleation, by seeding their gels with small (< 0.5 μm diameter) NaX crystallites. These authors propose that the linear rate of crystal growth is determined by reaction of nutrient species at the surface of the growing zeolite crystals and is given by:

$$\frac{dL}{dt} = k_s(c - c_s)$$

where dL/dt is the growth rate of the crystal diameter, k_s is the surface reaction rate constant, and c and c_s are the concentration of nutrients in the bulk liquid and in a solution saturated with respect to the zeolite, respectively [38]. The concentration of reactants, rate constant, and supersaturation, (c-c_s), are affected by the gel alkalinity.

The nutrients for zeolite crystal growth are envisioned to be aluminosilicate anions produced in the liquid phase of the gel through the reaction of silicate and aluminate anions. Inasmuch as Al is the limiting reactant, the concentration of nutrients should be proportional to the concentration of Al in the liquid phase. The observed increase in the Si/Al ratio of the zeolite product with decreasing alkalinity suggests that the Si/Al ratio of the associated nutrients also increases with decreasing alkalinity. This hypothesis is supported by a recent investigation of aluminosilicate solutions prepared using tetrapropylammonium hydroxide as the base [23]. ^{29}Si and ^{27}Al NMR spectra of these solutions shows that the connectivity of Al to Si increases with decreasing alkalinity. Similar conclusions may be drawn from [27]Al spectra of dissolved aluminosilicate species in solutions with potassium [39] and sodium [18] as the base.

In the synthesis gels, the nutrients to zeolite crystallization are provided by dissolution of the amorphous aluminosilicate solid. Initially, the consumption of nutrients by crystal growth is much slower than the rate of dissolution of the amorphous solid, as the surface area of the nuclei is small [40]. Under these conditions, the concentration of nutrients in solution should reflect the solubility of the amorphous aluminosilicate, since it is the primary solid phase (samples

F, H, and J). When all of the amorphous solid has been converted to zeolite (samples G, I, and K), further crystal growth consumes nutrients from solution, reducing the supersaturation. The solution eventually reaches saturation with respect to the zeolite ($c = c_s$), and growth stops [14,41,42]. The initial supersaturation in Al is 0.015 M for the gel with $x = 4.6$ ($[Al]_{sample\ H}$ - $[Al]_{sample\ I}$, see Table 2-3), which is about the same as that obtained after heating the seeded feedstock for 2 h ($[Al]_{sample\ F}$ - $[Al]_{sample\ G} = 0.017$ M, see Table 2-2) and much higher than the initial supersaturation for the gel with $x = 3.3$ ($[Al]_{sample\ J}$ - $[Al]_{sample\ K} = 0.005$ M, see Table 3).

Decreasing x (i.e., the gel alkalinity) from 4.6 to 3.3 decreases the supersaturation only by a factor of 3; however, the zeolite crystallization rate decreases by a factor of 10-20. This suggests that the decrease in alkalinity must also decrease the surface reaction rate constant, k_s. The surface reaction involves rearrangement of incoming nutrients to fit into the growing lattice structure [38]. There may be two reasons why the more siliceous nutrients take longer to rearrange at the lower alkalinity. The first is that rearrangement involves the breaking of Si-O-T bonds. This process is catalyzed by OH$^-$ ions, which are less abundant at the lower pH. The second reason that pH affects the value of k_s is that Si-O-Si bonds are cleaved at a rate that is orders of magnitude slower than that at which Si-O-Al bonds are cleaved [19,32]. As a consequence, the more siliceous nutrients are expected to take longer to react at the surface.

Our inability to crystallize NaY from a gel with low free base content ($x = 2.5$) agrees with the findings of Delprato et al. [43] who observed that NaY would not crystallize from gels with a SiO_2/Al_2O_3 ratio of 10 and Na_2O/Al_2O_3 ratios below 2.8. The problem in synthesizing NaY with Si/Al ratios above 3 may lie in the fact that at lower pH's, the silanol groups associated with dissolved nutrients are more fully protonated [44]. Under such circumstances, fewer Na$^+$ ions will be brought to the surface of the growing zeolite, and this, in turn, will reduce the availability of waters of hydration required to fill and stabilize the supercages. The net consequence would be that any zeolite formed is likely to be more dense than NaY. Delprato et al. [43,45] have recently overcome the difficulties in making high Si/Al NaY by adding crown ethers to complex with the Na$^+$ ions and serve as space fillers in the supercages, and using F$^-$ ions to accelerate the surface reaction. By this means, NaY crystals with Si/Al ratios up to the stoichiometric limit of the batch (5 in this case) were made by direct synthesis.

CONCLUSIONS

NaY nucleation from sodium silicate solutions is a consequence of the formation of hydrous aluminosilicate precursors during aging of a Na-rich seed gel. Uniform distribution of these precursors is necessary to avoid the formation of NaP zeolite.

Seed gels begin as clear aluminosilicate solutions after mixing the low-alkalinity sodium silicate solutions with the high-alkalinity sodium aluminate solutions. Oligomeric silicate anions depolymerize during aging to release low molecular weight silicate anions, which react with aluminate anions to produce aluminosilicate anions. The latter species then aggregate to form a gel with a solid phase that has a Si/Al ratio of 2 and in which the Si and Al atoms are intimately mixed. NMR and FTIR spectroscopies show that the solid phase contains NaY fragments with a Si/Al ratio of 2. These fragments contain a high concentration of Na^+ ions, which balance the negative charges associated with Al and the deprotonated silanol groups present at high pH. Heating the amorphous aluminosilicate present in the aged seed gel results in the formation of additional Si-O-Si and Si-O-Al bonds and the formation of crystalline NaY. The abundance of water in the amorphous material helps stabilize the open NaY structure.

The growth rate of NaY and the Si/Al ratio of the final zeolite are both affected by the alkalinity of the synthesis gel. A decrease in the alkalinity reduces the growth rate and increases the Si/Al ratio of the zeolite. The lower growth rate can be ascribed to a lower supersaturation of the liquid phase of the gel with aluminosilicate anions, produced by dissolution of an amorphous aluminosilicate solid phase in the gel, and to a lower rate of reaction of these anionic species at the surface of the growing zeolite. The increase in Si/Al ratio with decreasing gel alkalinity can be ascribed to the increase in connectivity of Al to Si in the aluminosilicate anions comprising the nutrients for zeolite growth.

ACKNOWLEDGMENTS

This work was supported by the Director, Office of Energy Research, Office of Basic Energy Sciences, Materials Sciences Division of the U.S. Department of Energy under contract DE-AC03-76SF00098.

References

1. Breck, D.W., US Patent 3,130,007, 1964.
2. Robson, H.E., US Patent 3,343,913, 1967.
3. Ciric, J. and Reid, L.J., Jr., US Patent 3,433,589, 1969.
4. McDaniel, C.V. and Duecker, H.C., US Patent 3,574,538, 1971.

5. Elliott, C.H., Jr. and McDaniel, C.V., US Patent 3,639,099, 1972.
6. Maher, P.K., Albers, E.W., and McDaniel, C.V., US Patent 3,671,191, 1972.
7. McDaniel, C.V., Maher, P.K., and Pilato, J.M., US Patent 3,808,326, 1974.
8. Vaughan, D.E.W., Edwards, G.C., and Barrett, M.G., US Patent 4,178,352, 1979.
9. Vaughan, D.E.W., Edwards, G.C., and Barrett, M.G., US Patent 4,340,573, 1982.
10. Breck, D.W., *Zeolite Molecular Sieves*, Wiley, New York, NY, 1974.
11. Milton, R.M., US Patent 3,008,803, 1961.
12. Robson, H., in *Zeolite Synthesis*, Occelli, M.L. and Robson, H.E., eds., ACS Symp. Ser. 398, Amer. Chem. Soc., Washington, DC, 1989, 436-447.
13. Barrer, R.M., in *Zeolite Synthesis*, Occelli, M.L. and Robson, H.E., eds., ACS Symp. Ser. 398, Amer. Chem. Soc., Washington, DC, 1989, 11-27.
14. Ginter, D.M., Bell, A.T., and Radke, C.J., submitted to *Zeolites*.
15. Li S., Li. L., and Xu R., Shiyou Xuebao, Shiyou Jiagong (Acta Petrolei Sinica, Petroleum Processing Section), 5 (1989), 30-36.
16. Kasahar S., Itabashi, K., and Igawa, K., in *Proceedings of the Seventh International Zeolite Conference*, Marakami, Y., Jijima, A. and Ward, J.W., eds., Elsevier, Amsterdam, 1986, 185-192.
17. Kacirek, H. and Lechert, H., *J. Phys. Chem.*, 80 (1976), 1291-1296.
18. Ginter, D.M., Went, G.T., Bell, A.T., and Radke, C.J., submitted to *Zeolites*.
19. Kinrade, S.D. and Swaddle, T.W., Inorg. Chem., 28 (1989), 1952- 1954.
20. Thangaraj, A. and Kumar, R., Zeolites, 10 (1990), 117-120.
21. Ginter, D.M., Bell, A.T., and Radke, C.J., J. Magn. Reson., 81 (1989), 217-219.
22. Englehardt, G., Zeigin, D., Jancke, H., Hoebbel, D., and Weiker, W., Z. *Anorg. Allg. Chem.* 418 (1975) 17-28.
23. Mortlock, R.F., Bell, A.T., and Radke, C.J., *J. Phys. Chem.*, 95 (1991), 372-378.
24. Klinowski, J., *Progress in NMR Spectroscopy*, 16 (1984), 237-309.
25. Englehardt, G., Hoebbel, D., Tarmak, M., Samoson, A., and Lippmaa, E., Z. *Anorg. Allg. Chem.*, 484 (1982), 22-32.
26. Flanigen, E.M., Khatami H., and Szymanski, H.A., in *Molecular Sieve Zeolites - I*, Adv. in Chem. Ser. 101, Gould, R.F., ed., Amer. Chem. Soc., Washington, DC, 1971, 200-229.
27. Fichtner-Schmittler, H., Lohse, U., Miessner, H., and Maneck, H.-E., Z. *Phys. Chem.* (Leipzig), 271 (1990), 69-79.
28. Engelhardt, G., *Trends in Anal. Chem.*, 8 (1989), 343-347.
29. Velde, B. and Couty, R., *Chem. Geol.*, 62 (1987), 35-41.
30. Gervais, F., Blin, A., Massiot, D., Coutures, J.P., Chopinet, M.H., and Naudin, F., J. *Non-Crys. Solids*, 89 (1987), 384-401.
31. Merzbacher, C.I. and White, W.B., *Am. Min.*, 73 (1988), 1089-1094.
32. McCormick, A.V., Bell, A.T., and Radke, C.J., *J. Phys. Chem.*, 93 (1989), 1741-1744.

33. Caullet, P., Guth, J.-L., Hurtez, G., and Wey, R., *Bull. Soc. Chim. Ser. I*, 1978, 253-257.
34. Gasteiger, H.A., and Frederick, W.J., Jr., AIChE Forest Products Division Sessions, Meeting Date 1988, TAPPI, Atlanta, 195-202.
35. Ting, D.C.-K., MS Thesis, Dept. of Chem. Eng., U. of Calif., Berkeley, CA, 1985.
36. Wijnen, P.W.J.G., Beelen, T.P.M., de Haan, J.W., Rummens, C.P.J., and van de Ven, L.J.M., *J. Non-Crys. Solids*, 109 (1989), 85-94.
37. Flanigen, E.M., in *Molecular Sieves*, Adv. in Chem. Ser. 121, Meier, W.M., and Uytterhoeven, J. B., eds., Amer. Chem. Soc.,Washington, DC, 1973, 119-139.
38. McCabe, W.L., Smith, J.C. and Harriott, P., *Unit Operations of Chemical Engineering*, McGraw-Hill, New York, NY, 1985, Ch. 28.
39. Harvey, G. and Dent Glasser, L.S., in *Zeolite Synthesis*, Occelli, M.L. and Robson, H.E., eds., ACS Symp. Ser. 398, Amer. Chem. Soc., Washington, DC, 1989, 49-65.
40. Ginter, D.M., Radke, C.J., and Bell, A.T., in *Zeolites: Facts, Figures, Future; Stud. in Surf. Sci. and Catalysis*, 49, Jacobs, P.A. and van Santen, R.A., eds., Elsevier, Amsterdam, 1989, 161-168.
41. Bodart, P., Nagy, J.B., Gabelica, Z., and Derouane, E.G., *J. Chim. Phys.*, 83 (1986), 777-790.
42. Ueda, S., Kageyama, N., and Koizumi, M., in *Proceedings of the Sixth International Zeolite Conference*, Olson, D. and Bisio, A., eds., Butterworth, London, 1984, 905-913.
43. DelPrato, F., Guth, J.-L., and Huve, L., *Zeolites*, 10 (1990), 546-552.
44. Iler, R.K., *The Chemistry of Silica*, Wiley, New York, 1979.
45. DelPrato, F., Guth, J.-L., Anglerot, D., and Zivkov, C., Eur. Pat. Appl. 364,352, 1990.

3

Synthesis of Zeolite Y in Gelatin Solution

Yoshiaki Goto, Hiroshi Saegusa[1], and Mitsue Koizumi
Department of Materials Chemistry, Ryukoku University, Otsu-shi 520-21, Japan
[1]Department of Chemistry, Gunma University, Kiryu-shi 376, Japan

An attempt to synthesize zeolite Y was made at 373 K in a gelatin solution with a concentration ranging from 0.1 to 10 wt%. After hydrothermal reaction for 0.5 to 10 h, zeolite Y was formed as a single phase. The crystallization rate decreased at the 7.5 to 10 wt% gelatin concentration. The crystals of zeolite Y were smaller in size and narrower in size distribution than those produced by the ordinary method. The Si/Al ratio in chemical composition of zeolite Y increased with the increment of gelatin concentration in synthesis. It appears that gelatin does not play a role as a template but supplies many crystallization sites and restrains diffusion of species.

INTRODUCTION

Many studies of zeolite synthesis in the system of Na_2O– Al_2O_3– SiO_2–H_2O have been carried out using oxides, hydroxides and gels as starting materials. In 1972, Argauer and Landolt synthesized ZSM–5 and ZSM–8 in the mixed base system, TPA–sodium and TEA–sodium. Up to now, novel zeolites have been successfully synthesized using various organic bases. It is considered that the organic base plays a role as a template in building the structure of zeolite. But it seems that no organic polymer has been used for the synthesis of zeolite Y.

The object of this study is to examine the possibility of zeolite formation in an $Na_2O-Al_2O_3-SiO_2-H_2O$ system with the addition of gelatin which is water–soluble polymer. It is suggested that the gelatin will be decomposed at 373 K in alkaline solution. But in actuality, synthesis of zeolite Y was possible in the system including gelatin. In this paper, the formation conditions and the characterization of synthesized zeolite Y will be reported.

EXPERIMENTAL

Synthesis of zeolite Y was attempted at 373 K in hydrothermal conditions by using a conical flask made of PMP(poly(4–methyl pentene–1))resin. Reagents used were sodium hydroxide (NaOH, Wako, 96%), sodium aluminate ($NaAlO_2$, Kanto Chemicals, Na_2O:33.00%, Al_2O_3:36.64%, H_2O:30.04%), colloidal silica (Ludox HS–40, Dupont), and gelatin (Gelatine weiß, Merck). The reaction mixture was prepared by adding slowly colloidal silica and gelatin to the solution of sodium hydroxide plus sodium aluminate. The mixture was stirred vigorously, and was put into a 300 ml flask, and then was digested for 1 h at room temperature. After heat treatment at 373 K for 0.5–10 h, crystallization occurred in the unagitated system. Solid product was immediately separated, and washed with cold distilled water until the pH of filtrate fell below 9.0. Hereafter, the ratio of gelatin to distilled water in amount is defined as the concentration of gelatin.

The solid product was identified by RIGAKU X–ray diffractometer, MODEL RAD–C (Cuka, graphite monochromator). The degree of crystallization of zeolite Y in the solid product was evaluated from the intensity of diffraction peaks; 111, 311, 331, 533 (Tatic and Drzaj 1989) with Shokubai Kasei zeolite Y (ZCP–50) as the reference. The unit cell parameter of zeolite Y was determined using the RAD–C computer program and the measurement by step scanning method.

Na_2O, Al_2O_3, and SiO_2 of the solid product were analyzed quantitatively by energy dispersion X–ray analysis (PHILLIPS, EDAX 9100) equipped with scanning electron microscopy

(JEOL, JSM–50A). Scanning electron micrographs were taken on Au–coated samples.

The particle size distribution of the solid product was measured by a centrifugal precipitation method using NIKKISO, Model Kl.

RESULTS AND DISCUSSION

The system of Na_2O–Al_2O_3–SiO_2–H_2O ($H_2O/Na_2O = 45$) and the same system to which gelatin was added were heated hydrothermally for 10 h at 373 K. The formation area of zeolite Y in the system including no gelatin is shown in Fig. 3–1 as a function of SiO_2 and Al_2O_3 molar fraction. This experiment is regarded as the ordinary method. The area where zeolite Y was found as a single phase is ranged from 2.45 to 4.15 mol% in SiO_2 and from 0.18 to 0.55 mol% in Al_2O_3. In the SiO_2 rich compositional range around the area of zeolite Y, no obvious reaction products were found, and the material was still amorphous. Herschelite coexisting with zeolite Y was formed from the SiO_2 poor composition.

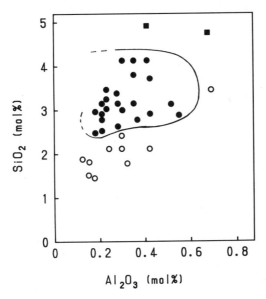

Fig. 3–1. Formation area of zeolites formed from batch compositions of Na_2O–Al_2O_3–SiO_2–H_2O(H_2O/Na_2O=45) system without gelatin after reaction for 10 h. (●)Zeolite Y, (■)Amorphous, (○)Zeolite Y + Herschelite.

The results of addition of 1% gelatin in the system are shown in Fig. 3–2. The addition resulted in the extension of the formation area of zeolite Y to low silica composition of 1.34 molar fraction. It is seen that hydroxysodalite and zeolite A forms with zeolite Y in the lower silica composition area. As shown in Fig. 3–3, the addition of a high amount of gelatin brought a more remarkable effect. Due to the addition of 10% gelatin, the formation area for SiO_2 moved toward the low silica composition area (0.54–3.15 mol%), and that for alumina was extended to the range of 0.08 mol%. This movement was due to the expending of NaOH during the partial decomposition of gelatin.

The effect of concentration change of gelatin on the crystallization of zeolite Y for 10 h at $8Na_2O \cdot Al_2O_3 \cdot 12SiO_2 \cdot 360H_2O$ batch composition is shown in Fig. 3–4. Although the degree of crystallization of zeolite Y is 86% for a standard sample (ZCP–50) in case when no gelatin is present, the degree of crystallization approaches 100% by the addition of 0.1% gelatin and the value was unchanged to 5%. The degree of crystallization decreases drastically when the gelatin content exceeds 5%. With a

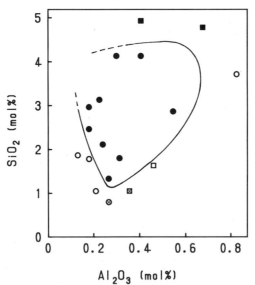

Fig. 3–2. Formation area of zeolites formed from batch compositions of Na_2O–Al_2O_3–SiO_2–H_2O($H_2O/Na_2O=45$) system including 1% gelatin for 10 h. (●)Zeolite Y, (■)Amorphous, (○)Zeolite Y + Herschelite, (□)Zeolite Y + Zeolite A, (⊠)Zeolite Y + Zeolite A + Hydroxysodalite, (⊗)Zeolite Y + Hershelite + Zeolite A.

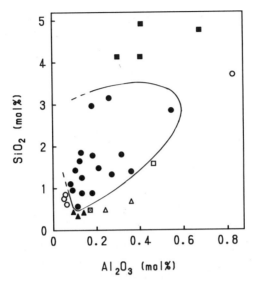

Fig. 3-3. Formation area of zeolites formed from batch compositions of $Na_2O–Al_2O_3–SiO_2–H_2O(H_2O/Na_2O=45)$ system including 10% gelatin for 10 h. (●)Zeolite Y, (■)Amorphous, (○)Zeolite Y + Herschelite, (□)Zeolite Y + Zeolite A, (⊠)Zeolite Y + Zeolite A + Hydroxysodalite, (▲)Zeolite Y + Hydroxysodalite, (△)Zeolite A + Hydroxysodalite.

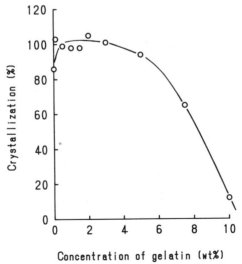

Fig. 3-4. Crystallization of zeolite Y from $8Na_2O·Al_2O_3·12SiO_2·360H_2O$ batch composition in various concentrations of gelatin for 10 h.

concentration of 10%, the crystallinity value is 12%. The effect of gelatin is that the degree of the crystallization of zeolite Y increases in low concentration of gelatin and decreases in high concentration.   The relationship between crystallization and reaction time is shown in Fig. 3–5. The behavior of crystallization in 1.0% and 7.5% gelatin is remarkably different. The tendency of crystallization in the case of 0.1% is similar to that of 0%, but the rate is greater. The rate of crystallization at 1% increases linearly and the degree of crystallization approaches 100%.   On the contrary, the rate at 7.5% is slow, and at 10% the induction period is elongated to 8 h.

The particle size distribution of zeolite Y sample is shown in Fig. 3–6. In the sample obtained from the batch composition of $8Na_2O \cdot Al_2O_3 \cdot 12SiO_2 \cdot 360H_2O$ which did not contain gelatin, the particle size distribution was widespread from 0.43 μm to 42.21 μm, and the particle size of cumulative 50% (median diameter) was 4.71 μm. In contrast, all the samples prepared from the batch compositions containing gelatin show narrower ranges of particle size distribution, as well as smaller median diameter of particle size. The samples in the case of gelatin contents of 0.1, 1.0, and

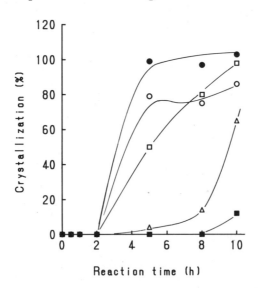

Fig. 3–5. Degree of crystallization of zeolite Y against reaction time in $8Na_2O \cdot Al_2O_3 \cdot 12SiO_2 \cdot 360H_2O$ batch composition. ($\bigcirc$)0% Gelatin, ($\bullet$)0.1%, ($\square$)1%, ($\triangle$)7.5%, ($\blacksquare$)10%.

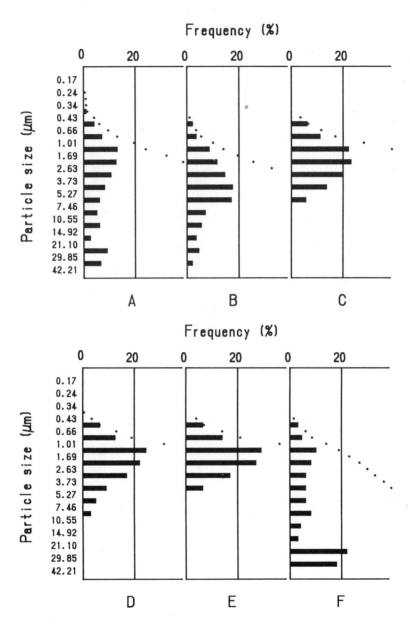

Fig. 3–6. Particle size distributions of samples formed from $8Na_2O \cdot Al_2O_3 \cdot 12SiO_2 \cdot 360H_2O$ batch composition in various concentrations of gelatin for 10 h. (A)ZCP–50, (B)0% Gelatin, (C)0.1%, (D)1%, (E)7.5%, (F)10%.

7.5% have median diameters of 2.18, 1.97, and 1.70 μm, respectively. This clearly shows that the samples of zeolite Y formed
using gelatin have a higher degree of uniformity in particle size
than that produced without using gelatin. The particle size of the
sample from 10% gelatin is 9.43 μm, and the distribution has two
peaks. This increment is due to the lower dispersion of sample
which contains a large amount of amorphous material. In this
way, the effect of addition of gelatin on the uniformity of particle
size and the size distribution was observed.

The scanning electron micrographs of zeolite Y prepared from
$8Na_2O \cdot Al_2O_3 \cdot 12SiO_2 \cdot 360H_2O$ batch composition for 10 h are
shown in Fig. 3–7. It is observed that zeolite Y synthesized
without gelatin shows octahedral morphology having sizes of 1–
3 um. The crystals from the batch composition including gelatin
exhibit much smaller particle size with a higher degree of uniformity, and the morphology is rounded polygonal configurations which are almost spherical. In 10% gelatin, amorphous
material and zeolite Y are observed.

The unit cell parameters of zeolite Y are shown in Table 3–1.
The unit cell parameters are within the range of those of zeolite Y
given by Breck 1974. It is noted that the unit cell parameter
decreases with the increment of gelatin concentration.

Table 3–1. Unit cell parameters of zeolite Y formed for 10 h.

Conc. of gelatin (%)	a axis (A)
0	24.834(2)
1	24.795(1)
5	24.773(2)
7.5	24.760(1)

Batch composition: $8Na_2O \cdot Al_2O_3 \cdot 12SiO_2 \cdot 360H_2O$

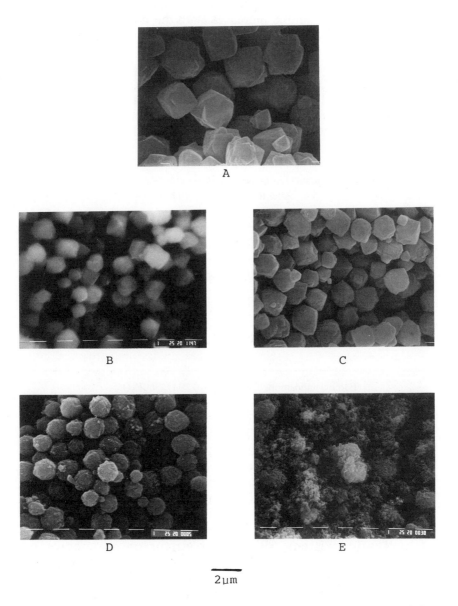

Fig. 3–7. Scanning electron micrographs of zeolite Y formed from $8Na_2O \cdot Al_2O_3 \cdot 12SiO_2 \cdot 360H_2O$ batch composition. (A)0% Gelatin, (B)0.1%, (C)1%, (D)7.5%, (E)10%.

The results of chemical analysis are shown in Table 3–2. The formulas are similar to that of zeolite Y reported by Breck 1974. The Si/Al ratio of sample increases with the increment of gelatin concentration. The amount of Na is insufficient for that of Al. It is considered that this deficiency is satisfied by H which was induced from the distilled water used for washing product or from the decomposition of gelatin. The results of analysis of C in samples indicated no existence of C. The linear relation between unit cell parameter and Si/Al ratio was confirmed to be same as the results by Dempsey, Kurl, and Olson 1969; Breck 1974.

Table 3–2. Chemical formulas and Si/Al ratios of zeolite Y formed after the reaction for 10 h.

Gelatin (%)	Chemical formula	Si/Al
0	$Na_{52.8}Al_{65.8}Si_{126}O_{384}$ $226H_2O$	1.92
1	$Na_{50.6}Al_{65.3}Si_{128}O_{384}$ $230H_2O$	1.94
5	$Na_{47.7}Al_{61.7}Si_{130}O_{384}$ $222H_2O$	2.11

Batch composition: $8Na_2O \cdot Al_2O_3 \cdot 12SiO_2 \cdot 360H_2O$

From the above results, it is found that the presence of gelatin has influence on the formation of zeolite. This is considered as mentioned below. Gelatin decomposes to a molecule of amino acid of low polymerization in alkaline solution and by heating at 373 K. This molecule has a large number of end groups such as $-NH_2$ and $-COOH$. In the alkaline solution, the end group COOH fixes Na^+ ion by exchange with H^+, and makes many sites of nucleation (Breck 1974). The diffusion of species of $SiO_2(OH)_2^{2-}$ and, especially, $Al(OH)_4^-$ (Wengin, Ueda, and Koizumi 1986; Dutta and Shieh 1986) are restrained in gelatin solution. As a consequence of the diffusion, the rate of crystal growth of zeolite decreased.

CONCLUSION

The presence of gelatin was made possible the synthesis of zeo–
lite Y having smaller size and narrower size distribution than that
produced by the ordinary method. It is expected that the addition
of other water–soluble polymers will give similar results.

REFERENCES

Argauer, R.J., and G.R.Landolt. 1972. U.S. Pat.3 ,702,886.

Breck, Donald. 1974. *Zeolite Molecular Sieves.* New York : John
Wiley & Sons.

Dempsey, E., G.H.Kurl, and D.H.Olson. 1969. *Variation of the
Lattice Parameter with Aluminum Content in Synthetic   Sodium
Faujasites. Evidence for Ordering of the Framework Ions.* J. Phys.
Chem. 73 (12): 387–90.

Dutta, P.K., and D.C.Shieh. 1986. *Crystallization of Zeolite A: A Spec–
troscopic Study.* J. Phys. Chem. 90 (11):2331–34.

Tatic, M., and B.Drzaj. 1985. *A Contribution to the Synthesis of the
Low–Silica X Zeolite.* In Zeolites–Synthesis, Structure, Technology,
and Application. Amsterdam: Elsevier Science Publishers.

Wengin, R., S.Ueda, and M.Koizumi. 1986. *The Synthesis of Zeolite
NaA from Homogeneous Solutions and Studies of its Properties.* In
New Developments in Zeolite Science and Technology. Tokyo:
Kodansha, Elsevier.

4

Synthesis and Characterization of Breck Structure Six (BSS)

Gary W. Skeels
UOP, Tarrytown Technical Center, Tarrytown, NY 10591.
Kathleen B. Reuter and Nancy K. McGuire
Union Carbide, Tarrytown Technical Center,
Tarrytown, NY 10591

Synthesis of pure Breck Structure Six (BSS) was recently reported by Delprato et al. Their basic synthesis method has been followed in the preparation of BSS. Basic characteri ation data are reported for this material including X-ray powder diffraction and unit cell, thermal analysis, hydroxyl region and framework region infrared spectra, and wet chemical analysis. In addition, electron diffraction and high-resolution electron microscopy were used to confirm whether the Delprato synthesis procedure produces pure hexagonal packing of sodalite cages (BSS) or whether some cubic packing (FAU) is retained. The relationship between this material and other members of the faujasite family is discussed.

INTRODUCTION

In the structure of the natural mineral faujasite, the polyhedral cages, or the sodalite or beta cages, are linked in a cubic array through double-six rings. The analogues of faujasite synthesized in a totally inorganic system are zeolites X or Y. Breck[1] proposed six theoretical structures based on various linkages of the sodalite cages. Two of the proposed structures involve linkage through the

42

double-six ring or hexagonal prism. The first structure was faujasite (FAU). The second proposed structure involved linkage of the sodalite cages through the double-six ring in a hexagonal array rather than cubic. This material has been known as Breck Structure Six since it was first proposed. No known natural mineral has this structure, and until recently, there were no synthetic analogues. Breck Structure Six (we have designated the material synthesized by the Delprato method as BSS) was once suggested as the framework structure of zeolites L[2] and ZSM-3[3], and was possibly related to zeolite beta. However, zeolite L consists of erionite cages rather than sodalite cages linked through double-six rings.[1] The suggestion has been made that ZSM-3 consists of random intergrowths of cubic- and hexagonal- packed sodalite cages linked through the double-six ring.[4] Zeolite beta is a disordered structure unrelated to faujasite (FAU) or BSS.[5]

Recently, two separate laboratories have claimed success in synthesizing a material called BSS. Delprato et al.,[6] prepared "hexagonal faujasite" using 18-Crown-6 as the structure directing agent (SDA), and Vaughan[4] prepared ECR-30 using triethyl-methyl ammonium hydroxide as the SDA. Both materials are represented as more or less pure BSS. The Delprato et al. material was compared to FAU synthesized with 15-Crown-5 as the SDA. They concluded that pure BSS was synthesized with the 18-Crown-6 SDA, but the slightly smaller 15-Crown-5 SDA produced only the cubic FAU.[7] The ECR-30 product was claimed to be relatively pure BSS.[4] However, extensive spectroscopic examination by Newsam has shown that ECR-30 is an intergrowth having approximately 85% hexagonal (BSS) character and about 15% cubic (FAU) character.[8]

The term hexagonal faujasite[7] has been used to describe BSS. Because faujasite has a totally cubic structure, the use of the term hexagonal to describe faujasite is incorrect. Vaughan[4] has labeled ECR-30 as BSS. Therefore, we have continued the notation BSS to describe the Delprato synthesis product. For purposes of simplifying the description of the entire family of materials containing sodalite cages connected through the double-six rings, the term <u>faujasite family</u> has been used in this report. The term should not be taken to describe a specific structural arrangement.

Other materials included in the faujasite family are ECR-4, ECR-32, CSZ-1, and ZSM-20. ECR-4 is FAU synthesized with bis-(2-hydroxyethyl)dimethylammonium cation as the SDA and having a SiO_2/Al_2O_3 ratio of greater than 6.[9] ECR-32 is FAU synthesized with tetrapropyl or tetra hydroxypropyl organic ammonium salt as the SDA and having a SiO_2/Al_2O_3 ratio of greater than 6.[10]   CSZ-1 is a distorted variant of the FAU framework. It is a cesium-containing, high-silica polymorph of FAU that crystallizes in hexagonal plates.[11] ZSM-20, synthesized by Ciric,[12] was conceived as a pure form of FAU. However, Newsam has recently shown that the structure is an intergrowth of BSS and FAU.[13] Factors affecting the synthesis of ZSM-20 have also been described recently by Dewaele et al.[14]

In this report, the synthesis of BSS with a product SiO_2/Al_2O_3 ratio of about 7.2 is described. The products have been characterized according to standardized procedures and are compared to LZ-210 with a SiO_2/Al_2O_3 of 8. LZ-210(8) is zeolite Y (FAU) with its framework SiO_2/Al_2O_3 ratio increased by chemical treatment with $(NH_4)_2SiF_6$ solution.[15]

EXPERIMENTAL

Synthesis of BSS

The BSS was synthesized according to the suggested procedure of Delprato et al.[6] The approximate gel ratio was as follows:

$$10 \; SiO_2 : Al_2O_3 : 2.4 \; Na_2O : 0.70 \; \text{18-Crown-6} : 140 \; H_2O$$

The 18-Crown-6 was reagent grade and was separately analyzed to determine the purity, particularly the absence of 15-Crown-5. The analysis showed the matrial to be pure 18-Crown-6. The reagents were mixed and aged at ambient temperature for 24 hours and then placed into an oven at 110^0C for 6 days. The crystallized solids were separated and washed according to standard procedures and characterized. Standard X-ray powder diffraction patterns, thermal analysis, hydroxyl region and framework region infrared spectra, and wet chamical analysis were obtained on all samples.

Electron diffraction was conducted on the BSS product. The samples were embedded using LRWhite acrylic and ultramicrotomed to obtain sections approximately 60 nm thick. The sections were supported on a standard 3 mm Cu grid and coated on one side with carbon to minimize charging. The ultramicrotomed samples were then examined on a JEOL 2000FX analytical transmission electron microscope operating at 200 kV.

To determine the lattice spacings from the diffraction pattern, a standard must be used. One of two methods was used: diffraction was conducted on a gold standard immediately after the diffraction information was obtained from the sample or the sample itself was partially coated with approximately 20 nm of gold.

Structure models for several molecular sieves whose structures are closely related to that of FAU have been constructed. The powder diffraction patterns generated by these models are being compared to the observed diffraction patterns from BSS, ZSM-3, and ZSM-20. Several new and interesting structural features that were not apparent from either X-ray diffraction or electron microscopy have been observed by model building.

Atomic coordinates for BSS have been established and a diffraction pattern has been simulated that compares well with ECR-30 and BSS. Using BSS and FAU as starting models, the intergrowths between these two structures have been modeled in an attempt to characterize the structures of ZSM-3 and ZSM-20.

RESULTS & DISCUSSION

Synthesis and Characterization of BSS

Characterization data for the as-synthesized BSS, the calcined and ammonium-exchanged BSS, and the ammonium-exchanged LZ-210 are shown in Table 4-1. A SiO_2/Al_2O_3 ratio of 7.2 is comparable

Table 4-1

COMPOSITION AND PROPERTIES OF BSS AND FAU, LZ-210

	BSS	NH$_4$-BSS	NH$_4$-LZ-210
Chemical Analyses:			
Na$^+$/Al	1.00	0.10	0.03
M$^+$/Al	1.00	0.92	0.95
SiO$_2$/Al$_2$O$_3$	7.2	7.2	8.0
Framework Infrared:			
Asymmetric Stretch, cm^{-1}	1027	1032	1038
Symmetric Stretch, cm^{-1}	1005	1015	1032
DTA Crystal Collapse Temp.,°C			
McBain-Bakr Adsorption, wt.%			
O$_2$, 100 torr, 80K	- - -	36.2	34.5
H$_2$O, 4.6 torr, 25°C	- - -	33.8	32.0

to the results reported by Delprato et al.[6] The Na^+/Al ratio of the as-synthesized BSS product is near unity, indicating that 18-Crown-6 does not balance any framework negative charges. The calcined and ammonium-exchanged BSS has a cation equivalent M^+/Al near unity and the SiO_2/Al_2O_3 ratio is essentially unchanged, indicating that most if not all of the aluminum remains in the BSS framework following the treatments.

X-ray powder patterns for the as-synthesized, calcined, and calcined-ammonium-exchanged BSS are compared in Figure 4-1. They show a well-crystallized zeolite. Relative intensities of the peaks are similar to the published data of Vaughan[4] and Delprato et al.[6] McBain-Bakr adsorption measurements confirm the X-ray crystallinity measurements. BSS adsorbed 36.2 wt% O_2 at 100 torr and 80 K. The value for LZ-210 (FAU) was 35.5 wt% O_2. Water adsorption values obtained at 4.6 torr and 25^0C values are also comparable: 33.8 wt% and 35.5 wt% respectively.

Hydroxyl region infrared spectra for the calcined-ammonium-exchanged form of BSS and the ammonium-exchanged LZ-210 (FAU) are compared in Figure 4-2. The calcined-ammonium-exchanged BSS shows the same two hydroxyl bands, at 3635 cm^{-1} and at 3550 cm^{-1}, found in FAU. The framework region infrared spectra are compared in Figure 4-3. The asymmetric stretch band at about 1032 cm^{-1} and the symmetric stretch band at about 803 cm^{-1} for BSS are in the same positions as the respective bands of FAU with a similar SiO_2/Al_2O_3 ratio.

Samples of NH_4-BSS were calcined in vacuum in 50^0C intervals from 500^0C to 650^0C, and the infrared spectrum of the hydroxyl region was obtained. The combined spectra are shown in Figure 4-4. Maximum hydroxyl strength (absolute absorbance) appeared to occur already at 500^0C. As the activation temperature was increased to 550^0C, a small loss occurred in hydroxyl strength. After 600^0C activation, the bands are small; and after 650^0C activation, the bands have nearly disappeared. The hydroxyl bands of LZ-210(8) behave similarly. The temperature at which the hydroxyl bands disappear agree quite well with the temperature at which a small endotherm is found in the DTA, and the occurrence

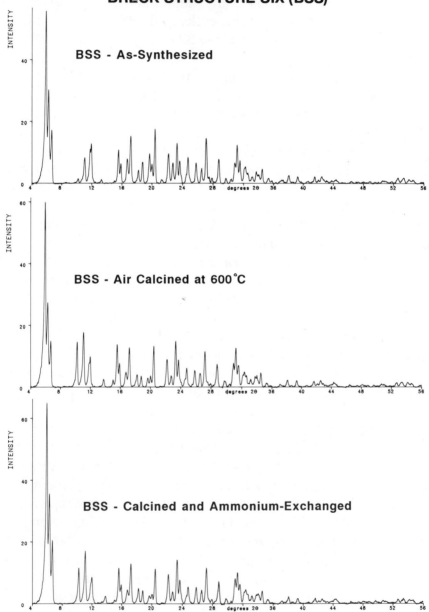

Figure 4-1
X-RAY POWDER PATTERN FOR BRECK STRUCTURE SIX (BSS)

Figure 4-2
COMPARISON OF THE
HYDROXYL REGION INFRARED SPECTRA
NH-BSS COMPARED WITH NH₄-LZ-210

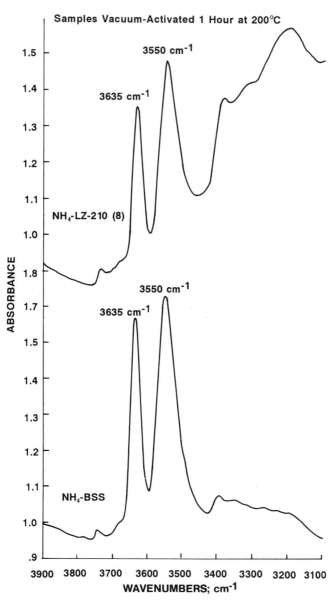

Samples Vacuum-Activated 1 Hour at 200°C

Figure 4-3
COMPARISON OF THE FRAMEWORK
REGION INFRARED SPECTRA
NH₄-BSS COMPARED WITH NH₄-LZ-210

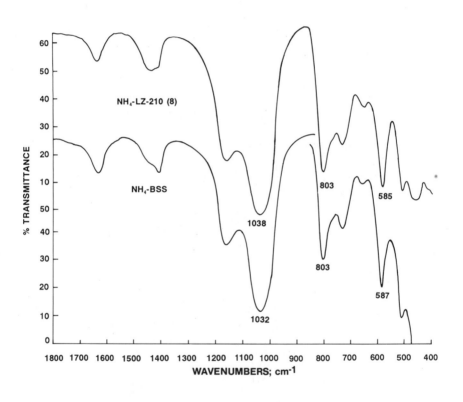

Figure 4-4
CHANGES IN THE HYDROXYL BANDS OF NH₄-BSS WITH ACTIVATION TEMPERATURE

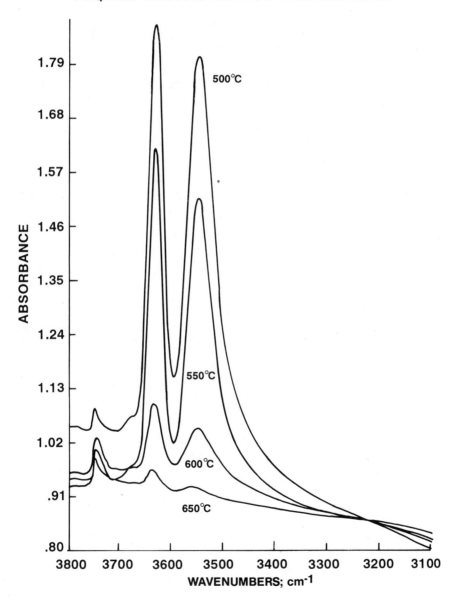

of a weight loss in the TGA, which are attributable to dehydroxylation.

Electron Diffraction

Typical electron diffraction results from BSS are shown in Figures 4-5A and 4-5B. Most of the patterns were indexed as the hexagonal phase with a = 17.3 and c = 28.3. However, the cubic phase was also observed. Evidence for the cubic phase in BSS is shown in Figure 4-5B. Electron diffraction patterns for both the cubic and hexagonal phase were obtained on crystals having hexagonal morphology. Of the 20 crystals examined using electron diffraction, 2 revealed evidence of the cubic phase.

High-resolution electron micrographs were also obtained from BSS. An example is shown in Figure 4-5C. The spacings on this micrograph correspond to the $\{\bar{1}011\}$ hexagonal spacings of the BSS lattice. Planar defects in this structure shown in Figure 4-5D, are most likely twins or stacking faults.

Calculated Powder Patterns

Faujasite has a structure that is composed of a cubic array of truncated cuboctahedral sodalite cages. Each sodalite cage is connected with four other sodalite cages at tetrahedral angles to one another. The connections are hexagonal prisms, which will be referred to hereafter as double-six rings. Two adjacent sodalite cages are related by a center of inversion at the center of the double-six ring connecting them. The sodalite cages are stacked along the cubic [111] direction in an abcabc sequence. This arrangement creates supercages that are connected with one another but no continuous linear channels. The supercages in FAU have four circular apertures. The diffraction pattern for FAU was calculated using the structure published by Olson and Dempsey.[16] This pattern is shown in Figure 4-6A.

The overall topology of the BSS structure has been known for many years.[1] This material is composed of sodalite cages that are

Figure 4-5
ELECTRON DIFFRACTION PATTERNS AND
HIGH RESOLUTION TEM IMAGES FROM BSS

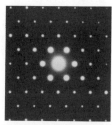

5a. Electron Diffraction Pattern from
BSS Hexagonal Phase - [0001]

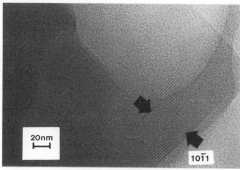

20nm

10$\bar{1}$1

5c. High Resolution TEM Image from BSS
Lattice Lines Correspond to [10$\bar{1}$1]

5b. Electron Diffraction Pattern from
BSS Cubic Phase - [$\bar{1}$14]

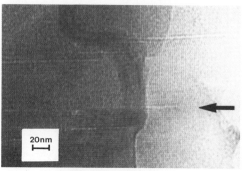

20nm

5d. High Resolution TEM Image from BSS
Planar Defects are Indicated

Figure 4-6
A,B,C, DIFFRACTION PATTERNS FOR FAUJASITE AND
STRUCTURALLY RELATED MATERIALS
A-Faujasite (FAU), Calculated from Structural Data in Reference 16.
B-Breck Structure Six (BSS), Calculated Using Distance Least Squares.
C-abcb Stacking of FAU/BSS,Generated by J. Newsam, Reference 17.

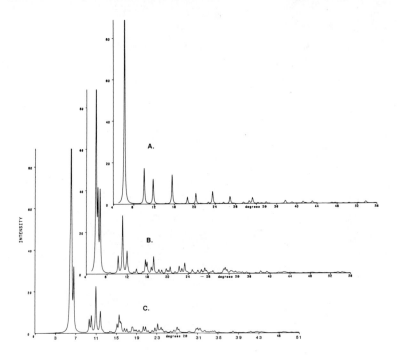

identical to that of FAU, but the two structures differ in their stacking sequence. The sodalite cages in BSS are stacked in an abab sequence, which gives rise to an overall hexagonal symmetry. Adjacent sodalite cages are related by a mirror plane bisecting the double-six ring between them. BSS has two types of supercage. One type has two circular apertures and three elliptical apertures, and the other type has three elliptical apertures. The circular apertures are aligned to form a continuous channel in one direction. At the time the present study was begun, the individual atomic coordinates for this structure had not been published. Therefore, in this study a plastic model was used in conjunction with the Insight II molecular modelling program (Biosym Technologies, San Diego CA) and the Distance Least Squares technique (Inst. fur Kristall. und Petrog., Eidg. Techn. Hochsch., Zurich, Switzerland) to determine the atomic coordinates for BSS. The program POWD10 (Dept. of Geosciences, Pennsylvania State U.) was used to create a simulated diffraction pattern, which is shown in Figure 4-6B.

One intergrowth model exhibits a system of intersecting channels that are continuous throughout the structure. This type of channel system is not present in FAU or BSS. Also becoming apparent is the fact that topological considerations constrain the number of intergrowth structures that are possible. This conclusion agrees well with electron microscopy studies showing that the FAU/BSS ratios in ZSM-3 and ZSM-20 fall within a limited range.

Electron microscopy studies have shown ZSM-20 to be an intergrowth of FAU and BSS.[13] Electron micrographs of ZSM-20 show coherent stacking of FAU and BSS blocks of varying thicknesses, but the overall BSS/FAU ratio is approximately 2.5:1. Diffraction patterns calculated by a recursion method[8,13] for FAU and BSS show that the range of BSS/FAU ratios that approximate the observed ZSM-20 diffraction pattern is fairly narrow. The abcb stackings of FAU/BSS suggested by Newsam generate the computed powder pattern for ZSM-20 shown in Figure 4-6C.[17]

In this study a plastic model has been constructed of a 1:1 coherent intergrowth of FAU and BSS. The model consists of sodalite cages linked in such a way that the six-rings composed of FAU and BSS sodalite cages share edges. A schematic of this model is shown in Figure 4-7. The FAU [111] direction is aligned with the BSS [001] direction. This model serves as a starting model from which other stacking sequences can be derived. A "shorthand" representation of the structure, in which a single tetrahedron represents the center of one sodalite cage, was used. This structure reduces the number of atoms in the model by a factor of 72. The resulting unit cell is 17.556 x 30.029 x 56.627 $Å^3$ and is in the orthorhombic space group $Cmc2_1$. The structure has continuous channels along the [110], [-1 -1 0], and [010] directions, a feature that is not observed in FAU or BSS.

A diffraction pattern has been calculated from this shorthand model. The pattern was obtained by reducing the dimensions of the unit cell equally in all directions until the distance between tetrahedra was approximately equal to the silicon-silicon distance observed in many zeolites. Dummy oxygen atoms were inserted to establish connectivity, and Distance Least Squares calculations optimized the atom positions and the unit cell dimensions. The refined unit cell was reexpanded until the distance between tetrahedra approximated the distance between sodalite cages in FAU-type materials. The dummy oxygen atoms were included in the diffraction pattern calculation so that even though not all the atoms in the structure were represented, the scattering factors represented a 1:2 mixture of silicon and oxygen. The preliminary calculated powder pattern generated from the shorthand model was qualitatively the same as the pattern generated by Newsam et. al.[8] However, not all of the atoms in the structure have been included in the shorthand model. The final calculated powder pattern for this model will have to await inclusion of all atoms in the structure.

The shorthand model was used as a guide for building the complete structure model using Insight II. Sodalite cages were brought into the correct orientation and "docked" using the real-time structure manipulation features of this program. Cartesian coordinates for the asymmetric unit have been obtained,

Figure 4-7
SCHEMATIC OF THE INTERGROWTH MODEL
**Each Vertex Represents the Center of a
Sodalite Cage. The Six-membered Rings Are
Labeled "B" or "C" Depending on Whether
They Are in a Boat Configuration (Eliptical Apertures)
or the Chair Configuration (Circular Apertures)**

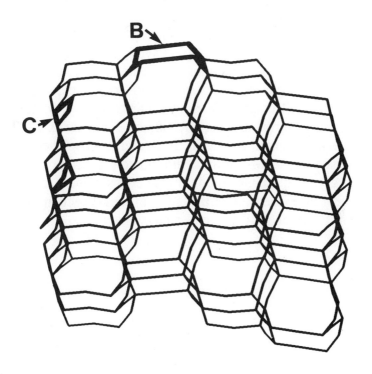

and a space group determination is in progress. Obtaining a repeat distance for ZSM-3 along the c axis has been difficult and the stacking is believed to be random in this direction. Development of the final structure model will have to await completion of the calculated powder pattern for the shorthand model. However, the preliminary results are encouraging and indicate that a number of BSS-FAU intergrowths are possible. ZSM-20 is a regular, ordered intergrowth while ZSM-3 may be a disordered 1:1 intergrowth of BSS and FAU.

CONCLUSIONS

The data would suggest that pure BSS has yet to be prepared. It was determined that ECR-30 is an intergrowth containing 85% BSS and 15% FAU character.[17] The present synthesis of BSS according to the other known method[4] contains a measurable amount of FAU, as demonstrated by electron diffraction and from calculated X-ray powder patterns. Structure modeling studies, coupled with X-ray diffraction and electron diffraction,[8] indicate that ZSM-20 is an approximately 2:1 intergrowth of BSS and FAU. These same studies indicate that ZSM-3 is an approximately 1:1 intergrowth of BSS and FAU. Preliminary modeling studies summaruzed in this report support these ideas. Confirmation of these speculations will have to await further synthesis and analysis work.

REFERENCES

1.   D.W. Breck, "Zeolite Molecular Sieves," Wiley, New York (1974), pp. 55-58.

2.   D.W. Breck and E.M. Flanigen, "Molecular Sieves," Society of the Chemical Industry, London (1968), p. 47.

3.   G.T. Kokotailo and J. Ciric, Adv. Chem. Ser. No. 102 (1971), p. 823.

4.   D.E.W. Vaughan, E.P. 0,315,461 (1988).

5.   M.M.J. Treacy and J.M. Newsam, Nature (London), 332, 249 (1988); J.M. Newsam, J.M.M. Treacy, W.T. Koetsier and C.B. DeGruyter, Proc. R. Soc. Ser. A. 420, 375 (1988).

6.   F. Delprato, J.L. Guth, L. Huve, J. Baron, L. Delmotte, M. Soulard, D. Anglerot, and C. Zivkov, "Zeolites for the Nineties," Recent Research Reports, presented at 8th International Zeolite Conference, Amsterdam (1989).

7.   F. Delprato, L. Delmotte, J.L. Guth, and L. Huve, Zeolites, 10, 546 (1990).

8.   M.J. Treacy, J.M. Newsam, and M.W. Deem, Proc. R. Soc., Ser. A 433, 499-520 (1991).

9.   D.E.W. Vaughan, U.S. 4,714,601 (1987).

10.   K.G. Strohmaier and D.E.W. Vaughan, E.P. 0,320,114 (1989).

11.   M.G. Barrett and D.E.W. Vaughan, U.S. 4,309,313 (1982).

12.   J. Ciric, U.S. 3,972,983 (1976).

13.   J.M. Newsam, M.M.J. Treacy, D.E.W. Vaughan, K.G. Strohmaier, and W.J. Mortier, J. Chem. Soc., Chem. Commun., 493 (1989).

14.   N. Dewaele, L. Maistriau, J.B. Nagy, Z. Gabelica, and E.G. Derouane, Applied Catalysis, 37, 273-290 (1988).

15.   G.W. Skeels and D.W. Breck, Proceedings of the Sixth International Zeolite Conference, Edited by D. Olson and A. Bisio, Butterworths Ltd., UK (1984), pp. 87-96.

16.   D. H. Olson and E. Dempsey, J. Catalysis, 13, 221-231 (1969).

17.   J. Newsam, Private Communication.

5

Synthesis and Characterization of VPI-6, Another Intergrowth of Hexagonal and Cubic Faujasite

Mark E. Davis

Department of Chemical Engineering, Virginia Polytechnic Institute and State University, Blacksburg, VA 24061
Current address: Department of Chemical Engineering, California Institute of Technology, Pasadena, CA 91125

We report here the synthesis procedures used to crystallize the zeolite VPI-6. VPI-6 is synthesized in the absence of organic materials at low temperatures (70-100°C) with short heating times (0.5-3 hours). This new zeolite is another intergrowth of cubic and hexagonal faujasite. The X-ray powder diffraction pattern from VPI-6 can be indexed on a hexagonal unit cell of dimensions a = 17.655(2) Å and c =.27.483(2) Å. Solid-state NMR and physical adsorption data from VPI-6 (SiO_2/Al_2O_3 = 2.4) are reported and compared to faujasite.

Cubic faujasite (FAU) has been known for quite some time (1). Recently, the hexagonal polytype of faujasite was synthesized by Delprato et al. (2). We have synthesized the hexagonal polytype of faujasite according to the procedures of Delprato et al. and have provided further characterization data (3). Also, we have dealuminated this material to SiO_2/Al_2O_3 = 53 by treatments with vapor phase silicon tetrachloride (4). The existence of the end members cubic and hexagonal faujasite allow for better understanding of existing intergrowth materials and lead us to speculate whether zeolites can be synthesized at any ratio of the two polytypes. By changing the ratio of the organic structure directing agents (15-crown-5 for FAU, 18-crown-6 for hexagonal polytype), we have recently synthesized intergrowth structures with varying contents of the cubic and hexagonal polytypes (5).

ZSM-20 has been shown to be comprised of intergrown cubic and hexagonal stackings of faujasite sheets (6). ZSM-3 (7), ZSM-2 (8) and ZSM-10 (9) may also be intergrowths of cubic and hexagonal faujasite. CSZ-1 is not an intergrowth structure but rather a rhombohedrally distorted faujasite (10). Also, ECR-30 (11) is claimed to be the hexagonal polytype of faujasite but it appears more closely related to ZSM-20.

ZSM-20, ZSM-10, ECR-30 and the true hexagonal polytype of faujasite require organic materials for their synthesis. ZSM-2, ZSM-3 and CSZ-1 use alkali other than sodium (Li, Li, Cs; respectively) for their preparation.

We have been attempting to synthesize either the hexagonal polytype of faujasite or intergrowth structures of the cubic and hexagonal faujasite in the absence of organic materials and cations other than sodium. Here we report on the synthesis and properties of an intergrowth phase synthesized from soda, alumina and silica.

Experimental Section

X-ray powder diffraction data were collected on a Scintag Θ - Θ diffractometer using CuKα radiation. Microscopy was performed on a JEOL 200CX microscope. Chemical analyses for silicon, aluminum and sodium were performed by Galbraith Laboratories, Knoxville, TN, USA. Argon adsorption isotherms were measured at liquid argon temperature (87K) on an Omnisorp 100 analyzer (Po: vapor pressure of argon at 87K).

The solid-state NMR measurements were carried out on a Bruker MSL 300 spectrometer. The ^{29}Si MAS NMR spectra were recorded at 59.6 MHz with a spinning speed of 3 kHz. ^{27}Al MAS NMR spectra were obtained at 78.2 MHz with a spinning speed of 7 kHz.

Results and Discussion

Synthesis of VPI-6. A representative synthesis procedure for crystallization of VPI-6 is given below. To 30 g of H_2O add 5 g of sodium aluminate (VWR Scientific) and 2.75 g NaOH and stir until dissolved. Next add 12.2 g sodium silicate (Fisher: 2 $Na_2O \cdot SiO_2 \cdot 5 H_2O$) and age for 24 hours with stirring at ambient conditions. This reaction mixture is then heated statically at 80°C. The product is recovered by filtration, washed with copious amounts of water and air dried.

Table 5-1 illustrates the effects of synthesis conditions on the final product crystallized. Aging time is important. Without aging or at short aging times, LTA is formed (see experiments A-C). If the aging time is extended past approximately 1 day, the purity of the final product declines (experiment D). Also, notice that there is a fairly narrow range of Na_2O/Al_2O_3, SiO_2/Al_2O_3 and

H_2O/Al_2O_3 which can be used to crystallize VPI-6 (experiments E-K) and that the heating times are very short. At 90°C, VPI-6 can be formed within 0.5 hours (experiment M).   Excessive heating times cause the loss of VPI-6 and thecrystallization of SOD. This feature makes the synthesis of VPI-6 autoclave size, shape, configuration etc. dependent. Thus, the optimum heating time will vary from laboratory to laboratory. Although the rapid crystallization of VPI-6 may cause problems in scale-up when using batch-type autoclaves, it does open the possibility of exploring continuous crystallization via flow processing.

Table 5-1. Variations in Synthesis Conditions

Exp.	$Na_2O/$ Al_2O_3	$SiO_2/$ Al_2O_3	$H_2O/$ Al_2O_3	aging time, hr	T, °C	heating time, hr	Result
A	5.0	2.5	90-95	24	80°	2	VPI-6
B	5.0	2.5	90-95	0	80	2	LTA
C	5.0	2.5	90-95	3	80	2	LTA
D	5.0	2.5	90-95	48	80	2	VPI-6 (low purity)
E	4.5	2.5	90-95	24	80	2	VPI-6 (med. purity)
F	6.0	2.5	90-95	24	80	2	VPI-6 (low purity)
G	9.0	2.5	90-95	24	80	2	SOD
H	5.0	2.0	90-95	24	80	2	LTA
I	5.0	4.0	90-95	24	80	2	VPI-6 (low purity)
J	5.0	2.5	60	24	80	2	SOD
K	5.0	2.5	120	24	80	2	LTA
L	5.0	2.5	90-95	24	70	3	VPI-6
M	5.0	2.5	90-95	24	90	0.5	VPI-6

We have found that sodium aluminate and sodium silicate are the preferred starting materials. The use of aluminum sulfate tends to yield FAU while Catapal B gives SOD. Colloidal silicas such as Ludox HS-40 have successfully been used but in general give a large amount of LTA in the VPI-6 product. The use of fumed silica tends to yield SOD.

The SiO_2/Al_2O_3 of the product VPI-6 is 2-3 (vide infra). In attempts to increase the silica content, further exploratory syntheses were conducted at higher temperatures. Table 5-2 gives the synthesis conditions and results for experiments involving temperatures of 90-110°C. At 100°C, VPI-6 can be crystallized from a reaction mixture with $SiO_2/Al_2O_3 = 4$. However, the product VPI-6 still has a $SiO_2/Al_2O_3=2.8$. Upon increasing further both Na_2O/Al_2O_3 and SiO_2/Al_2O_3, the product gives an X-ray powder diffraction pattern similar to CSZ-1. The X-ray powder diffraction pattern of this material can be indexed on a hexagonal unit cell of dimensions a = 17.71(1) Å and c = 28.59(2) Å.

Characterization. The X-ray powder diffraction data for VPI-6 are illustrated in Figure 5-1 along with those from FAU and the hexagonal polytype of faujasite for comparison. Table 5-3 gives a listing of the main diffraction lines from VPI-6. The X-ray powder diffraction pattern from VPI-6 can be indexed on a hexagonal unit cell of dimensions a = 17.655(2) Å and c = 27.483(2) Å. In general, the peak intensities and half-widths are low and broad, respectively. From electron micrographs (Figure 5-2), the crystal sizes are small. This observation is consistent with the X-ray peak broadening. The X-ray powder diffraction data reveal that VPI-6 is most likely yet another intergrowth of cubic and hexagonal faujasite.

Table 5-2. Synthesis Conditions at Higher Temperature and SiO_2/Al_2O_3*

Exp.	$Na_2O/$ Al_2O_3	$SiO_2/$ Al_2O_3	$H_2O/$ Al_2O_3	aging time, hr	T, °C	heating time, hr	Result
N	5.0	4	90-95	24	100	1.5	VPI-6 (small amount of LTA)
O	7.0	10	90-95	24	100	2.0	CSZ-1 (?)
P	3.5	4	90-95	24	90	3.0	CSZ-1 (?)

*Sodium aluminate and sodium silicate starting materials.

Figure 5-2 illustrates the morphology of VPI-6. Notice that some hexagonal shaped crystals are observed. Hexagonal morphology is apparent in ZSM-3 (7) and the hexagonal polytype of faujasite (2, 3) as well.

Figure 5-3 shows the ^{29}Si and ^{27}Al MAS NMR spectra from VPI-6 (Experiment N). The bulk chemical analysis of this sample gives SiO_2/Al_2O_3 = 2.4 (Na_2O/Al_2O_3=0.93) while SiO_2/Al_2O_3 = 2.2 from ^{29}Si NMR. Thus, within experimental error, the two techniques yield the same value. Since the $(SiO_2/Al_2O_3)_{gel}$ = 4, the product VPI-6 does not crystallize with the same SiO_2/Al_2O_3 as the synthesis mixture. (Syntheses with $(SiO_2/Al_2O_3)_{gel}$ = 2.5 give VPI-6 with SiO_2/Al_2O_3 = 2.2.) The ^{29}Si MAS NMR spectrum does not reveal a peak for amorphous SiO_2 (-109 ppm), only resonances for Si(4Al): -84.9 ppm and Si(3Al,1Si): -88.5 ppm. The ^{27}Al MAS NMR shows only a single resonance at 58.7 ppm which is the same as observed from both the cubic and hexagonal polytypes of faujasite (3). Thus, all NMR observable aluminum is Al(4Si).

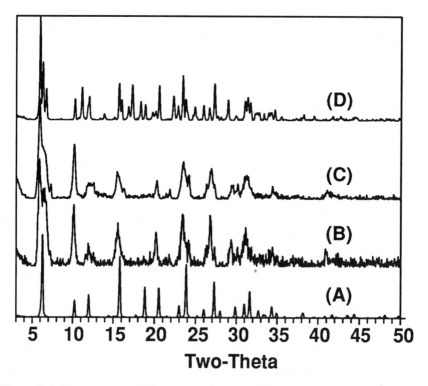

Figure 5-1. X-ray powder diffraction patterns. (A) FAU, (B) VPI-6 (contains small amount of LTA), (C) VPI-6 (contains LTA), (D) hexagonal faujasite.

Table 5-3. X-ray powder diffraction data for VPI-6

2 Θ	d($\overset{\circ}{A}$)	I/I$_o$	hkl
5.764	15.3320	100	100
6.425	13.7560	71	002
10.008	8.8379	54	110
11.580	7.6415	5	200
11.953	7.4037	16	112
15.324	5.7819	25	210
15.640	5.6658	14	211
16.131	5.4944	8	005
20.073	4.4234	24	214
21.706	4.0942	8	304
23.288	3.8195	36	400
26.225	3.3980	8	322
26.674	3.3418	35	404/410
27.176	3.2812	12	323
29.247	3.0534	23	500/009
29.987	2.9798	15	502
30.937	2.8904	25	420

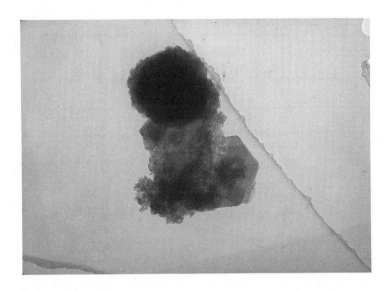

Figure 5-2. TEM of VPI-6. Magnification: 30,000X

Figure 5-4 shows the argon adsorption isotherms for VPI-6 and the two polytypes of faujasite. Notice that the initial uptake occurs at a lower P/P_o with VPI-6 and indicates that it contains a slightly smaller pore size (12). Also, the argon adsorption capacity declines in the order FAU, hexagonal faujasite, VPI-6. We have discussed previously possible reasons for the differences in argon capacity between FAU and the hexagonal polytype (3). It is not clear why the isotherm for VPI-6 is so different from faujasite since there appears to be little amorphous material present (SiO_2/Al_2O_3 from ^{29}Si NMR and chemical analysis nearly the same; no resonances in ^{29}Si or ^{27}Al NMR spectra to indicate amorphous material). However, we have noticed isotherms of this type for certain samples of ZSM-20. Recently, Anderson et al. (13) synthesized intergrowths of the cubic and hexagonal polytypes of faujasite using mixtures of 15-crown-5 and 18-crown-6 in a manner analogous to Arhancet and Davis (5). From high resolution TEM's, defect regions were observed in the samples prepared by Anderson et al. (13). Disrupted regions like those observed by Anderson et al. could certainly produce argon adsorption isotherms that vary significantly from those obtained from either the cubic or hexagonal polytypes of faujasite. It is unknown whether VPI-6 has these types of defects, but if so, they would have a major impact on its adsorption behavior.

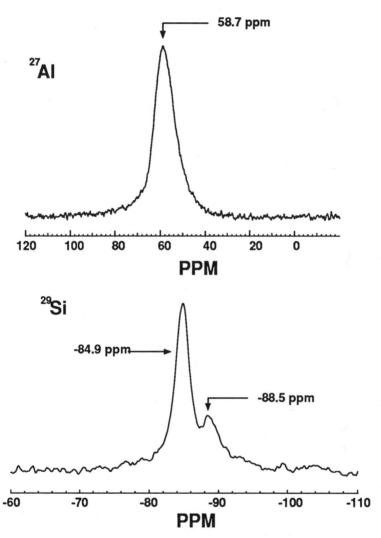

Figure 5-3. ^{29}Si and ^{27}Al MAS NMR Spectra from VPI-6.

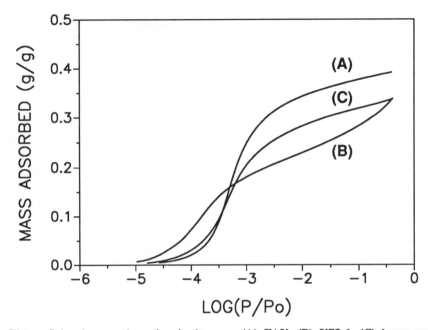

Figure 5-4.   Argon adsorption isotherms.   (A) FAU, (B) VPI-6, (C) hexagonal faujasite.

Summary
VPI-6 can be synthesized with inexpensive starting materials in very short times. Since the SiO_2/Al_2O_3 ratio is like that of zeolite X it is doubtful that this material will find application as a catalyst or catalyst support. Rather it may be useful as an adsorbant or ion exchanger. To that end, there has been much work on the synthesis of a $SiO_2/Al_2O_3 = 2$ zeolite X for these purposes. Unfortunately, when the reaction mixture composition is lowered to $SiO_2/Al_2O_3 = 2$ in this study, LTA is formed. Thus, at this time we are unable to crystallize VPI-6 with a framework composition of $SiO_2/Al_2O_3 = 2$.

Acknowledgment
We thank Dr. John B. Higgins of Mobil Oil Company for his help in analyzing the X-ray powder diffraction data and obtaining electron micrographs.

Literature Cited

1. Milton, R. M. U.S. Pat. 2,882,244 (1959).
2. Delprato, F.; Delmotte, L.; Guth, J. L.; Huve, L. Zeolites 1990, 10, 546 and references therein.
3. Annen, M. J.; Young, D.; Arhancet, J. P.; Davis, M. E.; Schramm, S. Zeolites 1991, 11, 98.
4. Li, H. X.; Annen, M. J.; Chen, C. Y.; Arhancet, J. P.; Davis, M. E. J. Mat. Chem. 1991, 1, 79.
5. Arhancet, J. P.; Davis, M. E. Chem. Mater. 1991, 3, 567.
6. Newsam, J. M.; Treacy, M. M. J.; Vaughan, D. E.; Strohmaier, K. G.; Mortier, W. J. J. C. S. Chem. Commun. 1989, 493.
7. Kokotailo, G. T.; Ciric, J. Adv. Chem. Ser. 1971, 101, 109.
8. Ciric, J. U.S. Pat. 3,411,874 (1968).
9. Ciric, J. U.S. Pat. 3,692,470 (1972).
10. Treacy, M. M. J.; Newsam, J. M.; Vaughan, D. E. W.; Beyerlein, R. A.; Rice, S. B.; deGruyter, C. B. MRS Symp. Proc. 1988, 111, 177.
11. Vaughan, D. E. W. Eur. Pat. Appl. 0315461 (1989).
12. Hathaway, P. E.; Davis, M. E. Catal. Lett. 1990, 5, 333.
13. Anderson, M.W.; Pachis, K.S.; Prebin, F.; Carr, S.W.; Terasaki, O.; Ohsuna, T.; Alfreddson, V. J. C. S. Chem. Commun. 1991, 1660.

6

Extra-Framework Al in Ultrastable Zeolite Y: Solid-State NMR Studies with ^{1}H - ^{27}Al Cross-Polarization and Quadrupole Nutation Under Fast Magic-Angle Spinning

João Rocha and Jacek Klinowski

Department of Chemistry, University of Cambridge, Lensfield Road, Cambridge CB2 1EW, U.K.

Fast (> 12 kHz) magic-angle-spinning NMR with ^{1}H - ^{27}Al cross-polarization and quadrupole nutation with fast magic-angle spinning quantitatively monitor several aluminous species in dealuminated zeolite Y (Si/Al = 4.0 - 14). The non-framework matter contains 4-, 5- and 6-coordinated Al; the ^{27}Al signal at 30 ppm is an independent resonance probably due to 5-coordinated non-framework Al and not part of a second-order quadrupole lineshape as previously reported.

INTRODUCTION

Brønsted acid groups in zeolites are associated with 4-coordinated framework aluminum, and their catalytic activity strongly depends on the concentration and location of Al in the structure. Upon hydrothermal treatment of zeolite NH_4-Y, the process known as "ultrastabilization",[1,2] part of the aluminum is ejected from the framework into the intracrystalline space, and the framework vacancies are reoccupied by silicon from other parts of the crystal. As a result, thermal stability of the zeolite is greatly increased, so that the product retains crystallinity at temperatures in excess of 1000°C. Ultrastable faujasitic catalysts are very important to the petroleum industry, and much effort has been devoted to the study of their properties.

Ultrastable zeolite Y has been extensively examined by ^{29}Si and ^{27}Al solid-state NMR.[3-12] ^{29}Si magic-angle-spinning (MAS) NMR can measure directly the framework Si/Al ratio of thermally treated samples,[3] while ^{27}Al MAS NMR shows how 6-coordinated non-framework Al species build up at the expense of the 4-coordinated framework Al as the calcination temperature is increased. However, the precise nature of non-framework aluminum entities, which are amorphous and subject to large second-order quadrupolar interactions, has not been established. Quadrupole nutation ^{27}Al NMR has also provided information about the coordination number of non-framework aluminum.[4-12] We demonstrate that ^1H - ^{27}Al CP/MAS can readily monitor Al^{NF} species in ultrastable zeolite Y. We have used samples of ultrastable zeolite Y with Si/Al ratios between 4.0 and 14 and ^1H - ^{27}Al CP/MAS with fast MAS (up to 12.5 kHz) and strong radio-frequency fields.

EXPERIMENTAL

Samples. Our starting material was 80% NH_4-exchanged zeolite Y with Si/Al = 2.41 prepared by three-fold contact of Na-Y with a saturated aqueous solution of NH_4NO_3 at 80°C followed by washing with water. The zeolite was heated in a tubular quartz furnace at 550°C for 2 hours under deep-bed conditions with water being slowly injected into the tube by a peristaltic pump so that the partial pressure of H_2O above the zeolite was 1 atm. The product (sample 2) was cooled down and re-exchanged with NH_4NO_3. Samples 3 and 4 were prepared by second calcination at 650-700°C. Sample 1 was steamed at 825°C for 2 hours, and is the same as sample A7 in ref. 3. The samples were finally hydrated in a desiccator with saturated NH_4NO_3. Powder XRD patterns were collected on a Phillips automatic diffractometer fitted with a vertical goniometer using Cu Kα radiation. Framework $(Si/Al)_{NMR}$ ratios of the samples were calculated from ^{29}Si MAS NMR^3 and are given along with unit cell parameters, a_o, in Fig. 6-1.^{6}

Techniques. It has been shown that special care needs to be taken to obtain quantitatively reliable spectra of non-framework aluminum in zeolites.^{5,6} ^{27}Al Bloch decays were recorded on a Bruker MSL-400 NMR spectrometer at 104.26 MHz with very short (0.6 μs, equivalent to $\pi/20$) radiofrequency pulses and 0.4 s recycle delays. Double-bearing MAS rotors 4 mm in diameter were spun in air at 12 - 13 kHz. ^{1}H - ^{27}Al CP/MAS spectra were recorded with a single contact, a contact time of 500 μs, a ^{1}H $\pi/2$ pulse of 3.5 μs, a recycle delay of 4 s and with spinning rates of 8 - 10 kHz. The Hartmann-Hahn condition was established in one scan on a sample of pure, highly-crystalline kaolinite using similar acquisition parameters. Because only the central $(+1/2 \leftrightarrow -1/2)$ transition is observed, excitation in the ^{1}H - ^{27}Al CP/MAS experiment is selective and therefore the Hartmann-Hahn condition is

$$3\gamma_{Al} \, B_{Al} = \gamma_H \, B_H \, .$$

3000 to 5000 transients were accumulated with a 0.4 s recycle delay. Chemical shifts are given in ppm from external $Al(H_2O)_6^{3+}$. Quadrupole nutation spectra were recorded with radiofrequency field of 120 kHz and MAS rates of 10.5 and 12.5 kHz. Typically, 64-70 data points (dwell times of 0.5-1.0 μs) were collected in the F_1 dimension and the FIDs were doubly Fourier transformed in the magnitude mode. 3000 to 5000 transients were accumulated with 0.4 s recycle delay.

RESULTS AND DISCUSSION

The ^{27}Al MAS NMR spectrum of zeolite Na-Y (not shown) consists of a single sharp signal at ca. 60 ppm due to 4-coordinated Al in the zeolitic framework (AlF). Hydrothermally treated samples (Fig. 6-1) give two further signals at ca. 0 and 30 ppm, associated with Al^{NF} species. The intensity of Al^{NF} peaks, particularly that at 30 ppm, increases with increasing degree of dealumination, while the intensity of the line at 60 ppm decreases. In addition, a very broad peak, extending from ca. 200 to ca. -180 ppm and responsible for the raised spectral baseline, grows with the degree of dealumination. At moderate spinning rates (< 8 kHz) the broad signal is more prominent, while the intensities of both Al^{NF} signals decrease relatively to the intensity of the peak at 60 ppm. This indicates that the broad signal is associated with Al^{NF} and that a quantitative treatment of the ^{27}Al MAS NMR spectra of dealuminated zeolite Y requires fast (> 12 kHz) MAS.$^{10-12}$ The CP/MAS spectrum of zeolite Na-Y

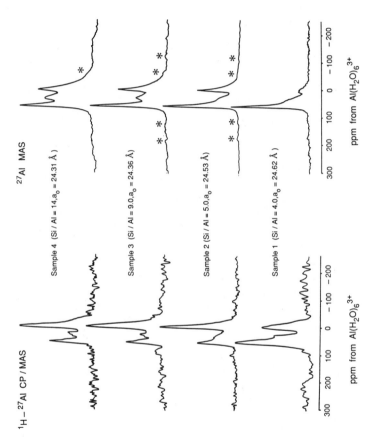

Fig. 6-1. ^{27}Al (MAS at 12 - 13 kHz) and ^{1}H - ^{27}Al CP/MAS (MAS at 8 - 10 kHz) NMR spectra of increasingly dealuminated (from bottom to top) zeolite HY. The Si/Al ratios calculated from ^{29}Si MAS NMR spectra and the unit cell parameters, a_o, are indicated. Asterisks denote spinning sidebands.

(not shown) contains two very faint signals at 61 and 9 ppm, due to a trace impurity. These signals are an order of magnitude weaker than the signals given by all the dealuminated samples examined (see Fig. 6-1) and will not be considered further. We believe that ^1H - ^{27}Al CP/MAS monitors only Al^{NF} species. A signal at ca. 60 ppm, associated with 4-coordinated Al^{NF}, is clearly seen in the CP/MAS spectra of dealuminated samples 1-4. The possibility that two (or more) ^{27}Al resonances overlap at 60 ppm must therefore be considered. Samoson et al.[4] concluded that the peak at 30 ppm is not an independent resonance but the low-frequency component of a second-order quadrupolar lineshape, the high-frequency counterpart of which overlaps with the Al^F signal at 60 ppm. However, other workers[8,9] assigned resonances in the range 27-30 ppm in aluminosilicates to 5-coordinated Al. Fig. 6-1 shows that when the spectra are recorded with CP the intensity of the signal at 0 ppm increases relative to that of the signals at 30 and 60 ppm. Although less important, there is also an increase of the signal at 30 ppm relative to that at 60 ppm. This suggests that the peak at 30 ppm is a separate ^{27}Al resonance. Furthermore, the position of NMR signals in ^{27}Al MAS NMR spectra reported in ref. 4, recorded at the higher field, B_o, of 11.7 T (as opposed to 9.4 T in the present study), is the same at 0, 30 and 60 ppm. Computer simulations indicate that increasing B_o should decrease the distance between the two components of a second-order quadrupolar lineshape. Since this is not observed, those features are likely to be independent resonances.

Fig. 6-2 shows F_2 cross-sections of quadrupole nutation spectra of samples 1 and 3 recorded with fast (12.5 kHz) MAS and strong (120 kHz) radiofrequency field. It is clear that: (a) four different resonances are resolved; (b) the two resonances at 62 and 56 ppm which overlap in the ordinary MAS spectrum (to produce the peak at 60 ppm) are resolved and the relative intensity of the 62 ppm component decreases with increasing degree of dealumination; (c) the signal at 30 ppm is an independent resonance which grows with increasing degree of dealumination. Because fast MAS can cause partial destructive interference of the spinning-dependent phase of the magnetization vector during during long radiofrequency pulses,[13] the quadrupole coupling constants, C_Q, can only be estimated from the F_1 cross-sections. In ref. 12 we discuss this problem in detail, and estimate the following C_Q values: 2 MHz (for the resonance at 62 ppm); 3.5 - 4 MHz (56 ppm); 4.5 - 5 MHz (30 ppm in sample 1), > 6 MHz (30 ppm in sample 3); 4 MHz (0 ppm).

The combined CP and quadrupole nutation evidence indicates that the resonances at 56 and 0 ppm are due to 4- and 6-coordinated Al^{NF}, respectively. Similar arguments allow the resonance at 62 ppm to be confidently assigned to Al^F. A striking feature shown by quadrupole nutation is the dramatic increase upon dealumination of C_Q for the resonance at 30 ppm. Our estimate that C_Q > 6 MHz for sample 3 is in line with the findings of Samoson et al.[4] for

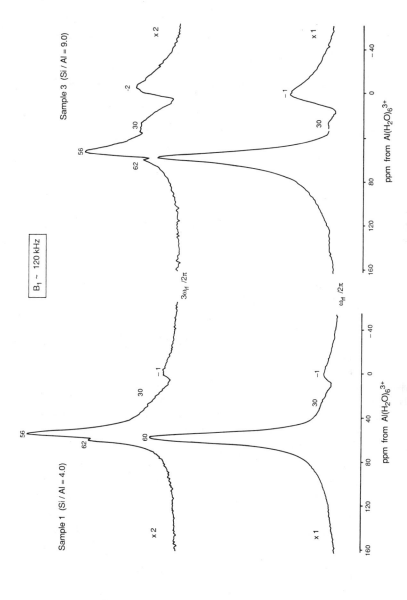

Fig. 6-2. F_2 cross-sections of the quadrupole nutation spectra of samples 1 and 3 recorded at $\omega_{rf}/2\pi = 120$ kHz, taken at $\omega_{rf}/2\pi$ and $3\omega_{rf}/2\pi$.

samples with Si/Al ratios in the range 5.2 - 20. The corresponding second-order quadrupolar shift of the centre of gravity for the central transition is 20 ppm ($\eta = 0$) to 27 ppm ($\eta = 1$) or larger.[14] If the centre of gravity is now assumed to be at 30 ppm, the isotropic chemical shift is at 50 to 57 ppm, which could be due to 4-coordinated aluminum.[14] When similar arguments are used for sample 1, one obtains a quadrupolar shift of 11 to 18 ppm ($C_Q = 4.5 - 5$ MHz) and an isotropic shift at 41 to 48 ppm which could well be due to 5-coordinated aluminum. We conclude that the non-framework species in ultrastable zeolite Y may contain 4-, 5- and 6-coordinated Al.

Acknowledgments. We are grateful to Unilever Research, Port Sunlight, for support.

References

1 . McDaniel, C. V.; Maher, P. K. *Molecular Sieves,* Society of Chemical Industry, London, 1968, p. 186; *Zeolite Chemistry and Catalysis,* ACS Monogr. **1976,** *171,* 285.

2 . Breck, D. W. *Zeolite Molecular Sieves: Structure, Chemistry and Use,* John Wiley, London 1974.

3 . Klinowski, J.; Fyfe, C. A.; Gobbi, G. C. *J. Chem. Soc., Faraday Trans. I* **1985,** *81,* 3003.

4 . Samoson, A.; Lippmaa, E.; Engelhardt, G.; Lohse, U.; Jerschkewitz, H.-G. *Chem. Phys. Lett.* **1987,** *134,* 589.

5 . Man, P. P.; Klinowski, *J. Chem. Phys. Lett.* **1988,** *147,* 581; *J. Chem. Soc., Chem. Comm.* **1988,** 1291.

6 . Man, P. P.; Klinowski, J.; Trokiner, A.; Zanni, H.; Papon, P. *Chem. Phys. Lett.* **1988,** *151,* 143.

7 . Hamdan, H.; Klinowski, J. *J. Chem. Soc., Chem. Comm.* **1989,** 240.

8 . Gilson, J.; Edwards, G. C.; Peters, A. W.; Rajagopalan, K.; Wormsbecher, R. F.; Roberie, T. G.; Shatlock, M. P. *J. Chem. Soc., Chem. Comm.* **1987,** 91.

9 . Alemany, L. B.; Kirker, G. W. *J. Am. Chem. Soc.* **1986,** *108,* 6158.

1 0 . Kellberg, L.; Linsten, M.; Jakobsen, H. *J. Chem. Phys. Lett.* **1991,** *182,* 120.

1 1 . Rocha, J.; Klinowski, J. *J. Chem. Soc., Chem. Comm.* **1991,** 1121.

1 2 . Rocha, J.; Carr, S. W.; Klinowski, J. *Chem. Phys. Lett.* **1991,** *187,* 401.

1 3 . Samoson, A.; Lippmaa, E. *J. Magn. Reson.* **1988,** *79,* 255.

1 4 . Engelhardt, G.; Michel, D. *High Resolution Solid-State NMR of Silicates and Zeolites,* John Wiley, New York 1987.

7

Crystallization Kinetics of the Zeolites Faujasite, P, Mordenite, L, Omega, Offretite, and Sodalite $(Na,K, TMA)AlO_2 n[(Na,K, TMA) mH_{4-m}SiO_4]pH_2O$

H. Lechert, *Institute of Physical Chemistry University of Hamburg, Bundesstr. 45, L 2000 Hamburg 13, Germany*
and
H. Weyda, *Fa. Süd - Chemie, Katalyse-Labor, Waldheimer Str. 13, D8206 Heufeld - Bruckmühl, Germany*

In the system
$$(Na,K, TMA)AlO_2 \; n[(Na,K,TMA)_m H_{4-m} SiO_4] \, pH_2O$$
the crystallization of the zeolites faujasite, P, mordenite, L, omega, offretite and sodalite is studied in dependence on the ratio of the cations. In the system with only Na ions faujasite, P and at higher dilutions and temperatures above 400 K mordenite is obtained. With K ions, L can be synthesized. For Na and TMA ions at 350 K faujasite and omega crystallize. At high TMA contents TMA sodalite is obtained. Offretite is formed in a wide range with all three cations in the batch. At 463 K and a composition
$$(Na,K,TMA)AlO_2 \; 12.0[(Na,K,TMA)_{0.85} H_{3.15} SiO_4] \; 200 \; H_2O$$
the crystallizing zeolite structures depend only on the ratio of the cations.

1 INTRODUCTION

In a system with the general batch composition
$$(Na,K,TMA)AlO_2 \; n[(Na,K,TMA)_m H_{4-m} SiO_4] \, pH_2O$$
a series of zeolites can be synthesized depending on the alkali-

nity of the batch given by m, the ratio of the cations Na, K, and TMA present in the batch, the water content p and the temperature. The Si/Al-ratio of the final product is determined mainly by the alkalinity, which has been thoroughly analyzed in two previous papers (Lechert and Kacirek 1991, Lechert, Kacirek and Weyda 1991). The cations present in the batch cause with rather strict regularity the formation of special structure elements in the crystallizing products, which may be combined to the final structure with some precision. The cations Na, K, and TMA have increasing radii and influence the water structure during the crystallization process and the possibility of the arrangement of the Si and Al in different building units.

Going through the literature summarized e.g. in the books of Breck (1974), Barrer (1982), and recently Occelli and Robson (1989) some of the mentioned regularities can be drawn from the synthesis procedures and the crystallization fields of the respective zeolites.

From batches containing only Na, faujasite, zeolite P and mordenite can be obtained. In the faujasite structure the Na ions may be responsible for the formation of the cubooctahedra as well as for the double six-membered rings. If K is added to the batch zeolite L crystallizes. K ions are often related to the formation of cancrinite cages.

With TMA the zeolites omega, offretite, sodalite and NA have been observed.

Omega contains gmelinite cages. Offretite crystallizes in the presence of all three cations and contains in its structure double six-membered rings, cancrinite cages and gmelinite cages. With the exception of mordenite, these structures contain mainly four- and six-membered rings forming the mentioned cavity structures.

In the different crystallization processes usually the zeolites with the largest pores are formed first as metastable phases which recrystallize in several cases to denser structures if the zeolite remains in the mother liquor.

Many of the very special manipulations of zeolite synthesis consist of attempts to separate the crystallization process of the wanted zeolite from recrystallization to unwanted products. In the present paper a series of systematic experiments will be done in the mentioned system to study the common features of the formation of the different structures.

2 EXPERIMENTAL

The starting materials for the batches were
- amorphous precipitated silicic acid (Merck) SiO_2 0.5 H_2O in several batches the silicic acid has been prepared from $Si(OC_2H_5)_4$
- NaOH pellets, KOH pellets, and TMA-OH solution
- Na- or K-aluminate prepared from $Al(O-iC_3H_9)_3$ 1000 g aluminate solution contained

$$2.08 \ NaAlO_2 \ 2.08 \ NaOH \ 41.43 \ H_2O$$

The NaOH or KOH pellets were dissolved in distilled water. Then the SiO_2 was added and finally the aluminate. After each step the solution was shaken for 30 minutes. To the gel formed after this procedure appropriate seeds (for the synthesis of Y) or the TMA-OH was added and the mixture was again shaken for 30 minutes. The crystallization was carried out in Teflon vessels which were enclosed into stainless steel autoclaves for reactions above 373 K. The reaction vessels were heated in an oven. After appropriate times the reaction was stopped and samples were taken. These samples were washed and dried overnight at 373 K. The phase composition and the crystallinity were checked by X-ray diffraction. The chemical composition of the samples was analyzed by X-ray fluorescence spectroscopy or by EDAX.

3 RESULTS

The starting point of our work was the synthesis procedure of the faujasite which crystallizes in batches containing only the Na ions

$$NaAlO_2 \ n(Na_m H_{4-m} SiO_4) p H_2O$$

Pure faujasites can be obtained for $n > 1.5$, $2.5 < m < 5$ at a water content of $p \approx 200$.

At lower n zeolite A is obtained. For $n < 2.5$ and without seeding zeolite P crystallizes. With seeding or by synthesis procedures ensuring the nucleation of X zeolite at the beginning of the process, Y zeolite can be grown down to about m = 0.65 giving $Si/Al \approx 3.5$ in the final product (Kacirek and Lechert 1975; Kacirek and Lechert 1976; Lechert 1984; Robson 1989).

If the batch is stirred for less than 1 hour before heating to the crystallization temperature zeolite Y is formed. Well homogenized batches which were stirred for more than 5 hours give P. For intermediate stirring times mixtures of both zeolites are obtained.

For alkalinities with m $>$ 5 at low Si/Al ratios n sodalite and at higher n analcime is obtained. The faujasites can be obtained only below a temperature of about 370 K. Above this temperature P crystallizes.

From many investigations it is known, however, that at higher temperatures in the described system mordenite can be crystallized. We have, therefore, studied the crystallization at different batch compositions and temperatures. Up to 403 K from batches within the concentration region of Y only P zeolite can be obtained.

Figure 1 shows the crystallization field at 423 K. Starting from the corner with the Si-rich samples at first only amorphous products were observed under the conditions indicated in the legend of the figure. Mordenite can be found as the substance with the most open lattice. Mordenite is followed by P zeolite, and by a comparatively broad region where P and analcime are crystallizing together. In the lower part of the diagram only analcime is observed.

To get further insight a series of experiments with varying water contents have been carried out. An example is given in Fig.2. Starting from a batch

$$NaAlO_2 \; 9.2(Na_{0.38}H_{3.62}SiO_4) \; p \; H_2O$$

at 423 K for p = 119 P crystallizes.

With increasing water content above p = 200 pure mordenite is obtained. At higher water contents the region of the crystallization of the template free ZSM-5 can be reached (Schwieger et al. 1989). Obviously the formation of five-membered rings is favored by higher dilutions. Whether this depends on kinetic effects or only on the lower alkalinity of the batches is not quite clear from the present experiments. The kinetic of

the formation of mordenite from a batch

$$NaAlO_2 \; 15.5(Na_{0.19}H_{3.81}SiO_4) \; 330 \; H_2O$$

at temperatures of 403, 423, and 463 K is shown in Fig.3.

It can be seen that an increase of the temperature influences the nucleation time as well as the growth time. Both are shortened with increasing temperature.

Kinetic experiments for the growth of mordenite are reported also by Culfaz and Sand (1973). By Thompson et al. (1985)

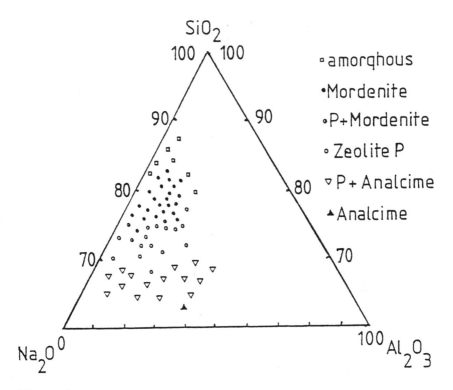

Figure 1. *Crystallization field of mordenite for a crystallization temperature of 423 K, a crystallization time of 24 hours and* $H_2O/SiO_2 = 20$.

and Ciric (1968) crystallization curves of this type have been analyzed by the Avrami theory describing the time dependence of the crystallinity Kr by

$$Kr = 1 - \exp(- kt^s)$$

k is the rate constant of the growth and s a characteristic constant for the kind of growth.

An analysis of the data from Fig. 3 shows that the exponent s has extremely high values indicating a complicated growth process which cannot explained by the Avrami theory without further assumptions. Looking at the crystal shapes obtained in the described experiments it can be seen that these differ for different temperatures. At 463 K columns with a star-shaped

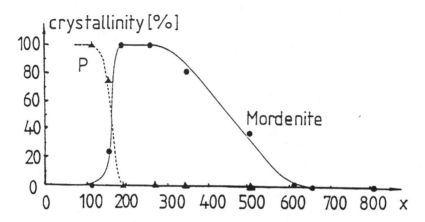

Figure 2. *Crystallization of mordenite in dependence on the* H_2O *content x of a batch with the composition*
  $NaAlO_2$ 15.5 ($Na_{0.38}$ $H_{3.62}$ SiO_4) x H_2O
Crystallization temperature 423 K, crystallization time 45 hours.

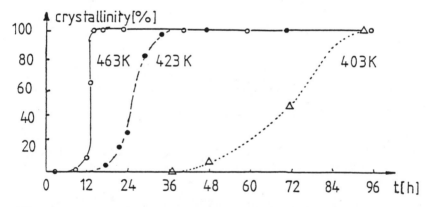

Figure 3. *Kinetics of mordenite crystallization at different temperatures. Batch composition*
  $NaAlO_2$ 15.5 ($Na_{0.38}$ $H_{3.62}$ SiO_4) 300 H_2O

basis can be observed. In other batches bundles of rods can be
obtained under apparently identical conditions. At lower temperatures plate-like crystals appear.

An explanation of the observed effects may be that the nucleation rate or also the k for the different crystal planes may

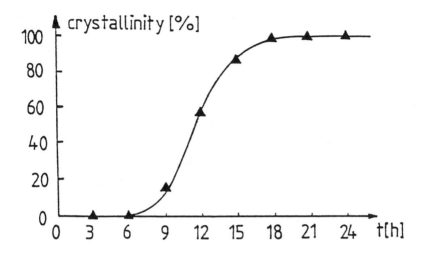

Figure 4. *Kinetic curve of the crystallization of L zeolite from a batch*
$(Na_{0.2}K_{0.8})AlO_2$ 14.5 $(Na_{0.16}K_{0.64}H_{3.2})$ SiO_4 200 H_2O

be time dependent in the course of the crystallization. Adding K ions to the batches from which Y and P zeolites crystallize at about equal amounts of Na and K zeolite L can be observed above 370 K (Breck 1974). The crystallization time of these samples is usually rather long. Therefore, optimization experiments have been carried out.

Finally, at 403 K pure L zeolite could be obtained within about 20 h from batch compositions near

$(Na_{0.2}K_{0.8})AlO_2$ 14.5 $(Na_{0.16}K_{0.64}H_{3.2}SiO_4)$ 200 H_2O

The time dependency of the crystallization of such a sample is demonstrated in Fig. 4. From the crystallization curve an Avrami exponent s = 2.2 and k $\approx$ 2×10^{-3} [h$^{2.2}$] is obtained.

For the crystallization of omega batches of

$NaAlO_4$ 4.35 $(Na_{0.28}H_{3.72}SiO_4)$ n TMA-Cl 80 H_2O

have been prepared after a series of preliminary experiments.

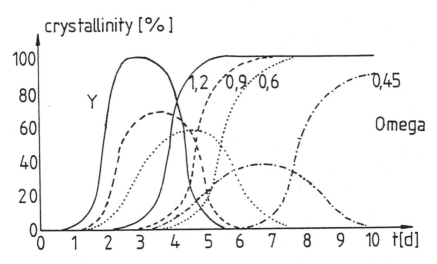

Figure 5a. *Kinetic curves for the crystallization of Y and omega zeolite from batches*

$$NaAlO_4 \ 4.35 \ (Na_{0.28} \ H_{3.72} \ SiO_4) \ n \ TMA\text{-}Cl \ 80 \ H_2O$$

for different TMA-Cl contents n . Crystallization temperature 358 K

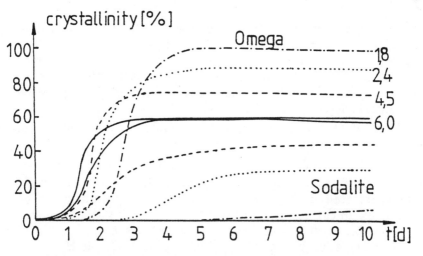

Figure 5b . *Kinetic curves for the crystallization of omega and Sodalite from batches*

$$NaAlO_2 \ 4.35 \ (Na_{0.28} \ H_{3.72} \ SiO_4) \ n \ TMA\text{-}Cl \ 80 \ H_2O$$

for different n . Crystallization temperature 358 K.

For the crystallization of omega batches of
$$NaAlO_4 \ 4.35 \ (Na_{0.28} \ H_{3.72} \ SiO_4) \ n \ TMA\text{-}Cl \ 80 \ H_2O$$
have been prepared after a series of preliminary experiments
and studied in dependence on the TMA content. TMA-Cl has
been used to keep the alkalinity of the batch constant. The results
for a temperature of 358 K can be seen in Figs. 5a and 5b.

At low TMA contents Y zeolite crystallized first and recrystallized then to zeolite omega. For $n = 1.2$ after about 10
hours 100% Y could be observed. Some of these experiments
have been reported by Dwyer and Chu (1979).

To study the mechanism of the recrystallization process
samples were taken from batches with $n = 1.2$ after 2 to 5

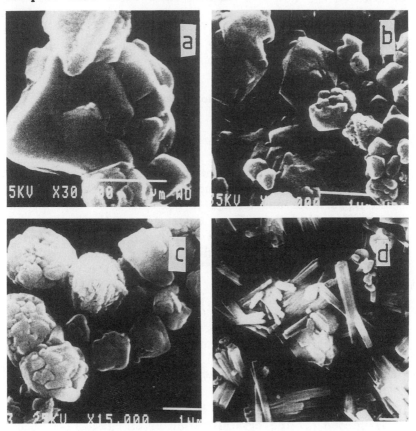

Figure 6. *Crystals from a batch according to Fig.5a and
the TMA-Cl content n = 1.2 after a. 3 hours, b. 4 hours,
c. 5 hours , d. 6 hours.*

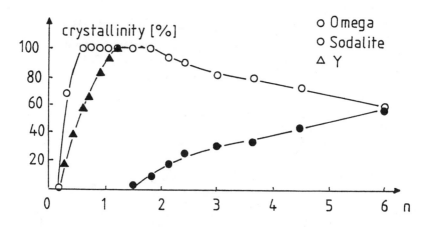

Figure 7. *Maximum crystallinity of Y zeolite, Omega and Sodalite from batches*
$NaAlO_2$ 4.35 ($Na_{0.28} H_{3.72} SiO_4$) n TMA-Cl 80 H_2O
for different n at a crystallization temperature of 358 K

hours and studied by the scanning microscope. Some of these pictures are shown in Fig. 6.

It can be seen that the formation begins on the surface of Y zeolite especially at the corners (Fig. 6a). Other crystals get brittled surfaces and change to the spherical shape which is seen in Figs. 6b and 6c. From these spheres the crystals of the Omega develop as it can be seen in Fig. 6d.

Increasing n further, increasing amounts of TMA sodalite can be observed in the products. A survey of the data for a batch with

$$NaAlO_2 4.35(Na_{0.28} H_{3.72} SiO_4) \text{ n TMA-Cl } 80 \text{ } H_2O$$

at 358 K and a broad range of n can be seen in Fig. 7.

The fact that at low TMA content n omega reaches 100 % inspite of an appreciable amount of Y means that Y recrystallizes completely to omega which is not the case with omega and sodalite.

Fig. 7 shows that at least 1 TMA ion on 20 $(Si+Al)O_4$ units is needed for the crystallization of omega.

For more than about 3 TMA ions on 20 $(Si+Al)O_4$ units beside omega sodalite can be found. For about 1 TMA ion on

1 $(Si+Al)O_4$ pure sodalite is obtained.

This means that for a cubooctahedron in the sodalite structure about 6 TMA ions are needed in the batch. In comparison with this the Y structure needs about 1 TMA ion for a sodalite cage and the omega structure about 1 to a maximum of 2 of these ions for a gmelinite cage.

For the TMA ion a size of about 60-100 $Å^3$ is usually reported in the literature. The sodalite cage has a space of 150 $Å^3$ and the gmelinite cage 170 $Å^3$ which agrees with our findings.

Thorough investigations on the kinetics of the growth of omega in the different directions of the crystal structure have been carried out in an excellent paper by Fajula et al. (1989)

Using a batch composition of

$$NaAlO_2 \; 4.1(Na_{0.29}H_{3.71}SiO_4) \; 0.8 \; TMA\text{-}OH \; 80 \; H_2O$$

which is almost identical with the compostion which was used in our experiments the authors found that the (001) face is proportional to $[Al]^{0.8}$ and the growth rate of the (hk0) faces proportional to $[Al]^{1.6}$. The activation energy was 96 kJ/mole in the first and 125.4 kJ/mole for the lateral surfaces.

The authors explain this result by the assumption that only the growth of the (100) surfaces occurs via the incorporation of the templating TMA ions. For increasing n the TMA ion causes first obviously the less stable sodalite cage and induces the crystallization of faujasite. Then, this structure recrystallizes to the omega structure.

Going to high n the sodalite structure is formed which consumes much more TMA ions than the more open structures.

Offretite may be synthesized with and without the TMA-ion. Without the TMA ion at 373 K within about six days almost pure offretite is obtained only in a quite narrow range around a batch composition of

$$(Na_{0.34}K_{0.66})AlO_2 \; 14.5[(Na_{0.24}K_{0.45})H_{3.3}) \; SiO_4] \; 200 \; H_2O.$$

The most important impurity is a steadily
increasing amount of erionite as has been reported already in the earlier literature. A typical kinetic run is demonstrated in Fig. 8. Pure offretite can be obtained in a broad range of compositions when TMA ions are added. This may be understood systematically from the fact that the offretite structure contains gmelinite cages the formation of which is favored by TMA ions, cancrinite cages which are preferred by K ions and double six-membered rings which are most probably caused by the pre-

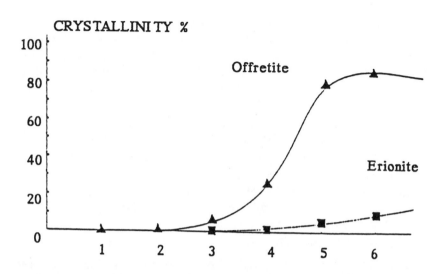

Figure 8. *Kinetic curve of the crystallization of Offretite and Erionite for a batch*
$(Na_{0.34}K_{0.66})AlO_4\ 12.5(Na_{0.24}K_{0.46}H_{3.3}\ SiO_4)\ 200\ H_2O$

sence of Na-ions.

To find a more systematic picture attempts have been made to find conditions under which all the mentioned zeolites can be obtained varying only the ratio of the cations in the batch
$(Na,K,TMA)AlO_2\ n\ [(Na,K,TMA)_m\ H_{4-m}SiO_4]\ p\ H_2O$

Varying systematically the temperature in the batches described above, it can be seen that at 463 K a large variety of zeolites can be found from a batch
$(Na,K,TMA)AlO_2\ 12.0\ [(Na,K,TMA)_{0.85}\ H_{3.15}SiO_4]\ 200\ H_2O$
in a crystallization time of 0 to 7 hours.

The ranges of cation ratios under which these zeolites crystallize are demonstrated in Fig. 9. The figure contains some peculiarities which shall be discussed briefly.

If only Na-ions are in the batch faujasite crystallizes after about 3 hours, then changes after 7 hours first to P zeolite and then to analcime which seems to be stable at the choosen temperature. P is found at the given temperature only for batches with only Na; otherwise the zeolites change directly into analcime at longer exposure to the mother liqour at 463 K. Only

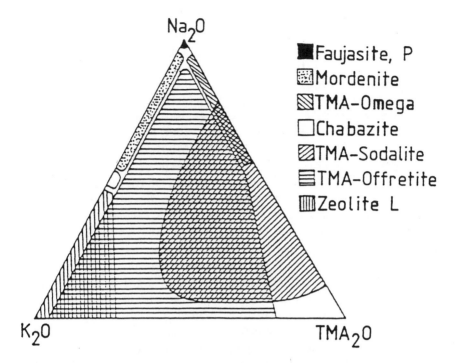

Figure 9. *Crystallization diagram of various zeolites at 463 K and a batch composition*
$(Na,K,TMA)AlO_2\ 12.0\ [\ (Na,K,TMA)_{0.85}\ H_{3.15}\ SiO_4\ ]\ 200\ H_2O$
in dependence on the ratios of the cations.

sodalite has a better stability.

A pure L phase is obtained if only K ions are in the batch. L is distinctly detectable, however, for appeciable amounts of Na and also of TMA ions present. Mordenite can be found for Na/(Na+K) = 0.6–0.9 in strongly varying low amounts. At Na/(Na+K) = 0.5 chabazite can be synthesized.

The formation of omega and, at higher TMA contents, of sodalite is at the higher temperature of 463 K similar to that described above. As mentioned, offretite can be observed in a rather broad region of ratios of the cation concentrations. Typical concentrations for the crystallization of pure offretite are   Na = 0 and K/(K+TMA) = 0.05 – 0.8

or $K/(K+TMA) = 0.05 - 0.2$ and $Na/(Na+K) = 0 - 0.9$

An increasing amount of K shortens the crystallization times.

Generally, in our experiments the appearance of the different zeolite types at the borders of the crystallization field is continuous. Exceptions are given with the intermediate formation of faujasite. Even with small amounts of K in the batch mordenite or offretite is found. Omega is found only with Na- and TMA-ions. Otherwise only omega does not change to P or analcime at longer exposure to the mother liquor.

Finally, the possibility of the synthesis of zeolite NA, an A zeolite with a Si/Al ratio greater than 1 shall be mentioned. This zeolite has been first synthesized by Barrer and Denny (1961). Later on systematic studies of the formation of this zeolite have been carried out by Kacirek and Lechert (1977).

NA crystallizes under the same conditions as zeolite omega, if care is taken that A nuclei are present in the batch. This may be done by adding TMA silicate to a gel of Si/Al = 1. Depending on the procedure under which the crystallization conditions are reached. A nuclei are formed already in the batch or may be added before heating.

Summarizing, the well known relations between the cations present in the solution and the structural elements in the crystalline products can be veryfied in our experiments with great detail under conditions where the batches differ only by the ratio of the cations present in the batch. This gives a broader basis for the controlled synthesis of zeolites with defined properties.

4 ACKNOWLEDGEMENTS

The authors thank the Deutsche Forschungsgemeinschaft for the support of their work.

5 REFERENCES

Barrer, 1982 R.M., *Hydrothermal Chemistry of Zeolites*, Academic: London.
Barrer, R.M., Denny, P.J., 1961, J. Chem. Soc., 971.
Breck, D.W.,1974, *Zeolite Molecular Sieves*, John Wiley and Sons, New York.
Ciric, J., 1968, Coll. Interf. Sci., 28, 315.
Culfaz, A. and Sand, L.B.,1973, in *Molecular Sieves*, Meier, W.M., and Uytterhoeven, J.B., Eds., Adv. Chem. Series, 121, Washington p. 140.

Dwyer, F.G., and Chu P., 1979, J. Catal., 50, 263.

Fajula, F., Nicolas, S., Di Renzo, F., Gueguen, C. and Figueras, F.,
1989 ; in *Zeolite Synthesis*, Occelli, M.L., Robson, H.E. Eds.,
ACS Symp. Series 398, Washington, p. 493.

Kacirek, H., Lechert, H., 1975, J. Phys. Chem., 79, 1589.

Kacirek, H., Lechert, H., 1976, J. Phys. Chem., 80, 1291.

Kacirek, H., Lechert, H., 1977; in *Molecular Sieves II* ,
Katzer, J.R. Ed., ACS Symp. Series 40, Washington , p. 244.

Lechert, H.Kacirek, H., 1991, Zeolites, 11, 720.

Lechert, H., Kacirek, H., Weyda, H., 1992, this issue.

Occelli, M.L., Robson, H.E. Eds., 1989, *Zeolite Synthesis*,
ACS Symp. Series 398, Washington.

Robson, H.E. 1989; in *Zeolite Synthesis* , Occelli, M.L., and
Robson, H.E. Eds., ACS Symp. Series, Washington, p. 436.

Schwieger, W., Bergk, K.-H., Freude, D., Hunger, M. and Pfeifer. H.
1989; in *Zeolite Synthesis*, Occelli, M.L., and Robson, H.E. Eds.,
ACS Symp. Series, Washington, p. 274.

Thompson, R.W., and Dyer, A., 1985, Zeolites, 5, 202 .

8

12-Ring Channel Templated Mazzite: A Stabilized Calcined Zeolite Structure

D. E. W. Vaughan and K. G. Strohmaier *Exxon Research Engineering Company, Annandale, NJ 08801*

Synthetic MAZ, made with tetramethylammonium cations trapped in gmelinite cages (ZSM-4, Omega), undergoes partial structural degradation when the template is burned out at high temperature (570°C), damage which is significantly increased in the gallium substituted form. If, however, bis-2-hydroxyethyl-dimethylammonium cations are used they locate in the 12-ring channel and are removed at a temperature below about 450°C, low enough to prevent major framework hydrolysis and structural breakdown. Structural degradation is shown by XRD peak broadening and lower hydrocarbon sorption capacities, the latter directly related to template burn-off temperatures. Extensive preparation results are presented and preferred synthesis properties defined.

INTRODUCTION

Mazzite is a 12-ring zeolite comprising six columns of connected gmelinite cages (Galli 1974), the adjacent columns offset by c/2. The natural mineral is rare (Gottardi and Galli 1985) but the synthetic forms, ZSM-4 (Ciric 1972) and "omega" (Flanigen and Kellberg 1980), have been extensively investigated for almost twenty years from the perspectives of synthesis compositional ranges and catalysis (Bowes and Wise 1971; Cole and Kouwenhoven 1973; Travers et al. 1991), particularly from the perspective of mordenite surrogacy in hydroisomerization and related processes. The commercial designation "Omega" should not be confused with the similarly labeled earlier (erroneous) proposed related theoretical structure for "Omega" (Barrer and Villiger 1969). The differences between these two structures is shown in Figure 8-1. The experimental problem of differentiating these structures in synthetic products is a difficult one, particularly as the expected differences in X-ray diffraction patterns are small. The question of whether one or the other, or intergrowths of the two, are preferred in particular templated systems is an unresolved issue in the absence of large single crystals of the synthetic products, or extensive analysis of synchrotron X-ray diffraction powder data.

Mazzite can be synthesized over a wide range of gel Si-Al-Ga compositions when sodium and tetramethylammonium cations (TMA) are present in the synthesis gel, including the pure end member aluminosilicate and gallosilicate forms. A wide range of silica source reactants have been successfully utilized, such as waterglass (Ciric 1972), colloidal silica (Flanigen and Kellberg 1980; Aiello and Barrer 1970), silica-alumina gels (Bowes and Wise 1971) and meta-kaolin (Fajula et al. 1989). It is frequently preceded by faujasite, but once formed it completely replaces faujasite and remains the stable phase for long times between 100° and 160°C, ultimately being replaced by GIS. Although wide ranges of Si/Al are claimed for its composition (Ciric 1972; Flanigen and Kellberg 1980), it is synthesized with composition ratios between about 2.7 and 3.2 with the TMA located mainly in the gmelinite cages of the structure. When the TMA is burned out of the structure to facilitate its use as an absorbent or

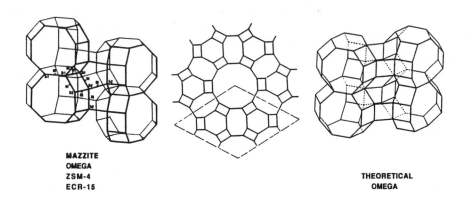

MAZZITE
OMEGA
ZSM-4
ECR-15

THEORETICAL
OMEGA

FIGURE 8-1. A Comparison of the MAZ mineral structure (Galli 1974) and an alternate theroretaical structure proposed for "Omega" (Barrer and Villiger 1969).

catalyst, some structural degradation and dealumination invariably result, causing a loss in sorption capacity and X-ray diffraction peak intensity, the latter sometimes showing significant peak broadening (Weeks et al. 1976). This structural instability has also been observed for the gallium form (Newsam, Jarman, and Jacobson 1985), in which it is a more acute problem. TMA, being trapped in a small cage, is not burned off until 570°C is attained, a temperature closely coincident with the dehydroxylation temperature for MAZ, facilitating the hydrolysis of Al or Ga T atoms with the resultant formation of lattice defects.

Recent synthesis studies with bis-2-hydroxyethyldimethyl ammonium (E_2M_2) and bis-2-hydroxypropyldimethyl ammonium (P_2M_2) cations have shown that, when traces of TMA are present in the gel, a mazzite derivative can be synthesized with the E_2M_2 or P_2M_2 cations located only in the 12-ring channel, from which they can be burned off at a lower temperature without T atom hydrolysis, preserving properties commensurate with a 12-ring channel structure. This material has been designated ECR-15 (Vaughan 1991). In the absence of TMA ECR-1 crystallizes (Vaughan and Strohmaier 1989). Presumably the TMA nucleates mazzite and once nucleated the structure is propagated by E_2M_2 inclusion in the channels. Pure mazzite has not been synthesized in the total absence of TMA , although it has been observed as an overgrowth on ECR-1 (Leonowicz and Vaughan 1987), possibly induced by E_2M_2 disproportionation to TMA in the highly basic synthesis mixture at a late stage in the crystallization of the ECR-1. Seeding using only TMA in a small portion of seeds and no other template (Albers and Vaughan 1976) is a crystallization route to low TMA-MAZ, but the same process does not work with E_2M_2 or P_2M_2 in place of TMA, producing no mazzite in this composition range. It is possible that syntheses of MAZ using choline or pyrollidine (Rubin, Plank, and Rosinski 1977) or DABCO (Rubin and Rosinski 1982) also require the presence of traces of TMA nucleant.

CATION INFLUENCES ON MATERIALS SYNTHESIS

Syntheses in the seeded (Vaughan, Barrett and Edwards 1982) composition range:
$$2 \text{ to } 3.5 \text{ } Na_2O: Al_2O_3: 6 \text{ to } 10 \text{ } SiO_2: 80 \text{ to } 180 \text{ } H_2O$$
produce faujasites of various ratios (silica source was PQ N brand Na silicate Si/Na = 0.33; alumina source was ALCOA C-31 $Al_2O_3 \cdot 3H_2O$ dissolved in NaOH). When Na is partly replaced (up to 0.6 moles) by Li, K, Rb, Cs, TMA or E_2M_2 in stirred (S) and unstirred reactions, the products shown in Table 8-1 result. In some preparations various faujasite forms may appear as weak transient phases at 125°C. At Si/Al$_2$ ratios over about 12 mordenite begins to appear and predominates with increasing silica levels in the group 1A cation systems.

TABLE 8-1 Initial Synthesis Products as a Function of Cation Combinations

Cryst. temp(°C) /2nd cation	100°(S)	100°	125°	140°
Li	ABW	FAU	ABW	ABW
Na	GIS	FAU/GME	GIS(P)	GIS(P)
K	MER	pFAU	MER	MER
Rb	MER	pFAU/MER	MER	MER
Cs	ANA	CSZ-1*/CSZ-3*	ANA	ANA
TMA	+	FAU/MAZ	MAZ	MAZ
E_2M_2	+	FAU/ECR-4*	ECR-1	ECR-1
(TMA)+E_2M_2	+	FAU/ECR-4*	FAU/MAZ	MAZ

p = platelet morphology; * various forms of FAU; for codes see (Meier and Olson 1987).

MATERIALS PREPARATION

TMA-MAZ Synthesis

Our standard comparison material was made at a composition:

$$2.75Na_2O: Al_2O_3: 9SiO_2: 0.5TMA_2O: 140H_2O$$

by first dissolving 75gm $Al_2O_3 \cdot 3H_2O$ in a solution of 59gm NaOH in 100gm H_2O at 100°C, cooling it to room temperature, and further diluting with H_2O to a total weight of 250gm. To 256.5gm sodium silicate solution (PQ N brand waterglass) in a blender were added, sequentially with mixing, 18.2gm of a nucleant seed solution($13.3Na_2O:Al_2O_3:12.5SiO_2:267H_2O$), 51.1gm 25% TMAOH, 45.2gm of the above sodium aluminate solution, and finally (slowly) 65.3gm 50% wt. solution of $Al_2(SO_4)_3 \cdot 17H_2O$ and 63.8gm of H_2O. After crystallizing the gel in Teflon lined Parr autoclaves at 150°C in an air oven for 1 day, the product was filtered, washed and dried at 110°C. The product was analyzed (Leeman Labs Plasma Spec ICP 2.5) as having a composition:

$$0.81Na_2O: Al_2O_3: 5.66SiO_2$$

The morphology of the products are shown in Figure 8-2a and the X-ray diffraction pattern in Figure 8-3a for the template occluded form and 8-3b for the "burnt-out" form. Changing the crystallization temperature from 140°C to 125°C or 160°C does not change the Si/Al ratio of the product.

ECR-15 Synthesis

The preferred (not exclusive) silica source for ECR-15 is colloidal silica (DuPont Ludox HS-40), and because E_2M_2 stabilizes FAU(ECR-4) at lower

FIGURE 8-2 Scanning electron micrographs of mazzite samples: (a) upper - TMA-MAZ described in text; (b) lower - E_2M_2-ECR-15 sample described in the text. Bar = 1 μ.

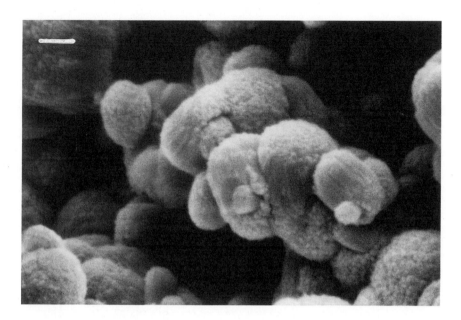

FIGURE 8-2(c) Scanning electron micrograph of mazzite sample P_2M_2-ECR-15. Bar = 1 μ.

crystallization temperatures, 140° to 160°C is the preferred crystallization temperature. The comparison sample of ECR-15 was made at a composition:

$$2.4E_2M_2I:\ 1.95Na_2O:\ Al_2O_3:\ 7.5SiO_2:\ 0.05TMA_2O:\ 120H_2O$$

by reacting components in the manner described above, except that the template was in the iodide form rather than the hydroxide, and the use of colloidal silica eliminated the need for aluminum in the form of alum (Vaughan 1991).The morphology of the products are shown in Figure 8-2b and the X-ray diffraction pattern in Figure 8-3c for the template occluded form and 8-3d for the "burnt-out" form. Substitution of P_2M_2I for E_2M_2I yielded the ECR-15 products shown in Figure 8-2c. The influences of different variables in this series are shown in Table 8-2.

At the same synthesis gel composition, the Si/Al ratios increase from about 2.7 to 3.7 as a function of template volume in the sequence:

$$P_2M_2> E_2M_2 > \text{choline} > \text{TMA}$$

A similar sequence also applies to the Ga substituted forms. With all other variables kept constant, the more aluminous syntheses yielded marginally lower Si/(Al,Ga) products. Pyrrolidine syntheses (Rubin and Rosinski 1982), made from higher Si/Al gels (Table 8-2) produce the highest Si/Al ratio MAZ, but in highly uneconomic gel chemical compositions.

TABLE 8-2
ECR-15 Synthesis Gels and Products

R_2O:Al_2O_3	R	TMA_2O:Al_2O_3	Na_2O:Al_2O_3	SiO_2:Al_2O_3	H_2O:Al_2O_3	Seeding	Si Source	Al/Ga Source	Temp. °C	Time days	Si/Al
1.1	$(PrOH)_2Me_2NI$	0.10	1.95	7.5	120	2.0%	HS-40	C-31	150	2.0	3.61
1.2	$(EtOH)_2Me_2NI$	0.00	1.95	7.5	120	2.0%	HS-40	Ga_2O_3	150	7.0	3.26
1.2	$(EtOH)_2Me_2NI$	0.00	1.95	7.5	120	2.0%	HS-40	C-31/Ga_2O_3	150	7.0	3.40
		0.60	2.40	9.0	135	1.0%	N	C-31/Ga_2O_3	100	3.0	2.78
1.2	$(PrOH)_2Me_2NI$	0.00	1.95	7.5	120	2.0%	HS-40	Ga_2O_3	150	10.0	3.70
0.6	$(PrOH)_2Me_2NI$	0.00	1.95	7.5	120	2.0%	HS-40	Ga_2O_3	150	10.0	3.27
1.2	$(EtOH)_2Me_2NI$	0.04	1.95	7.5	120	1.0%	HS-40	C-31	150	4.0	3.63
1.2	$(EtOH)_2Me_2NI$	0.05	1.95	7.5	120	1.0%	HS-40	C-31	160	5.0	3.76
1.2	$(EtOH)_2Me_2NI$	0.04	1.95	7.5	120	1.0%	HS-40	C-31	160	4.0	3.42
1.2	$(EtOH)Me_3NCl$	0.00	1.95	7.5	110	1.0%	HS-40	C-31	150	3.0	3.19
1.8	$(EtOH)_2Me_2N$	0.00	0.60	9.0	180	2.0%	N	C-31	150	7.0	4.30
16.5	$(EtOH)_2Me_2N$	0.00	2.67	15.0	1150	0.0%	Silica gel	Na alumin.	130	31.0	4.25
14.0	Pyrrolidine	0.00	3.28	16.8	574	0.0%	N	C-31	100	46.0	5.42
		0.50	2.75	9.0	140	2.0%	N	C-31	150	1.0	2.83

Silica gel = Grace 923; Na alumin. = Fisher sodium aluminate; C-31 = ALCOA $Al_2O_3 \cdot 3H_2O$; Ga_2O_3 = Ingal; seeding = %$(Al,Ga)_2O_3$ from seeds.

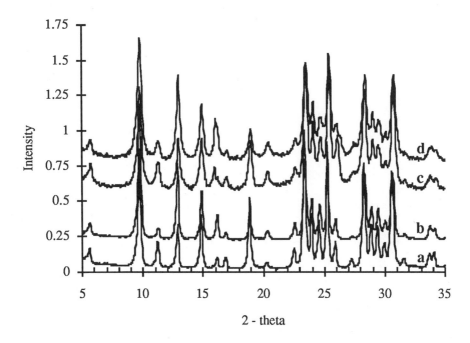

FIGURE 8-3 X-ray Diffractograms for TMA-MAZ (Omega, ZSM-4) befor (a) and (b) after calcination to remove template, and E_2M_2-ECR-15 before (c) and after (d) calcination.

CHARACTERIZATION

X-ray diffractograms (Siemens D500; Cu K radiation; 2°-60°) show few relative intensity differences between the template containing varieties, as shown in Figure 8-3. Unit cell differences are minor (~1%) and even significant changes in Si/Al ratio, template size and loading induce little change, as illustrated in Figure 8-4. After removal of the template major reductions in intensities are observed and peak broadening may occur, particularly at higher angles. These are more pronounced with increasing Ga substitution for Al, and with increasing TMA contents.

Thermogravimetric analyses (absolute and derivative, Figure 8-5) were obtained using DuPont 951 and 2950 instruments at a heating rate of 10°C/min. in air. They show the major differences in template decomposition temperatures - about 435°C for E_2M_2 and 575°C for TMA, distinctly differentiating the former in the channel site and the latter in the gmelinite cage. TMA spiked ECR-15 syntheses (Figure 8-5c) show a progressive ancillary weight loss up to about

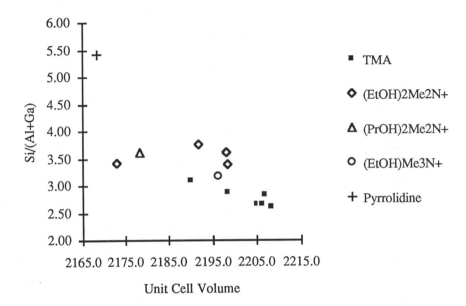

FIGURE 8-4 Plot of hexagonal unit cell volume ($\mathring{A}^3$) for various differently templated MAZ samples against Si/(Al+Ga) ratio.

550°C. The tailing in the ECR-15 TGA (Figure 8-5b) may be indicative of polymeric degradation products, or impurities from the custom manufactured E_2M_2 template solution, trapped in gmelinite cages (these are probably TMA or trimethylamine). However, analyses (^{13}C-NMR and liquid chromatography) failed to detect these in the E_2M_2 solution.

Hexane sorption experiments (obtained using an automated Cahn 2000 combined with Baratron PDR-C-1B pressure gauges, ASCO solenoids and controlled by a Hewlett-Packard 86B/3497A computer data acquisition system) are the preferred method to evaluate "accessible sorption capacity," and depending on the specific mode of synthesis and calcination, these results may be highly variable. (Weeks et al. 1976 discuss in detail the desirability of controlling the mode of calcination to retain structural integrity.) Evaluating the sorption of hexane before and after burn off shows the interesting characteristics demonstrated in Figure 8-6 for post calcination n-hexane isotherms at room temperature. When calcined at 300°C the TMA-MAZ shows a high capacity as the channels are cleared of water and are not blocked by TMA which is located in the channels, but after calcination at 600°C to burn off the TMA, the hexane capacity is reduced 30% (Figure 8-6a). The E_2M_2-MAZ on the other hand shows lower initial sorption capacity but higher capacity after template "burn-off"

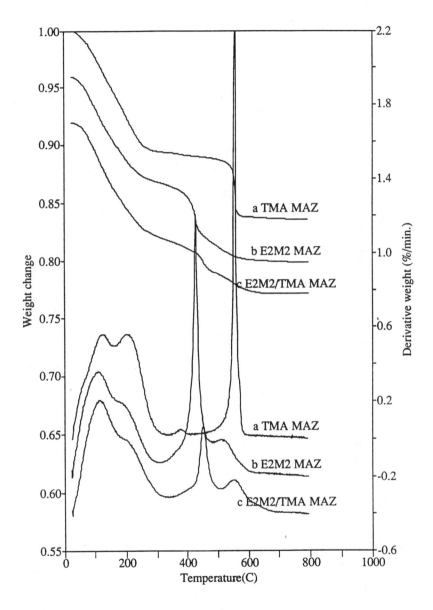

FIGURE 8-5: TGA (upper) and DTA (lower) scans of a) TMA-MAZ (Omega, ZSM-4) showing a "burn-off" at 557°C; b) E_2M_2-ECR-15 made without "spiked" TMA, and showing tailing in the "burn-off" at 435°C (presumably derived from TMA impurity in the E_2M_2); c) E_2M_2-ECR-15 made with a TMA "spike," showing an extended "burn-off" to 550°C.

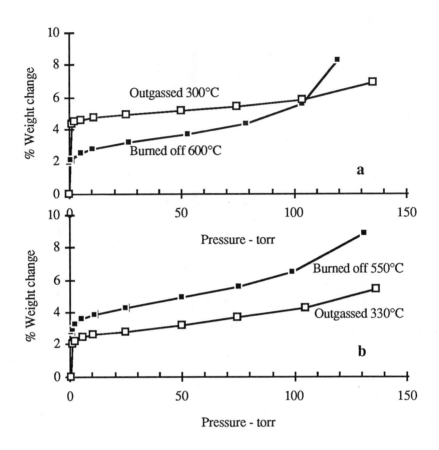

FIGURE 8-6 (a) TMA-MAZ n-hexane sorption isotherms at room temperature before and after template removal. (b) Same isotherms for E_2M_2-MAZ.

(Figure 8-6b), and similar to that for the TMA-MAZ before "burn-off." The E_2M_2 products retained higher hexane sorption capacities than the equivalent TMA products calcined under similar conditions being in the range 5 to 9% wt. (No sample has been made with a hexane sorption capacity comparable to high quality LTL or MOR products.) However, a wide range of calcined MAZ materials made with different templates all gave water sorption capacities between 14% and 15% wt., indicating that pore volume accessible to small molecules is largely retained even if partial channel blockage occurs during template "burn-off."

CONCLUSIONS

Synthetic MAZ zeolites have varying properties depending on the template used in the synthesis and its corresponding "burn-off" temperature. The most siliceous products are obtained in the absence of gallium, using highest levels of the largest template in a more siliceous gel composition. Crystallization times are optimized by seeding the precursor gel. For catalyst and sorbent uses a lower "burn-off" temperature is desirable, in addition to optimization of other recognized variables which may significantly influence structural integrity and accessibility. These include Si/(Al,Ga) ratios (high Si, low Ga preferred), morphology (small, short crystals to minimize channel blocking), and low template content. A zero template route to MAZ is highly desirable for property optimization.

The issue of the existence of the alternate proposed structure for "omega" in different templated synthetic systems is an unsolved problem worthy of further investigations.

References

1. Aiello, R., and R. M. Barrer. 1970. *J. Chem. Soc. A* :1470.
2. Albers, E. W., and D. E. W.Vaughan. 1976. *U.S. Patent* 3,947,482.
3. Barrer, R. M., and H. Villiger. 1969. *J. Chem. Soc. Chem. Commun.* :659.
4. Bowes, E., and J. J. Wise. 1971. *U.S. Patent* 3,578,723.
5. Ciric, J. 1972. *U.S. Patent* 3,923,639; 1968. *British Patent* 1,117,568.
6. Cole, J .S., and H .W. Kouwenhoven. 1973. In *Molecular Sieves (Adv. Chem. Ser.* **121**), ed. W. M. Meier and J. B. Uytterhoeven, p. 583. Washington: Amer. Chem. Soc.
7. Fajula, F., S. Nicolas, F. DiRenzo, C. Gueguen and F. Figueras. 1989. In *Zeolite Synthesis (A. C. S. Symp. Ser.* **398**), ed. M. L. Occelli and H. E. Robson, p. 493. Washington: Amer. Chem. Soc.
8. Flanigen, E. M., and E. R. Kellberg. 1980. *U.S. Patent* 4,241,036; also 1970 *British Patent* 1,178,186.
9. Galli, E. 1974. *Crystal Structure Commun.* **3**: 339.
10. Gottardi, G., and E. Galli. 1985. Natural Zeolites, p. 160. Berlin: Springer-Verlag.
11. Leonowicz, M. E., and D. E. W. Vaughan. 1987. *Nature* **329**:819.
12. Meier, W. M., and D. H. Olson. 1987. *Atlas of Zeolite Structure Types*, London: Butterworths/IZA.
13. Newsam, J. M., R. H. Jarman and A. J. Jacobson. 1985. *Mater. Res. Bull.*
14. Nicolas, S., P. Massiani, M. Vera-Pacheo, F. Fajula and F. Figueras. 1988. In *Innovations In Zeolite Materials Science, (Stud. Surf. Sci. and Catal.* **37**), ed. P. J. Grobert et al., p. 115. New York: Elsevier.

15. Rubin, M. K., and E. J. Rozinski. 1982. *U.S. Patent* 4,331,643.

16. Rubin, M. K., C. J. Plank and E. J. Rozinski. 1977. *U.S. Patent* 4,021,447.

17. Travers, Ch., Ch. Marcilly, F. Raatz, Th. DesCourieres, A. Perrard, F. Fajula and M. Boulet. Spring 1991. A. I. Ch. E. Natl. Mtg., *Petrochem. Ref.* #30E.

18. Vaughan, D. E. W. 1991. *U.S. Patent* 5,000,932.

19. Vaughan, D. E. W., and K. G. Strohmaier. 1989. In *Zeolite Synthesis (A. C. S. Symp. Ser.* **398**), ed. M. L. Occelli and H. E. Robson, p. 506.

20. Vaughan, D. E. W., M. G. Barrett and G. C. Edwards. 1982. *U.S. Patent* 4,340,573.

21. Weeks, T. J., D. G. Kimak, R. L. Bujalsic and A. P. Bolton. 1976. *J. Chem. Soc. Faraday 1* **72**:575.

9

Common Factors in the Synthesis of Na-Chabazite and K-Mazzite

Francesco Di Renzo, François Fajula, François Figueras, M.Agnès Nicolle and Thierry Des Courieres[1]

Laboratoire de Chimie Organique Physique et Cinétique Chimique Appliquées, URA 418 du CNRS, Ecole Nationale Supérieure de Chimie, 8 rue de l'Ecole Normale, 34053 Montpellier, France, fax 33-67144353

[1]Centre de Recherche ELF-France Solaize, BP 22, 69360 St. Symphorien d'Ozon, France

This paper emphasizes the influence of the rate of aluminum incorporation to reactive gels on the orientation of zeolite synthesis. The faujasite/chabazite selectivity can be reversed by aging of the silica source before the addition of aluminum. In a second example, Na,K,TMA-mazzite is produced instead of offretite when aluminum is provided through the slow dissolution of a crystalline source. The changes of the microenvironments of the nucleation are related to modifications of the degree of polymerization of silicoaluminate species in solution and the surface reactivity of amorphous precursors.

INTRODUCTION

It is well known that reactive gels of the same global composition can give rise to different zeolitic phases. For instance, the way of aluminum incorporation in the reagents has been shown to exert a directing role on the nucleation of zeolites X and NaP (Freund 1976). Zeolite X and zeolite B have been prepared from the same reagents by changing the rate of mixing (Kerr 1968). The gmelinite/faujasite and gismondine/faujasite selectivity can be completely reversed by room temperature aging of the crystallization gels (Polak and Cichocki 1973, Rollmann 1984). These results are not surprising, if we consider that history-dependent factors exert a primary influence on the characteristics of the reactive gel. Both the properties of the colloid surface and the partition coefficients between solid and liquid phase may be modified. In this way, the nature of the silicoaluminate units in solution and the availability of heterogeneous nucleation sites, namely the two main factors governing zeolite formation, are affected.

More correlations between the properties of the reactive gel and the kind of zeolite formed may prove useful tools in the understanding of the nucleation mechanisms. In this communication, we want to emphasize two examples of selectivity control through modification of the microenvironment of the nucleation.

EXPERIMENTAL

The reagents used were precipitated silica (Zeosil 175MP, BET surface area 175 m^2/g, pore volume 0.08 ml/g, grain size 2-20 μ, H_2O 12%, Na 0.9%, Al 0.4%), silica gel (Rhône Poulenc, BET surface area 660 m^2/g, pore volume 0.62 ml/g, grain size 20-200 μ, H_2O 21%, Na 0.1%, Al 200 ppm), sodium aluminate (Carlo Erba RLE), potassium and sodium hydroxide (Prolabo RP Normapur), tetramethylammonium hydroxide pentahydrate (TMA, Fluka purum grade), deionized water. Zeolite Y, used as a reagent, was prepared from a reactive gel of composition $4(0.80Na \cdot 0.20TMA)_2O \cdot Al_2O_3 \cdot 9.2SiO_2 \cdot 160H_2O$. The gel was stirred for 24 hours at room temperature and heated at 50°C for 3 weeks. The solid obtained, whose composition was $0.96Na_2O \cdot 0.03TMA_2O \cdot Al_2O_3 \cdot 4.6SiO_2 \cdot 19H_2O$, featured the XRD diagram of well-crystallized zeolite Y.

Hydrothermal reactions at a temperature lower than 100°C were carried out in sealed polypropylene bottles, without stirring. Syntheses at higher temperature were carried out in stirred stainless steel autoclaves.

CHABAZITE CRYSTALLIZATION FROM SODIUM GELS

Synthetic chabazite is readily obtained from gels containing relevant amounts of potassium. Pure chabazite of composition $K_2O \cdot Al_2O_3 \cdot (2.30-4.15)SiO_2 \cdot (3.4-4.6)H_2O$ was obtained by hydrothermal treatment at 150°C from gels of composition $K_2O \cdot Al_2O_3 \cdot (3-6)SiO_2$ (Barrer and Baynham 1956). When sodium was used instead of potassium, chabazite-gmelinite intergrowths were obtained (Barrer et al. 1959, Il'in et al. 1966). The XRD results published suggest gmelinite as the prevailing phase. When the synthesis was carried out in mixed sodium-potassium systems, gmelinite-free chabazite was obtained only for $K/(K+Na)$ ratios higher than 0.8 (Colella and Aiello 1975). Nevertheless, these experiments in the sodium system were carried out from gels with Na/Al ratio higher than 2 and Si/Al ratio higher than 4. The hydrothermal treatment of a gel of composition $2Na_2O \cdot Al_2O_3 \cdot 4SiO_2$ led to the formation of zeolite R, featuring a chabazite-like XRD diagram (see Table 9-1) and a composition of $Na_2O \cdot Al_2O_3 \cdot 3.5SiO_2 \cdot 7H_2O$ (Milton 1962; Breck and Flanigen 1968).

According to literature data, reactive gels of constant OH^-/SiO_2 ratio, between 0.5 and 0.65, should give rise to chabazite for a Si/Al ratio between 2 and 3, gmelinite for a Si/Al ratio between 3.5 and 7, and faujasite for a Si/Al ratio higher than 7 (Breck and Flanigen 1968). That seems to preclude the obtaining of more silicic chabazite by increasing the Si/Al ratio of the synthesis medium. It should be possible to overcome this difficulty by promoting the formation of reactive gels of non-uniform composition. Once the most critical step -nucleation- is overcome in the most aluminic part of the gel, zeolite crystals should be able to grow in a more silicic environment. In order to modify the scale of homogeneity of the gel, we tested different procedures of mixing and aging of the reagents.

TABLE 9-1. X-ray diffraction data of Na-chabazite and reference phases.

Na-chabazite		Natural chabazite (Gottardi and Galli 1985)		Zeolite R (Milton 1962)	
d (A)	I/I_o	d (A)	I/I_o	d (A)	I/I_o
11.7	9				
9.38	20	9.36	73	9.51	88
6.88	38	6.90	12	6.97	35
				5.75	16
5.53	11	5.56	33	5.61	26
5.03	50	5.02	39	5.10	45
		4.68	15	4.75	12
4.30	64	4.33	95	4.37	78
4.10	13			4.13	12
3.98	28	3.980	13	4.02	14
3.90	21	3.875	38	3.92	35
				3.80	16
3.61	17	3.594	47	3.63	41
3.44	55	3.446	18	3.48	25
3.28	18			3.34	12
3.18	15	3.183	13	3.21	18
				3.13	12
2.934	100	2.928	100	2.95	100
		2.891	55	2.89	16
				2.80	14
		2.694	19	2.71	14
		2.687	19	2.66	10
2.606	41	2.606	27	2.62	25
2.508	10	2.507	32	2.53	22
				2.39	10
2.297	12		2.298	10	
2.090	20	2.089	14	2.10	14

Reactive gels of composition $4Na_2O \cdot Al_2O_3 \cdot 9.2SiO_2 \cdot 160H_2O$, when heated under hydrothermal conditions immediately after mixing, are known to give rise to gmelinite (Aiello and Barrer 1970). In our experiment, a gel with the above composition was prepared by adding precipitated silica to the alkaline aluminate solution and aged 24 hours at room temperature under stirring. After 2 days at 50°C in quiescent conditions traces of zeolite Y could be detected by XRD. Crystallization was complete after 14 days of synthesis. The final product consisted of pure, well-crystallized zeolite Y with a Si/Al ratio of 2.3.

In another experiment, precipitated silica was added to the alkaline aluminate-free solution and stirred at room temperature for 24 hours. Sodium aluminate was added under stirring to the silicate clear solution. The gel obtained, again of composition $4Na_2O \cdot Al_2O_3 \cdot 9.2SiO_2 \cdot 160H_2O$, was aged at room temperature for 24 hours under stirring. After 15 days at 50°C the gel was still XRD-amorphous. After 20 days at 50°C chabazite was present in large amounts. Crystallization was complete after 27 days of synthesis. The XRD diagram of the final product, reported in Table 9-1, corresponds to well-crystallized chabazite with traces of gmelinite intergrowths. The composition of the final product was $1.03Na_2O \cdot Al_2O_3 \cdot 4.2SiO_2 \cdot 7.9H_2O$.

Synthesis temperature can be varied in a reasonable field without affecting the results. At 75°C the same phases were obtained after a shorter synthesis time.

A small fraction of the reactive gel was withdrawn and filtered just before the hydrothermal treatment. The filtrate contained 440 mmol/l Si and 18 mmol/l Al in the case of the gel giving rise to faujasite, and 840 mmol/l Si and 13 mmol/l Al in the case of the gel giving rise to chabazite. The solid fraction, washed to remove excess alkalinity and dried, presented a composition of $1.04Na_2O \cdot Al_2O_3 \cdot 6.5SiO_2 \cdot 6.2H_2O$ in the case of the gel giving origin to faujasite, and $1.02Na_2O \cdot Al_2O_3 \cdot 6.0SiO_2 \cdot 7.2H_2O$ in the case of the gel giving origin to chabazite. The chabazite parent system is hence characterized by the highest Si/Al gradient between solid phase and solution.

The pH was 12.9 and 13.6 for parent gels of chabazite and faujasite, respectively. The pH values suggest that tetrahedra are more connected in the faujasite parent system, fewer hydroxyls being immobilized at the solid surface (Lowe 1983). As a consequence, we can visualize the system issued from the rapid addition of aluminate as formed of smaller particles than the faujasite parent system.

MAZZITE CRYSTALLIZATION FROM SODIUM-POTASSIUM-TMA GELS

Synthesis

Mazzite (zeolite omega, ZSM-4) was first synthesized in a mixed sodium-tetramethylammonium system (Ciric 1966; Flanigen and Kellberg 1967). Attempts to use potassium instead of sodium were not successful, and no pure mazzite could be obtained for K/Na ratios higher than 0.1, offretite being the most favored phase for higher potassium contents (Aiello and Barrer 1970; Cole and Kouwenhoven 1973). A fairly silicic synthesis mixture (Si/Al$\geq$6) was indicated as necessary to favor the formation of mazzite instead of offretite in the mixed potassium-sodium systems (Cole and Kouwenhoven 1973).

On this basis, we supposed that a still more silicic reactive gel could allow mazzite to form at a higher potassium content. Higher alkalinity was probably needed to compensate for the lower reactivity of a less aluminic system. A gel of composition $17(0.67Na \cdot 0.28K \cdot 0.05TMA)_2O \cdot Al_2O_3 \cdot 38SiO_2 \cdot 900H_2O$ was prepared by adding a ground silica gel to the solution containing the other reagents. The gel, stirred for 1 hour at room temperature, was heated at 115°C for 24 hours. The final product was well-crystallized offretite of composition $0.15Na_2O \cdot 0.57K_2O \cdot 0.29TMA_2O \cdot Al_2O_3 \cdot 5.8SiO_2 \cdot 11H_2O$, in good agreement with the literature data for less silicic systems.

In order to further decrease the activity of the aluminate species in the nucleation medium, we had recourse to a less reactive source of aluminum. Zeolite Y is known to be metastable towards mazzite (Plank, Rosinski, and Rubin 1968; Dwyer 1969; Dwyer and Chu 1979), hence we considered it an appropriate delayed-action source of aluminum. A gel of the same composition as the former one was prepared by adding the ground silica gel and the appropriate amount of faujasite Y to the alkaline solution. The gel, stirred for 1 hour at room temperature, was heated at 115°C for 24 hours. The final product was pure, well-crystallized mazzite of composition $0.33Na_2O \cdot 0.38K_2O \cdot 0.29TMA_2O \cdot Al_2O_3 \cdot 8.7SiO_2 \cdot 9H_2O$ (Di Renzo et al. 1989).

The incorporation yield of potassium in mazzite is higher than the sodium one. As a consequence, potassium can be hardly considered as a poison for mazzite crystallization. More probably, the rationale for the selectivity shift from mazzite to offretite at increasing K/Na ratio has to be based on the competition between the nucleation rates of the two zeolites. Mazzite is clearly favored by more silicic environments. Could this effect be related to the prevalence of different silicoaluminate units in the synthesis solution? It can be observed that five-tetrahedra rings are secondary building units for mazzite, and not for offretite.

The domains of formation of the phases obtained in a wider range of cation ratio and alkalinity are represented in Figure 9-1. In the whole domain of crystallization offretite and mazzite were the predominant phases, the field of stability of mazzite being broadened when faujasite (Y) was used as source of aluminum. Phillipsite (zeolite ZK-19) was always accompanied by offretite or mazzite. TMA-sodalite and gismondine were detected only as trace contaminants of mazzite.

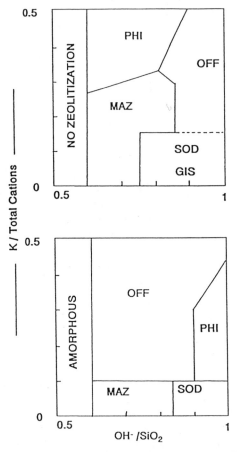

Figure 9-1: Fields of formation of zeolite phases after 24 h reaction at 115°C from reactive gels of composition $y[xK \cdot (0.95-x)Na \cdot 0.05TMA]_2O \cdot Al_2O_3 \cdot 38SiO_2 \cdot 900H_2O$. Source of aluminum: top, faujasite; bottom, sodium aluminate.

Properties of K-Na-TMA Mazzite

The synthesis of mazzite from sodium-potassium-TMA gels produces a silica-rich zeolite (Si/Al = 4.2 to 5.5) appearing as hexagonal single crystals 1-2 μ long, 0.3-0.5 μ wide. Typical specimens contain approximately two potassium, two sodium and two TMA cations per unit cell. Thermal analysis profiles feature two zones with equivalent weight losses for the decomposition of the TMA cations, at 360 and 560°C, suggesting that one TMA is located in the main channel and the second in the gmelinite cages (Aiello and Barrer 1970). A peculiarity of the present system is that, whereas a single ion exchange at reflux temperature in 2M ammonium nitrate removes nearly all the sodium ions, the zeolite retains between 1.5 and 2 wt% potassium (about one cation per unit cell) after two such treatments. These non-exchangeable cations are probably located inside the gmelinite cages, where they substitute for TMA ions. A location in the main channel can be ruled out in view of the sorption properties presented below.

Protonic forms of mazzite were prepared by calcination in air of the parent material at 550°C followed by two ion exchanges with ammonium nitrate and further calcination. In Table 9-2 some relevant properties of the protonic forms

TABLE 9-2. Properties of protonic forms of K,Na-TMA mazzite. Comparison with Na-TMA mazzite.

Sample[1]	Treatment[1]	Sorption Capacity (ml/g)		XRD Cryst.
		$CyC_6{}^2$	TMB^2	(%)
H(K)-MAZ	Calc. 550°C	0.09	0.058	100
H(K)-MAZ	Calc. 750°C	0.07	0.04	90
H(K)-MAZ	Steam Deal.	0.06	0.03	75
H-MAZ	Calc. 550°C	0.08	0.064	100
H-MAZ	Calc. 750°C	0.06	0.045	80
H-MAZ	Steam Deal.	0.02	0.01	30

[1] See text.
[2] CyC_6 = cyclohexane; TMB = 1,3,5-trimethylbenzene.

of mazzite samples synthesized in the presence (H(K)-MAZ) and in the absence (H-MAZ) of potassium are compared. Both zeolites were obtained as hexagonal single crystals of comparable size. The H(K)-MAZ sample contained 0.01 wt% Na and 1.7 wt% K and had a Si/Al ratio of 5. The H-MAZ specimen contained 0.03 wt% Na with a Si/al ratio of 3.2. The sorption capacity for cyclohexane in both materials corresponded to the calculated void space of the main channel (Breck and Grose 1973). Calcination in a dry atmosphere had little effect on the sorption characteristics and crystallinity, confirming the excellent thermal stability of the structure (Vera-Pacheco et al. 1986). After a dealumination treatment, consisting of a steaming at 750°C followed by an acid attack in 1M nitric acid at reflux temperature, the H(K)-MAZ was still highly crystalline whereas H-MAZ suffered considerable loss in sorption capacity and XRD crystallinity. This improved hydrothermal stability of H(K)-MAZ is likely to be due to the presence of residual cations, which would prevent the extensive hydrolysis of Al-O bonds (Chauvin et al. 1990), combined with a high silica content.

CONCLUSION

These two examples provide a new illustration of how the degree of polymerization of the silicoaluminate species in solution and the surface reactivity of the amorphous precursors may direct the zeolite nucleation. In both cases selectivities have been reversed by changing the rate of incorporation of aluminum to the reactive gel.

In the case of the faujasite-chabazite system, a sodium aluminosilicate gel of the composition usually resulting in gmelinite formation may give origin to different zeolites when suitably prepared. Plain aging leads to the formation of faujasite, usually obtained from more silicic systems. Preparation of a more heterogeneous gel by rapid addition of aluminate to predissolved silica results in the formation of chabazite, normally prepared from more aluminic systems.

In the case of mixed sodium-potassium-tetramethylammonium systems, the use of a delayed-dissolution source of aluminum allows the formation of mazzite in a range of K/Na ratio that usually produces offretite.

There is no doubt that a better knowledge of the complex gel chemistry involved in the precrystallization stage will result in a better understanding of the mechanisms of zeolite nucleation. From a more practical point of view, the fact that pure zeolitic phases may be obtained from gels whose composition falls beyond their preferential field of formation may prove useful for modifying the properties of the final material. Very often the conditions for crystal growth are imposed by the composition of the initial reaction medium, which is primarily designed so as to produce a required structure type. Broader nucleation fields should allow a given phase to grow under conditions yielding crystals with improved composition or size. Nucleation engineering is the long term challenge.

REFERENCES

Aiello, R., and R.M. Barrer. 1970. Hydrothermal Chemistry of Silicates. Part XIV. Zeolite Crystallization in Presence of Mixed Bases. J. Chem. Soc. A: 1470.

Barrer, R.M., and J.W. Baynham. 1956. Hydrothermal Chemistry of the Silicates. Part VII. Synthetic Potassium Aluminosilicates. J. Chem. Soc. (London): 2882.

Barrer, R.M., et al. 1959. Hydrothermal Chemistry of Silicates. Part VIII. Low-temperature Crystal Growth of Aluminosilicates, and of Some Gallium and Germanium Analogues. J. Chem. Soc. (London): 195.

Breck, D.W., and E.M. Flanigen. 1968. Synthesis and Properties of Union Carbide Zeolites L, X and Y. In Molecular Sieves, p. 49. London, Society of Chemical Industry.

Breck, D.W., and R.W. Grose. 1973. A Correlation of the Calculated Intracrystalline Void Volumes and Limiting Adsorption Volumes in Zeolites. Adv. Chem. Ser. 121: 319.

Chauvin, B., et al. 1990. Factors Affecting the Steam Dealumination of Zeolite Omega. Zeolites. 10: 174.

Ciric, J. 1966. Synthetic Crystalline Aluminosilicate Catalyst. Fr. pat. 1,502,289 to Mobil Oil.

Cole, J.F., and H.W. Kouwenhoven. 1973. Synthesis and Properties of Zeolite Omega. Adv. Chem. Ser. 121: 583.

Colella, C., and R. Aiello. 1975. Sintesi idrotermale di zeoliti da vetro riolitico in presenza di basi miste sodico-potassiche. Rend. Soc. Ital. Mineral. Petrol. 31: 641.

Di Renzo, F., et al. 1989. Nouvelle zéolithe de la famille de la mazzite, son procédé de synthèse et son utilisation comme catalyseur. Fr. appl. 89 11383 to Société Nationale ELF-Aquitaine.

Dwyer, F.G. 1969. Production of synthetic faujasite. U.S. pat. 3,642,434 to Mobil Oil.

Dwyer, F.G., and P. Chu. 1979. ZSM-4 Crystallization via Faujasite Metamorphosis. J. Catal. 59: 263.

Flanigen, E.M., and E.R. Kellberg. 1967. Crystalline Zeolite Molecular Sieves. Neth. pat. 6,710,729 to Union Carbide.

Freund, E.F. 1976. Mechanism of the Crystallization of Zeolite X. J. Crystal Growth 34: 11.

Gottardi, G., and E. Galli. 1985. Natural Zeolites, p. 333. Berlin: Springer.

Il'in, V.G., et al. 1966. Some Peculiarities of the Crystallization and Properties of High-silica Faujasites. Dokl. Akad. Nauk SSSR 166 (3): 604.

Kerr, G.T. 1968. Chemistry of Crystalline Aluminosilicates. IV. Factors Affecting the Formation of Zeolites X and B. J. Phys. Chem. 72 (4): 1385.

Lowe, B.M. 1983. An Equilibrium Model for the Crystallization of High Silica Zeolites. Zeolites 3:300.

Milton, R.M. 1962. U.S. pat. 3,030,181.

Planck, C.J., E.J. Rosinski and M.K. Rubin. 1968. Zeolite Crystallization. Fr. pat. 69,45544 to Mobil Oil.

Polak, F., and A. Cichocki. 1973. Mechanism of Formation of X and Y Zeolites. Adv. Chem. Ser. 121: 209.

Rollmann, L.D. 1984. Synthesis of Zeolites, an Overview. In Zeolites: Science and Technology, eds. F. Ramôa Ribeiro et al., NATO ASI Series 80, p. 109. The Hague: Martinus Nijhoff.

Vera-Pacheco, M., et al. 1986. Synthesis of a Thermally Stable Form of Zeolite Omega. In New Developments in Zeolite Science and Technology, Preprints of Posters Papers, The 7th International Zeolite Conference, p. 21. Tokyo: Japan Association of Zeolite.

10

Precursor Phases and Nucleation of Zeolite TON

Francesco Di Renzo, Françoise Remoue[1], Pascale Massiani[2],
François Fajula, François Figueras and Thierry Des Courières[3]

Laboratoire de Chimie Organique Physique et Cinétique Chimique Appliquées,
URA 418 du CNRS, Ecole Nationale Supérieure de Chimie, 8 rue de l'Ecole
Normale, 34053 Montpellier, FRANCE, fax (33)67144353

[1]present address: Centre de Recherches de Pont à Mousson, BP 109, 54704 Pont
à Mousson, FRANCE

[2]present address: Laboratoire de Réactivité de Surface et Structure, URA 1106
du CNRS, Université Pierre et Marie Curie, 4 Place Jussieu, 75252 Paris Cedex
05, FRANCE

[3]Centre de Recherche ELF-France Solaize, BP 22, 69360 St. Symphorien
d'Ozon, FRANCE

The evolution of the synthesis medium of zeolite TON has been studied under
various experimental conditions. The induction time for zeolite nucleation is related to the coalescence kinetics of the precursor amorphous solid. Zeolite crystals appear at the surface of the gel, suggesting a mechanism of heterogeneous
nucleation. Diffusion barriers inside the gel are responsible for non-uniform nucleation rates and broadening of the crystal size distribution.

INTRODUCTION

The formation of an amorphous phase is a nearly general feature of the first steps
of zeolite crystallization procedures (Barrer 1982). A high concentration of reactive silica and the presence of a condensing agent (organic or inorganic) are required for the ordered crystallization of the zeolite. The same factors promote a
rapid disordered polymerization of the silicoaluminate units. The colloid formed
is metastable towards zeolite formation in exactly the way zeolites are metastable
towards the crystallization of denser phases (Breck 1974).

This colloid is usually referred to as a gel. The high contact surface between solid and liquid phases is its most important feature, and it is the basis of its universally accepted property: the gel is a reservoir of reactants for zeolite formation, its solubility controlling the composition of the crystallization solution. Another possible role of the gel has been at the focus of a long debate between zeolite researchers: do zeolites form in the synthesis solution or inside the gel (for reviews see Guth and Caullet 1986; Szostak 1989)? Nowadays it is widely accepted that zeolite crystal growth is fed by the surrounding solution, but the location and mechanism of the zeolite nucleation are still unclear (Jansen 1991).

The openness of the debate suggests that it is worth collecting more data about the characterization of synthesis gels, paying special attention to those evolution patterns which may be correlated to zeolite formation. This paper relates our results about the evolution of the synthesis medium of zeolite TON.

EXPERIMENTAL

The influence of the synthesis conditions on the crystallization kinetics of zeolite TON has been detailed in a previous paper (Di Renzo et al. 1991). Among the examples of this paper we chose the synthesis systems corresponding to the shorter and to the longer induction times, in order to characterize gels following fairly different evolution patterns: respectively, a gel A of composition $3.6Na_2O \cdot 56TETA \cdot Al_2O_3 \cdot 54SiO_2 \cdot 1050H_2O \cdot 50NaCl$ (where TETA stands for triethylenetetramine) and a gel B of composition $4Na_2O \cdot 48DEA \cdot Al_2O_3 \cdot 70SiO_2 \cdot 1780H_2O$ (where DEA stands for diethanolamine).

Crystallization experiments were carried out in a 0.5 liter stainless steel reactor equipped with an anchor-shaped stirrer. A sampling outlet allowed specimens to be withdrawn without disturbing the crystallization process. The sampling pipe could slide from the sampling point to a rest point above the liquid level. After the withdrawal of each specimen a vapor purge under autogeneous pressure prevented the slurry from accumulating inside sampling pipe and valves. The reactor was heated by an electrical furnace controlled by a thermocouple sunk in the crystallization batch. The reaction temperature (160°C for gel A and 170°C for gel B) was attained in about 20 minutes and the stirring rate was 300 rpm.

The reagents were silica sol (Cecasol 30, SiO_2 25%, Na 0.2%, Al 60 ppm, pH 8.8, grain size 12-18 nm), sodium aluminate (Carlo Erba RLE), diethanolamine and triethylenetetramine (Prolabo), sodium hydroxide and sodium chloride (Prolabo RP Normapur), and deionized water. ^{13}C NMR of the triethylenetetramine used indicates the presence of a fraction of the branched triethylenetetramine isomer and traces of $R_2N(CH_2)_2OH$ groups. The reagents were mixed under stirring in the order: alkaline solution, aluminate, organic agent, silica sol and, when required, aqueous chloride solution. The mixture was stirred for 12 hrs at room temperature before the beginning of the synthesis.

Samples of 10 ml were periodically withdrawn. The slurry was collected in 100 ml of cold deionized water in order to optimize the separation of the solid and liquid fractions (Iler 1979, p. 233). The solid fraction was recovered by filtration. The clear solution, containing the diluted mother liquor, was preserved for elemental analysis. The solid was washed with deionized water up to pH 9 and dried at 70°C in air.

X-ray powder diffraction (CGR Theta 60 diffractometer, Cu $K\alpha$ radiation) was used to identify the phases present in the solid fraction. The degree of crystallinity of the specimen was evaluated from the area of selected diffraction lines. Physical mixtures of well-crystallized zeolite and amorphous solid withdrawn at the beginning of the synthesis were used as crystallinity standards. Texture of the amorphous gel and morphology of the zeolite crystals were determined by scanning electron microscopy (Cambridge Instruments Stereoscan 260).

PHYSICAL ENVIRONMENT OF THE NUCLEATION

The silica sol used consists of a stable suspension of silica particles 15-25 nm in size. When this sol is added to the alkaline aluminate solution its particles coalesce to form roughly shaped spheres 5-50 μ in diameter. The observation of some spheres broken by stirring indicates that they are indeed hollow shells. A possible mechanism for the formation of such aggregates is based on the formation of an emulsion of silica sol droplets inside the organic solution when reagents are mixed. The silica particles begin to form a three-dimensional network at the interface with the alkaline solution, stabilizing the drop shape. Other silica particles coalesce and the thickness of the shell increases as soon as aluminate and cations diffuse through the previously formed outer layer. The high viscosity of the organics used (Reid, Prausnitz and Sherwood 1977; Sand et al. 1987) probably contributes to the formation of a stable emulsion when silica sol is added.

The agglomeration process does not affect the morphology of the primary sol particles. The surface of the shell consists of a loose packing of well-defined particles about 20 μ in size, very similar to the one depicted in Figure 10-1b.

The pH of systems A and B at room temperature was 11.5 and 12.2, respectively. The pH of the synthesis systems depends on the interplay of several buffer equilibria. The protonation of the organic amines and the hydroxylation of the silica surface are among the reactions involved (Lowe 1983). In order to evaluate the influence of the organics used we measured the pH of several synthesis mixtures, with and without addition of sodium chloride (see Table 10-1). To weigh the buffer effect of silica we measured the pH of solutions whose compositions were the same, except for the absence of silica (see Table 10-2). In every case a much lower pH was observed in the presence of silica, indicating that the hydroxylation of the silica surface is the main buffer reaction in these systems, at least at room temperature. The formation of anionic species by silica dissolution probably contributes to the decrease of pH.

Table 10-1. The pH of synthesis mixtures of composition $3.8Na_2O\cdot(0-60)$amine$\cdot Al_2O_3\cdot 60SiO_2\cdot 950H_2O\cdot(0-50)NaCl$ after 24 hours of stirring at 25°C.

pH	without NaCl	with NaCl
No organic	12.3	11.7
Diethanolamine	12.0	11.1
Triethylenetetramine	12.6	11.4

Table 10-2. The pH of solutions of composition $3.8Na_2O\cdot(0-60)$amine$\cdot Al_2O_3\cdot 950H_2O\cdot(0-50)NaCl$ at 25°C.

pH	without NaCl	with NaCl
No organic	13.2	12.9
Diethanolamine	13.2	12.9
Triethylenetetramine	14.0	13.6

The addition of sodium chloride is another pH-lowering factor. The decrease of the activity of the hydroxyl ions at increasing ionic strength can account for a part of the pH lowering. The larger pH decrease when sodium chloride is added in the presence of silica can be accounted for by a higher level of adsorption of sodium ions and charge-compensating hydroxyls by the silica surface. Indeed the surface charge density of silica rapidly increases with the electrolyte concentration (Iler 1979, p. 356).

The pK_a of the amines used is 8.9 for diethanolamine and 9.9 for triethylenetetramine. At the higher pH of the synthesis systems the protonation of both amines should be negligible, and no detectable bufffer effect is expected, at least at room temperature. Indeed some effect of diethanolamine on the pH is only observed when alkalinity has been lowered by silica. The addition of triethylenetetramine always leads to a large increase of pH, suggesting a reagent contamination by free alkali.

GEL EVOLUTION AND INDUCTION TIME

The surface morphology of some gels is deeply affected by heating in hydro-thermal conditions. After 30 minutes at 160°C the shells of gel A feature a continuous hunchbacked surface, appearing sintered (Figure 10-1a). Such a morphology indicates that the primary particles of silica are dissolved to some extent. Dissolved silica precipitates again forming a less soluble flat surface (Iler 1979, p. 227). On the contrary, the surface of gel B is not affected to such an extent by hydrothermal conditions. In that case, the primary silica particles are still well-defined after more than 11 hours at 170°C (Figure 10-1b).

Figure 10-1. Surface morphology of the gels before zeolite nucleation. (a) Gel A; (b) gel B.

a

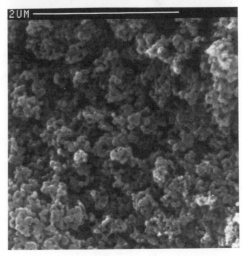

b

Compositions of the solid and liquid fractions of systems A and B during the synthesis are reported in Figures 10-2 and 10-5, respectively. The crystallinity-versus-time curves are also reported in the Figures.

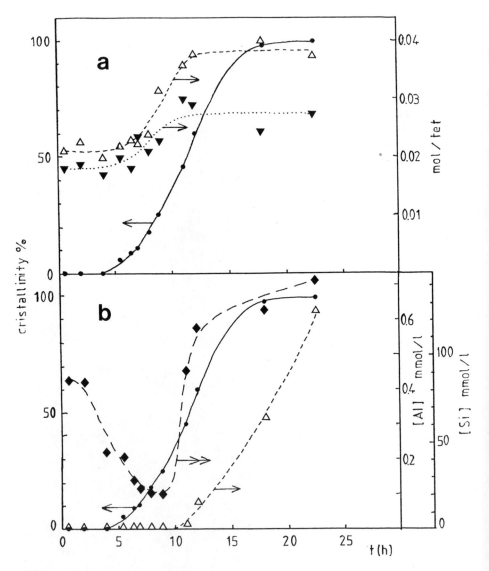

Figure 10-2. Composition of the synthesis system and crystallinity versus time, experiment A. (a) Mole fraction (referred to Si+Al) of aluminum (Δ) and triethylenetetramine ($\blacktriangledown$) in the solid. (b) Concentration of silicon ($\blacklozenge$) and aluminum (Δ) in solution. Crystallinity curve in full line.

System A evolves more rapidly. The actual induction time is significantly shorter than 4 hours. Electron microscopy of the sample withdrawn after 4 hours of synthesis already shows tiny zeolite crystals embedded in the outer surface of the amorphous shells (Figure 10-3a). The amount of crystallized material is still too low to be detected by XRD. No crystals can be detected on the concave inner surface of broken shells. This surface appears coated by an incomplete, tattered layer of vitreous luster (Figure 10-3b). The coating has probably been deposited by evaporation of the silica-rich solution trapped inside the shell and apparently unaffected by filtering and washing. Probably the less-volatile components of the shell-locked solution have not been affected by drying. This effect could account for the high organic content measured for the amorphous solid (Figure 10-2). This finding suggests that most of the broken shells have been fractured during post-sampling treatments, especially during drying and solid

Figure 10-3. Morphology of the outer (a) and inner (b) surface of the hollow shells of gel A at the appearance of the first zeolite crystals.

a

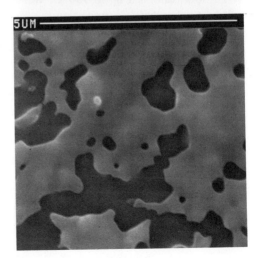

b

recovery. As a consequence, the differences observed by SEM may be attributed to actual differences between the inner and outer surfaces during the hydrothermal treatment.

During the first phases of the synthesis the silica concentration in solution steadily decreases (Figure 10-2b), in agreement with a lowering solubility of the amorphous solid. Gel sintering is likely to still be occurring. At a later stage of the synthesis silica concentration grows again and maintains the highest level reached also after the end of the crystallization. A similar phenomenon has already been observed in the synthesis of less silicic zeolites (Katovic et al. 1989). In that context, the increase of silica concentration in solution could be justified by a preferential incorporation of aluminum in the crystals. Such an explanation hardly seems to apply to the present system. The Si/Al ratio of the final solid is 25.3, fairly near to the Si/Al ratio of the reagent mixture. Moreover, the concentration of aluminum in solution also increases in the last phase of the synthesis (Figure 10-2b). It is likely that the phenomenon depends on the accumulation in solution of silicic and silicoaluminate units unfit for incorporation in the zeolite lattice. Moreover, a pH rise related to the zeolite formation (Lowe 1983, Araya and Lowe 1984) may account for an increased gel solubility, a possible concomitant cause of the upward concentration trend.

When the composition of the solid fraction of system A is considered (Figure 10-2a), it can be observed that the triethylenetetramine content in the early stages of synthesis is only slightly lower than the organic content in the final zeolite product. A lower extent of amine incorporation would be expected for an amorphous silica (Araya and Lowe 1984). It is likely that the analysis results include the organic content of the solution trapped inside the hollow shells. The TG curves of the initial gel and of the final zeolite of experiment A are reported in Figure 10-4. Weight loss above 200°C was in good agreement with the organic content of the sample (Araya and Lowe 1984). In the case of the zeolite the weight loss continued up to 750°C, whereas for the initial gel the weight loss ended up below 550°C, suggesting that in the last case the amine is not locked inside a microporous structure.

System B presents a fairly longer induction time. No zeolite crystals are detected, either by XRD or SEM, after more than 11 hours of synthesis (Figure 10-5). At that moment, no detectable sintering of the primary silica particles has yet taken place (Figure 10-1b). As a consequence, the silica aggregates are permeable to washing and the inner solution can be removed from the solid. A low organic content of the solid fraction at that stage of the synthesis is indeed observed (Figure 10-5a).

No significant change in the composition of the liquid fraction occurs during the first 11 hours of synthesis (Figure 10-5b). When the first zeolite crystals are detected, slightly before 20 hours of synthesis, the silica concentration in solution has significantly decreased. Needle-like zeolite crystals appear to be embedded in a tattered vitreous matrix (Figure 10-6). In this case crystals appear

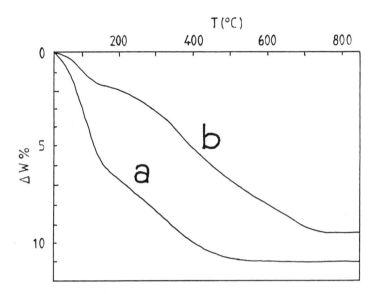

Figure 10-4. TG curves of the initial gel (a) and final zeolite product (b) of experiment A.
Heating rate 5°/min under air flow.

at both the outer and the inner surface of the hollow-shell silica aggregates. Both
the decrease of silica concentration and the presence of a sintered solid indicate
that dissolution of the primary silica particles and reprecipitation in a less soluble
form takes place, but several questions remain open. Did the reorganization of
the gel precede or accompany the nucleation of the zeolite? Were the crystals
formed at the surface of a sintered solid or throughout all the thickness of a still
macroporous shell?

As far as the slope of the crystallinity curves is concerned, the crystallization
of system A is slower than the crystallization of system B (Figures 10-2 and
10-5). This result is probably related to the number of crystals which grow in
the two systems, as suggested by the crystal size data presented in the next
section.

CRYSTAL SIZE

The final products obtained in experiments A and B are depicted in Figure 10-7. In both cases the crystals of zeolite TON still hold together according to the hollow-shell morphology of the initial gel (Figure 10-7b). Crystals
of sample A form smaller radial hedgehog-like clusters (Figure 10-7a); (100)
and (110) twinnings commonly occur. Crystals of sample B are randomly entan-
gled (Figure 10-7c). Twinning seems to be les common.

Crystal size distributions, evaluated by SEM, are reported in Figures 10-8 and 10-9 for samples A and B, respectively. We calculated separate statistics for crystals observed at the inner and outer surface of the hollow shells. Both

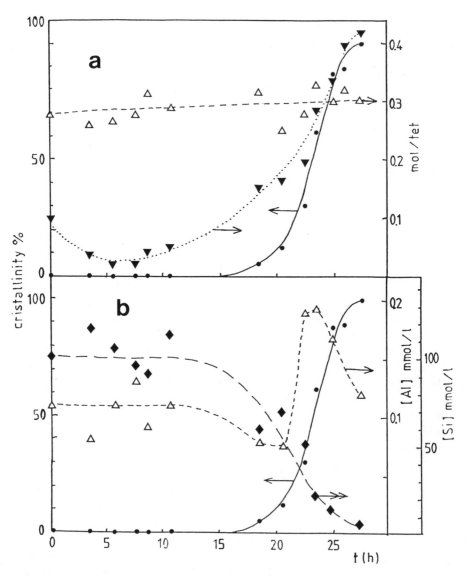

Figure 10-5. Composition of the synthesis system and crystallinity versus time, experiment **B.** (a) Mole fraction (referred to Si+Al) for aluminum (△) and diethanolamine (▼) in the solid. (b) Concentration of silicon (◆) and aluminum (△) in solution. Crystallinity curve in full line.

Figure 10-6. Morphology of system **B** at the appearance of the zeolite crystals.

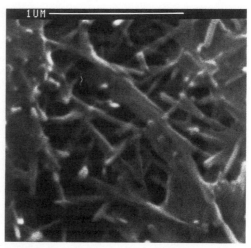

radial and random orientations of the crystals made it difficult to evaluate their length in a statistically significant way. Hence we measured the crystal width at the main beam. The average width of the crystals of sample A was 0.55 ± 0.14 μm at the inner and 0.37 ± 0.09 μm at the outer surface of the aggregate. In the case of the crystals of sample B, the average width was 0.042 ± 0.009 μm at the inner and 0.036 ± 0.009 μm at the outer surface of the hollow shell. The aspect ratio of the crystals (length/width ratio) was about 5 for sample A and 12 for sample B.

Larger crystals correspond to a lower number of nucleation sites. The observation of larger crystals in the case of sample A confirms that the nucleation occurred on the limited surface of a non-macroporus solid. Radial clusters suggest that several crystals originated from a small number of nucleation sites. The crystal size gradient through the thickness of the shell confirms that nucleation was easier on the outer than on the inner surface. In contrast, the crystals of sample B present a smaller size, random distribution and a lesser size gradient through the thickness of the shell. These factors suggest that nucleation takes place on every available surface throughout a still macroporous solid.

CONCLUSION

Zeolite TON forms in an environment which is heterogeneous from many points of view. Hollow silica aggregates create diffusional barriers and concentration gradients inside the liquid phase. As a consequence, the evolution of different portions of the solid phase depends on their location in the reaction medium as well as on the synthesis conditions.

Figure 10-7. Morphology of the final
zeolite product. (a) Experiment A,
(b, c) experiment B.

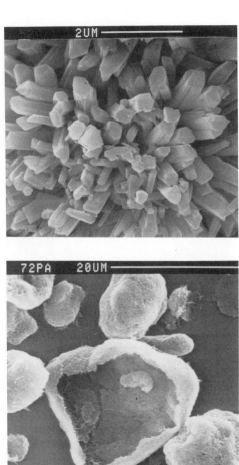

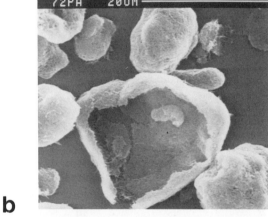

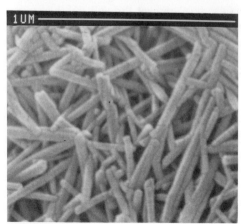

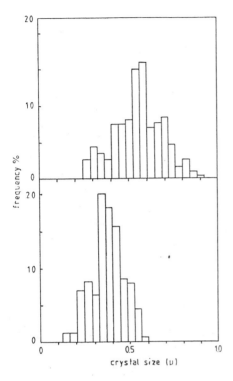

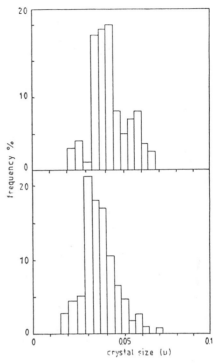

Figure 10-8. Experiment A. Crystal size distribution of the final zeolite product at the inner (top) and outer (bottom) sufaces of the hollow-shell aggregates.

Figure 10-9. Experiment B. Crystal size distribution of the final zeolite product at the inner (top) and outer (bottom) surfaces of the hollow-shell aggregates.

The zeolite crystallization is not the only phenomenon affecting the phase composition of the system. Sintering of the amorphous silica by dissolution and reprecipitation under a less soluble form always precedes, or accompanies, the nucleation of the zeolite. Earlier evolution of the gel corresponds to a shorter induction time for zeolite nucleation. Can any causal correlation be established between these two phenomena?

The first zeolite crystals appear at the outer surface of the gel aggregates, and the number of crystals formed seems to be proportional to the area of gel surface accessible to solution. Surface and solution co-operate to the crystallization, hence a mechanism of heterogeneous nucleation should be logically postulated. The question of the role of the gel in this mechanism remains. Does it intervene directly, by epitaxial nucleation of the zeolite on its surface, or in a less direct way, by modifying the silicic species present in the surrounding solution?

REFERENCES

Araya, A., and B.M. Lowe. 1984. Synthesis and Characterization of Zeolite Nu-10. Zeolites 4: 280.

Barrer, R.M. 1982. Hydrothermal Chemistry of Zeolites, p. 119. London: Academic Press.

Breck, D.W. 1974. Zeolite Molecular Sieves, p. 248. New York: Wiley.

Di Renzo, F., et al. 1991. Crystallization Kinetics of Zeolite TON. Zeolites 11: 539

Guth, J.L., and Ph. Caullet. 1986. Synthèse des Zéolites, Perspectives d'Avenir. J. Chim. Phys. 83 (3): 155.

Iler, R.K. 1979. The Chemistry of Silica. New York: Wiley.

Jansen, J.C. 1991. The Preparation of Molecular Sieves. In Introduction to Zeolite Science and Practice, ed. H. Van Bekkum, E.M. Flanigen, and J.C. Jansen, p. 77. Amsterdam: Elsevier.

Katovic, A., et al. 1989. Crystallization of Tetragonal (B_8) and Cubic (B_1) Modifications of Zeolite NaP from Freshly Prepared Gel. Part 1. Mechanism of the Crystallization. Zeolites 9: 45.

Lowe, B.M. 1983. An Equilibrium Model for the Crystallization of High Silica Zeolites. Zeolites 3: 300.

Reid, R.C., J.M. Prausnitz and T.K. Sherwood. 1977. The Properties of Gases and Liquids, 3rd ed., p. 636. New York: McGraw-Hill.

Sand, L.B., et al. 1987. Large Zeolite Crystals: their Potential Growth in Space. Zeolites 7: 387.

Szostak, R. 1989. Molecular Sieves, Principles of Synthesis and Identification, p. 190. New York: Van Nostrand Reinhold.

11

Zeolite Synthesis in the Na-TMA System at 100°C

P. Donald Hopkins *Amoco Oil Company, MS H-9, P.O. Box 3011, Naperville, IL 60566-7011*

The synthesis of zeolites in systems related to those that produce FAU (both zeolite Y and zeolite X) was studied both in the presence and absence of tetramethylammonium (TMA) cations. The synthesis of zeolite Y requires the Na/(Si + Al) ratio in the reactants to be at least 0.75. Zeolite Y was observed in low crystallinity as the initial product in the synthesis of GME (zeolite S) and SOD (hydroxysodalite) just as previously reported for MAZ (omega or ZSM-4) syntheses. Addition of various salts to the zeolite Y synthesis revealed that bromide prevents crystallization of FAU as it does for at least two other zeolites. Substitution of TMA for sodium in zeolite X syntheses caused the product to change to LTA at lower levels of substitution than found previously for zeolite Y syntheses. Results are discussed in the context of one possible synthesis mechanism.

INTRODUCTION

Zeolite Y, structure code FAU in the IUPAC system (Meier and Olson 1987), is an important component of cracking and hydrocracking catalysts. A knowledge of the synthesis mechanism of zeolite Y could lead to methods for synthesizing Y in various crystallite sizes or Si/Al ratios, or to more cost-effective syntheses. Additional benefits might be insights into methods for synthesis of other related structures such as the hexagonal analog of cubic FAU known as Breck structure six (Breck 1974, pp. 54-7). Our studies of the synthesis mechanism of zeolite Y have included studies of two other closely related structures, zeolite A (LTA) and hydroxysodalite (SOD); all three structures include a common structural component, variously called the sodalite unit, Type I 14-hedron or beta cage.

In a previous communication (Hopkins 1989) the effect of the tetramethylammonium cation (TMA) on the syntheses of FAU, LTA and SOD was discussed. In addition to verifying the well known structure directing effect of TMA to zeolites containing one of the two 14-hedra, the gmelinite cage or the sodalite cage (Kerr 1966; Aiello and Barrer 1970; Whyte et al. 1971; Jarman and Melchior 1984), several additional observations were made. In reactant mixtures designed to produce FAU or LTA, high TMA/Na ratios led to the synthesis of LTA and low ratios directed the synthesis toward FAU. As TMA content was increased in some series, zeolites, such as MAZ or EAB, that contain gmelinite cages were synthesized at lower ratios than required for the synthesis of the more stable LTA. This is somewhat surprising because TMA appears to be a better fit for the spherical sodalite cage of LTA than for the non-spherical gmelinite cage, and, therefore, one would expect TMA to preferentially template zeolites containing sodalite cages. Sodalite cages in ZK-4, a high-silica LTA, and SOD were completely occupied by TMA ions as shown previously (Jarman and Melchior 1984) but in FAU the sodalite cages were occupied approximately statistically by either one TMA or two sodium ions. Atomic ratios of Si/Al in products appeared to be unaffected by the TMA/Na ratio of the reactant mixture or whether the product was FAU or LTA. This paper reports the continuation of this experimental program and its extension to include syntheses in the absence of TMA and syntheses of zeolite X, which has the FAU structure but higher aluminum content than zeolite Y.

EXPERIMENTAL

The experimental procedures and reagents used were identical to those reported previously (Hopkins 1989). The primary reactant mixtures that were used are summarized in Table 11-1. All of the mixtures in Table 11-1 are variations of zeolite Y recipes except for mixture 4, which is based on a zeolite X recipe (Breck 1974). In zeolite X (FAU) type syntheses the silica source was sodium silicate (PQ N Brand, 27.9% SiO_2 and 8.9% Na_2O) rather than the colloidal silica (Ludox HS-40) used in the other syntheses. The syntheses were carried out in Pyrex flasks, with stirring, or in Teflon bottles, without stirring; the temperature for all syntheses was 373 K. Most reactant mixtures were aged at room temperature for at least 24 hours before heating. TMAOH was used when TMA was substituted for NaOH in order to

TABLE 11-1   Reactant Ratios for Zeolite Synthesis

Reactant Mixture	Moles/Moles of Al_2O_3				
	SiO_2	Na_2O	$(TMA)_2O$	M_2O(a)	H_2O
1	10.6	3.95	0	--	161
2	10.2	15.1	0	--	446
3	17-27	13.5	0	--	400
4	3.0	--	--	3.6	144
5	27.3	13.5	0	--	440

(a) M = TMA + Na

minimize pH changes. The products discussed here are the first major products which appeared, usually in the first 24 hours of heating. Product phases and purities were identified by powder X-ray diffraction.

RESULTS

Synthesis of Zeolite X

The effect of TMA on zeolite X synthesis was investigated using variations of reactant mixture 4, substituting TMA for sodium but maintaining the sum of the two constant. Results are shown in Table 11-2. Substitution of less than one-half of the sodium by TMA caused the product to change from FAU to LTA. A similar change was observed before in syntheses of zeolite Y and ZK-4 (high silica LTA) but at higher TMA substitutions (Hopkins 1989), even though the total concentration of TMA and sodium ions was about the same in both the Zeolite X and zeolite Y synthesis mixtures, about 1.4 cations per T atom.

The Effect of TMA/Na on Product Structure in Zeolite Y Type Syntheses

Studies of the effect, reported previously (Hopkins 1989), of the TMA/Na ratio on the product derived from alteration of zeolite Y (FAU) type syntheses have been extended to lower overall amounts of total base as shown in Table 11-3. Initial sampling was done at various times between 24 and 72 hours. All reactant mixtures with Na/T atom (where T, or tetrahedral atom, stands for both Si and Al) above 0.75 initially produced FAU even if some TMA was present. Some mixtures with Na/T atom below 0.75 also produced FAU if sufficient TMA was present. (The importance of the Na/TMA ratio of 0.75 will be emphasized in a later section of this paper).

TABLE 11-2 Effect of TMA on the Synthesis of Zeolite X
(Variations of Reactant Mixture 4)

TMA/Na	Age	Stir	Major Product
0.00	Yes	Yes	FAU
0.34	"	"	"
0.79	"	"	LTA
0.00	Yes	No	FAU
0.14	"	"	"
0.48	"	"	LTA

At the low levels of base employed no LTA was synthesized but two other
structures, MAZ and SOD, that contain 14-hedra were produced in
preference to FAU under some conditions. TMA was always present when
MAZ or SOD were formed but their synthesis appears to be a complex
function of the TMA/Na and Si/Al ratios.

The Effect of Anions on Zeolite Synthesis

The effect of anions on the synthesis of zeolite Y (FAU) was determined by
addition of salts to reactant mixture 5. Salts were added in the amount of
one equivalent per sodium equivalent in the basic synthesis mixture. The
results, listed in Table 11-4, show that bromide has a deleterious affect on the

TABLE 11-3 Initial Products of Syntheses in the
Al-Si-Na-TMA System as a Function of the Reactant
Mixture Composition
(H_2O/Al_2O_3 = 400, Mixtures Aged and Stirred)

TMA/Al	Na/Al	SiO_2/Al_2O_3			
		25	20	15	10
0.0	5.0	Amorph	Amorph	Amorph	FAU
2.5	2.5	Amorph	Amorph	Amorph	FAU
0.0	7.0	Amorph	Amorph	FAU	FAU
2.0	5.0	SOD	Amorph	FAU,MAZ	FAU
3.5	3.5	SOD	Amorph	FAU	FAU
2.0	7.0	MAZ	MAZ	FAU	FAU

TABLE 11-4 Initial Products of Syntheses in the
Al-Si-Na System as a Function of Added Salts
(1 Equivalent of Salt per Sodium Equivalent in
Synthesis Mixture 5 - Mixture Aged and Stirred)

Added Salt	Initial Product
--	FAU
NaCl	FAU
NaBr	Amorphous
Na_2SO_4	GME + CHA
KCl	Amorphous
TMABr	Amorphous
TMACl	FAU

reaction; both sodium bromide and TMA bromide prevented formation of crystalline products in the first 24 hours. Chloride, on the other hand, appeared to have no effect as either the sodium or TMA salt; KCl inhibited crystal formation but this appears to be the effect of potassium alone or a combination of potassium and chloride. The presence of sulfate caused the product to change from FAU to a mixture of GME and CHA.

FAU as a Precursor to GME-CHA and SOD Formation

FAU was detected, in small amounts which existed for only a few hours, as a precuror to highly crystalline GME (zeolite S) when using reactant mixture 1. This mixture can produce highly crystalline Y under some conditions (Breck 1974). In some instances this synthesis produced a mixture of two closely related structures, CHA (zeolite R) and GME following the short-lived FAU. SOD (hydroxysodalite) was also produced following a brief appearance of FAU when reactant mixture 2 was employed. There have been reports of FAU as a precursor to zeolite omega (MAZ) which is synthesized in the presence of TMA (Perrotta et al. 1978; Dwyer and Chu 1979). The presence of FAU in the early stages of the synthesis of these three zeolites indicates a strong tendency for this structure to form over a wide range of conditions, with and without TMA and at various Na/Al ratios in the reactants (compare the Na_2O contents of reactant mixtures 1 and 2).

The Effect of Na/(Si + Al) Ratio the on Synthesis of FAU

The effect of sodium concentration in the the reactant mixture was investigated using variations of reactant mixture 3, which normally produces zeolite Y in the SiO_2 range of 17-27. Both the sodium and silicon contents were varied as shown in Table 11-5. If any crystalline phase appeared within 24 hours it was invariably FAU (zeolite Y) which frequently transformed into GIS within 72 hours; minor amounts of ANA also appeared eventually in some syntheses. The designation "amorph" means that no crystalline phase formed in the first 72 hours. As the Si/Al ratio in the reactant mixture was reduced the demarcation between FAU and amorphous products moved to lower Si/Al ratios. If product structure is plotted as a function of Na/T atom in the synthesis mixture, as in Figure 11-1, a sharp demarcation between amorphous products and FAU occurs at about 0.75. Thus, a minimum sodium concentration appears to be required for formation of FAU. The pH at the time of mixing and at the time the first product was observed, or at the end of the synthesis if no crystalline product appeared, are listed in Table 11-5.

TABLE 11-5 Initial Products of Syntheses in the
Al-Si-Na System as a Function of the Reactant Mixture
Composition with Initial (Top) and Final (Bottom) pH
(H_2O/Al_2O_3 = 400, Mixtures Aged and Stirred)

$SiO_2/$ Al_2O_3	Na/Al				
	5.0	7.5	10.0	12.5	15.0
30	--	--	Amorph 13.8 12.1	FAU 13.8 12.5	FAU 13.8 --
25	--	Amorph 13.6 12.0	Amorph 13.7 12.2	FAU 13.7 12.8	FAU 13.7 13.2
20	--	Amorph 13.7 12.0	FAU 13.8 12.9	FAU 13.8 13.4	FAU 13.6 13.4
15	Amorph 13.5 12.1	FAU 13.5 13.0	FAU 13.3 13.3	FAU 13.7 13.5	--
10	FAU 13.5 13.0	FAU 13.3 13.2	FAU 13.3 13.3	--	--

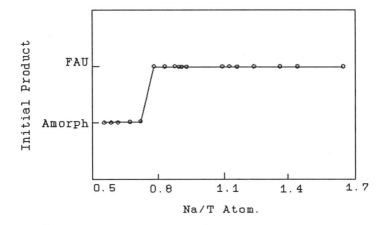

FIGURE 11-1 Effect of Na/T Atom in the Synthesis Mixture on the Initial Product.
See Table 11-5 for Reactant Compositions.

The initial pH varied only slightly over the range of compositions studied, but
the pH remained higher in the reactions in which FAU formed than in the
ones in which it did not. In a similar reaction, not listed in the table, FAU
was the primary product, highly crystalline and with only a trace of GIS, as
the pH dropped to 12.2.

DISCUSSION

As NaOH is added to synthesis mixtures there are increases in sodium ion
concentration, pH and ionic strength. However, as seen in Table 11-5, the
initial pH varied only slightly over the whole range of reactant compositions
studied. Also FAU was observed over nearly the entire final pH range given
in Table 11-5. Therefore, we may dismiss pH change as a significant
variable in controlling the synthesis of FAU in the system used here, and
only the sodium ion concentration and ionic strength need be considered.
The present experiments do not provide enough information to determine
conclusively which of the factors is operable and each is consistent with
possible synthesis mechanisms.
    Synthesis results cannot, by themselves, prove or disprove a particular
synthesis mechanism. Nor can they distinguish between reactions that occur
in solution or in the solid ("gel") phase. However, one can speculate on

possible mechanisms that are consistent with the experimental results. The results are consistent with FAU being formed by a linking of double six rings of T atoms (D6R) in the solution or gel phases or on the surface of growing crystallites. Some of the postulated steps also would hold for single six ring involvement. It should be pointed out that there is no conclusive evidence for the existence of D6R in aluminosilicate solutions or gels. Refinement and proof of the mechanisms must await additional instrumental and modeling studies.

If sodium cations were to serve as a template for FAU formation it would almost certainly be the D6R that are templated. Previous work demonstrated that sodalite cage formation is not a controlling step in FAU synthesis because TMA, which templates sodalite cages, fosters the synthesis of LTA rather than FAU. The D6R of FAU appear to be completely filled with sodium cations as about 16 ions per unit cell (one per D6R) are difficult to exchange at room temperature (Herman and Bulko 1980). In a D6R a Na^+ would be surrounded by six lattice oxygens at 0.262 nm in a distorted octahedron, by six more at 0.344 nm in another distorted octahedron and six more at 0.363 nm in a ring; the sum of the ionic radii of Na^+ and O^{2-} is about 0.244 nm. This should be a stable environment for Na^+ both electrostatically and sterically.

The increase in ionic strength at higher NaOH concentrations is also consistent with D6R being involved in the controlling step in FAU synthesis. Approach of two D6R in a geometry suitable for growth of sodalite cages, and therefore of FAU, requires that six oxygens of each ring, a number of which are undoubtedly protonated, approach the second ring closely. Two D6R joined, by the elimination of two oxygens, in the proper geometry for FAU are shown in Figure 11-2a. Increased ionic strength would assist this approach by supplying more cations to ameliorate the high concentration of negative charge in this region. It is interesting to note that at lower ionic strengths, related synthesis mixtures produce GME or CHA, structures that consist of D6R joined in parallel rows. Approach of two D6R to join in the configuration required for these two structures, as shown in Figure 11-2b, does not concentrate as many oxygens in one region. Also, the nearest approach of non-bonded oxygens in the latter configuration is about 0.70 nm (A-A in Figure 11-2), not as close as the A-A distance of 0.50 nm in the FAU configuration. The next nearest approach of non-bonded oxygens (A-B) is 0.92 nm in the GME configuration but only 0.74 nm in the FAU configuration.

The widespread appearance of FAU in the early stages of the synthesis of a variety of zeolites suggests that raw reaction mixtures, after short aging periods, contain species that easily join to form this structure. The rapid disappearance of FAU demonstrates its instability, consistent with its relative

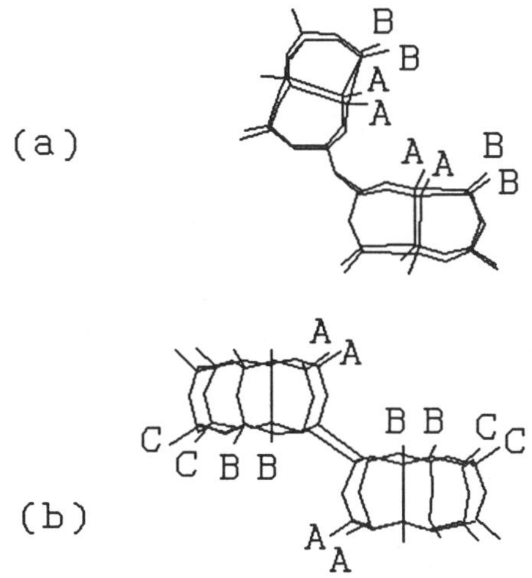

FIGURE 11-2 The Two Methods of Joining Two D6R: (a) FAU Precursor,
(b) GME or CHA Precursor. A, B, and C indicate oxygen locations.

scarcity in natural zeolite deposits. At least part of the tendency for FAU to
form is consistent with the possible mechanism discussed above. The
tendency for FAU to form initially indicates that this reaction is facile,
possibly aided by the high sodium ion concentration in raw synthesis
mixtures. GME (and CHA) would replace FAU as the free sodium ion
concentration becomes depleted by incorporation in D6R and sodalite cages
and would remain the primary species due to greater stability. MAZ would
replace FAU as TMA ions encourage the growth of gmelinite cages.
Formation of FAU in the early stages of SOD synthesis is a bit more difficult
to explain in this context but it does support the strong tendency for FAU
formation even at the high sodium hydroxide concentrations, under which
SOD is generally synthesized. Either the high stability of SOD or the low
stability of FAU could cause the growth of SOD at the expense of FAU.
The arrangement of four sodium ions and one hydroxyl ion in each sodalite
cage of SOD (Smeulders et al. 1987) may be particularly stable and direct the
structure towards SOD at very high sodium ion concentrations.
    The inhibiting effect of bromide ions on FAU formation is difficult to
explain. Other similar effects of anions are known in zeolite synthesis. For
instance, tetraethylammonium (TEA) hydroxide is the only good template for

synthesis of zeolite beta; if TEABr is substituted either MOR or MFI is the major product (Hopkins 1991). Also, in the course of a related study (Hopkins, et al. 1988, 1990) we have found that ZK-4, a high silica form of LTA, can be synthesized using TMAOH but not TMABr. No explanation is available as yet and more work on anion effects in zeolite synthesis should be pursued.

REFERENCES

Aiello, R. and R. M. Barrer. 1970. J. Chem. Soc. (A): 1470-5.

Breck, D. W. 1974. Zeolite Molecular Sieves. New York: John Wiley.

Dwyer, F. G. and P. Chu. 1979. J. Catal. 59: 263-71.

Herman, R. G. and J. B. Bulko. 1980. In Adsorption and Ion Exchange with Synthetic Zeolites, ed. W. H. Flank, pp. 177-86. Washington: American Chemical Society.

Hopkins, P. D. 1989. In Zeolite Synthesis, ed. M. L. Occelli, H. E. Robson, pp. 152-60. Washington: American Chemical Society.

Hopkins, P. D. 1991. Unpublished data.

Hopkins, P. D. et al. 1988. In IPNS Progress Report 1986-1988, Section 10, p. 61. Argonne, IL: Argonne National Laboratory.

Hopkins, P. D. et al. 1990. In IPNS Progress Report 1988-1990, Section 10, p. 75. Argonne, IL: Argonne National Laboratory.

Jarman, R. H. and M. T. Melchior. 1984. J. Chem. Soc. Chem. Commun. 414-6.

Kerr, G. T. 1966. Inorg. Chem. 5: 1537-41.

Meier, W. M. and D. H. Olson. 1987. Atlas of Zeolite Structure Types. London: Butterworths.

Perrotta, A. J. et al. 1978. J. Catal. 55: 240-8.

Smeulders, J. B. A. F. et al. 1987. Zeolites 7: 347-52.

Whyte Jr., T. E. et al. 1971. J. Catal. 20: 88-96.

12

Low Temperature Crystallization of MFI Zeolites in an Alkaline Fluoride Medium

M. J. Eapen, S. V. Awate, P. N. Joshi, A. N. Kotasthane
and V. P. Shiralkar,
National Chemical Laboratory, Pune 411 008, India

Low temperature (369-373 K) hydrothermal crystallization of zeolite ZSM-5(MFI, SiO_2/Al_2O_3 80-100) was performed using different templates viz. TPA-Br (tetrapropylammonium bromide), TEBA-Br (triethylbutylammonium bromide) and/or EDA (ethylenediamine) in the presence of potassium fluoride between the pH range 10.5-12.5. The accelerated crystallization rate during ZSM-5 formation with the addition of KF is thought to be caused by the increased solubility of the reactive hydrogel mixtures. The products were characterized by XRD, TG/DTA, IR and the sorption measurements.

INTRODUCTION

Hydrothermal crystallization at elevated temperature (413-473 K) of MFI zeolite from an aluminosilicate gel in an alkaline medium covering a broad range of SiO_2/Al_2O_3 (10 - 10,000) is well established (Argauer and Landolt 1972; Marosi, Stabenow and Schwarzmann 1980; Rubin, Plank and Rosinski 1979; Kouwenhoven and Stork 1980; Ratnasamy et al. 1983; Kulkarni et al. 1982). Production in tons of ZSM-5 by these processes demands huge investment in installing high pressure autoclaves. On the other hand, crystallization at lower temperature (368 - 373 K) for significantly longer periods (up to a month) also becomes rather uneconomical. Earlier attempts (Dwyer 1976; Plank, Rosinski and Rubin 1979) utilizing alkali fluoride in addition to the template in the hydrogel have lowered considerably the crystallization period; thereby providing an incentive for using autoclaves suited for atmospheric pressure. Very high concentration of the templating species makes these processes extremely expensive. Low temperature hydrothermal crystallizations are carried out both in the alkaline (Gubitosa and Gherardi 1983) and non-alkaline (Guth et al. 1986) medium. Non-alkaline fluoride medium facilitates larger crystallites, ease of incorporation of isomorphous species and use of alkylammonium salts instead of hydroxide. Alkaline fluoride medium provides smaller crystallites in reasonable yields with comparatively faster rates of crystallization. In view of this we have systematically studied the optimization of the amount and nature of the templating species; alkali fluoride consumption with comparatively lower crystallization periods and higher product yields. We have also studied the effect of dilution of the gel on the product quality and yield. The results are reported in the present communication.

EXPERIMENTAL

The reagents used to prepare the gel mixture were: sodium silicate (28.0% SiO_2, 8.4% Na_2O and 63.6% H_2O), aluminum sulfate, hexadecahydrate (G.R. Loba-Chemie), sulfuric acid (98%, SRL), tetrapropylammonium bromide (TPA-Br, Aldrich), triethylbutyl-ammonium bromide (TEBA-Br, SRL), ethylenediamine (Loba-Chemie) and seed crystals of ZSM-5 ($SiO_2/Al_2O_3 \simeq 80$) from previously synthesized batches (Shiralkar et al. 1991). In a typical synthesis experiment 41.5 g sodium silicate was diluted with 50 g of distilled water. Added to that with stirring was a solution of 2 g TPA-Br dissolved in 50 g distilled water. To the resultant mixture an acid slurry (prepared by dissolving 1.2 g aluminum sulfate 16 H_2O, 2 g of potassium fluoride, 2.1 g sulfuric acid and 90 g of distilled water) was added with vigorous stirring. Finally, a finely powdered 0.6 g (5 wt% of the estimated product yield on a dry basis) seed crystals of ZSM-5 were added. It was then stirred until it became homogeneous with its pH in the range 11.1 ± 0.2 and its oxide mole composition to be:

$$30 \text{ Na}_2\text{O}:1.0 \text{ Al}_2\text{O}_3:100 \text{ SiO}_2:X \text{ K}_2\text{O}:Y \text{ R}_2\text{O}:5700\text{-}7300 \text{ H}_2\text{O}$$
where $X = 9\text{-}36$ and $Y = 2\text{-}80$ depending upon template used.

The gel mixture was transferred and sealed in a stainless steel autoclave (250 ml capacity) and then was crystallized under static conditions (unstirred) at 369 K for 4-15 days depending upon the composition. At the end of the crystallization, the autoclave was quenched under cold water and the solid product was separated by suction filtration or centrifugation. The solid product was washed with distilled water until it was free from soluble occluded species including potassium fluoride. Filtrates along with the washings were analyzed for their silica and alumina contents in order to estimate solid product yield and its silica to alumina ratio. The solid product was dried at 393 K for 6 h and was then moisture equilibrated at room temperature prior to further characterization. Sorption properties were measured on the samples in the sodium form (obtained by decomposing the template at 743 K).

Chemical compositions of the solid products were determined by wet chemical methods including atomic absorption (Hitachi, Z-8000 Japan) and inductively coupled plasma (Jobin Yuon-JY-38 VHR) spectrometers and Flame photometry (Toshniwal, Model). The percent crystallinity and crystalline phase purity was studied by scanning XRD patterns on a Philips diffractometer (PW-1730) using CuK_x ($\lambda = 1.54017$ Å) radiations. The zeolite product with highest crystallinity without amorphous impurities was used as a reference standard. The framework IR spectra (400-1250 cm^{-1}) of the sample were obtained on a Pye-Unicam SP-300 spectro-photometer. Thermoanalytical curves (TG/DTA) were recorded simultaneously on a Netzsch automatic derivatograph (STA-490) using pre-calcined alpha-alumina as a reference material. Crystal morphology was studied by scanning electron microscope (Stere-oscan-150, Cambridge) by coating the samples with Au-Pd evapo-rated film. Sorption properties were measured at 298 K on a McBain type gravimetric unit connected to a high vacuum system. Prior to the sorption measurements the sample was degassed at 673 K under vacuum (10^{-6} torr). Sorption kinetics were deter-mined for water, n-hexane and cyclohexane at 298 K and $P/P_o = 0.8$.

RESULTS AND DISCUSSION

It is well established now that during the hydrothermal synthesis of zeolite in the non-alkaline fluoride medium, the role of solubilization, usually played by OH⁻ (in conventional alkaline medium), is played by F⁻ ions. Therefore when F⁻ is used in addition to OH⁻, the enhanced solubilization effect lowers the temperature of zeolite crystallization. Compared to the conventional method reported (Kulkarni et al. 1982), a large amount of excess templating agent in addition to a fluoride salt is used during this process (Gubitosa and Gherardi, 1983). In an attempt to minimize the consumption of both the templating species and the fluoride salt, we have carried out some synthesis experiments following the procedure reported by Gubitosa and Gherardi (1983) and Shiralkar et al. (1991), modified by using sodium silicate (instead of silica gel) with or without seed crystals. Table 12-1 summarizes the synthesis experiments carried out to optimize the potassium fluoride for the crystallization of ZSM-5 zeolite. Experiment No. 3 (Table 12-1) was

Table 12-1

Optimization of potassium fluoride in the synthesis of ZSM-5 (SiO_2/Al_2O_3 = 100) in fluoride medium
H_2O = 6608 moles; TPA-Br = 39.0 moles;
sulfuric acid = 11.0 moles

Expt. No.	KF (moles)	Crystal- lization time(days)	Crystal- linity (%)	Solid yield (%)
1.	48	12	100	45
2.	38	8	100	50
3.	29	4.5	100	54
4.	18	4.5	100	82
5.	9	4.5	20	10

almost identical to that reported elsewhere (Gubitosa and Gherardi 1983) except that sodium silicate was used in place of silica gel as a source of silica. The XRD pattern (Fig. 12-1) shows that the fully crystalline ZSM-5 was obtained in four and half days. This crystallization period is somewhat higher than that reported earlier (Gubitosa and Gherardi 1983) with the same oxide mole composition. Perhaps the use of sodium silicate in place of silica gel may be responsible for the slower crystallization. Fig. 12-2A shows the framework I.R. spectrum of a representative sample obtained during the present study and it confirms the product to be fully crystalline ZSM-5. Fig. 12-2B shows the DTA curves of the as-synthesized sample No. 4 (Table 12-1). Curve 1 (Fig. 12-2B) is the DTA curve for the sample after its regular washing with distilled water. In addition to the usual (Kotasthane and Shiralkar 1986) exothermic effect (573-773 K) on account of the template decomposition, two additional exotherms are observed at 502 K and 862 K. The first exotherm in DTA (corresponding to only 2 wt% loss in TG) may probably be due to a decomposition of the physically occluded template and/or potassium fluoride. However, there was no loss in weight in TG corresponding to an exotherm at 862 K. It was then suspected that some of the potassium fluoride may also be occluded. Therefore, the sample was washed again thoroughly and the filtrate was found to contain F⁻ ions. The DTA of the sample washed twice (Curve 2, Fig. 12-2B) shows the

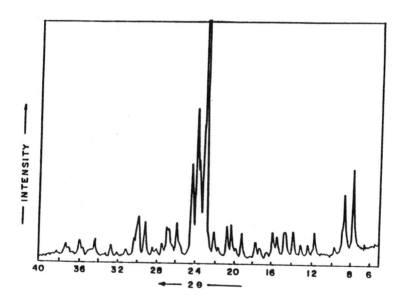

Fig. 12-1. XRD pattern of ZSM-5 synthesized in alkaline fluoride medium at 369 K.

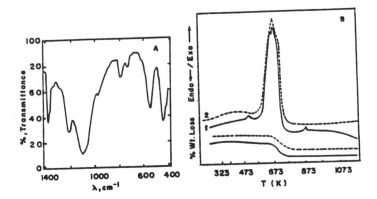

Fig. 12-2.(A) Infrared spectrum of zeolite ZSM-5 synthesized in KF medium.
(B) TG/DTA curves of as-synthesized zeolites (1) before washing (2) after washing.

absence of both the exotherms at 502 K and 862 K. Therefore, when as-synthesized sample was not washed thoroughly, traces of the template and potassium fluoride were occluded and an exothermic effect in DTA may be due to the decomposition of the template and transformation of KF into some complex, respectively.

In view of the above observations, the amount of potassium fluoride needed for the low temperature crystallization was optimized and its influence on the crystallization period was studied. Sample numbers 1 and 2 of Table 12-1 show that as the moles of KF were increased from 29 to 48 the crystallization, the process became slower and the crystallization period increased from four and a half days to eight days and also the percent solid yield of ZSM-5 decreased from 54 to 45. When KF moles were decreased to 18, the crystalline ZSM-5 was obtained in four and a half days but with the increased solid product yield up to 80%. On further decreasing KF moles to 9 it was observed that at the end of four and a half days, only ZSM-5 of 20% crystallinity was obtained, indicating slower crystallization. This shows that 18 moles KF (per mole of alumina) gives crystalline ZSM-5 in as many as four to five days with maximum solid yield to be 82%. Usually in case of high silica zeolites like ZSM-5, variataion in silica reactivity during hydrothermal crystallization affects the percent solid yield and so also the product SiO_2/Al_2O_3 ratio. However, in the present studies the initial four experiments (Table 12-1) produced fully crystalline products whose SiO_2/Al_2O_3 ratios were found to be dependent on the percent solid yield. This indicates that as a result of variation in KF, the crystallization rate of the aluminosilicate gel increases up to certain concentrations (18 moles KF per mole of alumina) of KF, below and above which, it then decreases. Sample no. 3 and 4 indicate that KF concentrations of 29 and 18 moles yielded fully crystalline product in four and a half days but the latter concentration yielded a higher percent of solid product.

Once the concentration of the potassium fluoride (in moles) in the starting gel composition was optimized, it was then thought to optimize template concentration, the most cost-bearing factor in the zeolite synthesis. During these experiments the main aim was to get fully crystalline ZSM-5 at 369 K in a maximum of four to

Table 12-2

Optimization of template TPA-Br in the synthesis of ZSM-5 at 369 K in fluoride medium ($SiO_2/Al_2O_3 = 100$)
KF = 18 moles; H_2O = 6608 moles;
sulfuric acid = 11 moles

TPA-Br (moles)	Seed (wt%)	Crystal- lisation time(days)	Crystal- linity	Solid yield
39	nil	4.5	100	82
21	nil	4.5	100	82
10	5	4.5	100	82
4	nil	7.5	100	82
4	5	6.0	100	82
4	10	4.5	100	82
4	15	4.5	100	82
2	10	6.0	50	-
nil	15	11.0	20	-

five days with the highest synthesis efficiency. The SiO_2/Al_2O_3 ratio was fixed at 100 with 18 moles of KF, 6608 moles of water and 11 moles of sulfuric acid. Table 12-2 clearly demonstrates that as the TPA-Br concentration was decreased from 39 to 4 moles, the seed concentration up to 10% was found to be helpful in obtaining fully crystalline ZSM-5 in four and a half days with 82% solid yield. On further decreasing TPA-Br and increasing seeds up to 15 wt% it was found that fully crystalline product was not obtained even by keeping it for 11 days. Therefore, it is concluded that the optimum concentration of templating species (Rao et al. 1990) does give a fully crystalline zeolite. Now by using 18 moles of potassium fluoride and 4 moles of TPA-Br using only 5 wt% seeds, a fully crystalline zeolite ZSM-5 (SiO_2/Al_2O_3 = 100) was obtained in six days. By increasing the seed crystals up to 10 wt% the crystallization period was reduced to four and a half days, but further increase in seed crystals did not have any positive influence in terms of crystallization period. The effect of gel dilution or alternatively that of variation in OH^-/H_2O on the percent yield of

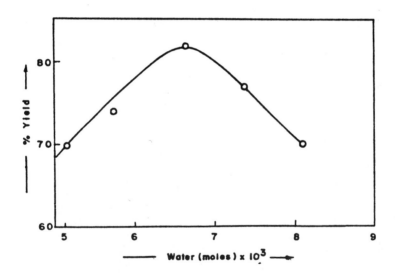

Fig. 12-3. The effect of gel dilution in the crystallization of ZSM-5 from alkaline fluoride medium at 369 K.

the solid product is shown in Fig. 12-3 for a gel with SiO_2/Al_2O_3 100, KF = 18 moles, TPA-Br = 4 moles and sulfuric acid = 11 moles. The highest yield of 82% of fully crystalline product with a four and a half days crystallization at 369 K was obtained at 6608 moles of water in the gel composition. Either an increase or a decrease in the moles of water lowers the percent yield of the solid product. An increase in moles of water leads to a decrease in OH^- /H_2O or F^-/H_2O; on the other hand, a decrease in moles of water imparts the higher F^-/H_2O and OH^-/H_2O ratios. From Fig. 12-3, it appears that 6608 moles of water is optimum with respect to OH^- /H_2O ratio or F^-/H_2O, thereby producing the optimum solid product yield.

So far, it has been demonstrated that concentrations of template (TPA-Br), KF and water are optimized to give fully crystalline ZSM-5 zeolite at 369 K with highest percent solid product yield. It has been also demonstrated in the literature (Kulkarni et al. 1982; Kotasthane et al. 1986) that zeolite ZSM-5 was synthesized at elevated temperature (up to 473 K) by replacing comparatively expensive template TPA-Br by commercially economical templates such as triethyl-propylammonium bromide (TEPA-Br) and triethylbutylammonium bromide (TEBA-Br). It was, therefore, thought of interest whether the hydrothermal synthesis of ZSM-5 in a fluoride medium at 369 K could be achieved by using TEBA-Br in place of TPA-Br. The results are summarized in Table 12-3. Preliminary experiments showed that using only TEBA-Br in place

Table 12-3

Optimization of template TEBA-Br in the synthesis of ZSM-5 (SiO_2/Al_2O_3 = 100) in fluoride medium
KF = 18 moles; water = 6608 moles; sulfuric acid = 2.0 moles

TEBA-Br (moles)	Crystal- lization time(days)	Seed (wt.%)	Crystal- linity (%)	Solid yield (%)
36.6	21	10	100	80
14.3	18	10	100	78
11.0	15	10	100	78
6.6	12	10	100	78
3.3	12	10	10	-

of TPA-Br without seed crystals, could yield only partly crystalline ZSM-5 even after 30 days of crystallization. This indicates that some factors, such as the geometry of the TPA-Br molecule, favor the faster crystallization as compared to that with TEBA-Br. This was also reflected in lower apparent energy of crystallization E_c (83 KJ mole⁻) with TPA-Br as template (Erdem and Sand 1979) as compared to that (90 KJ mole⁻) with TEBA-Br as template (Rao et al. 1990). Taking into consideration previous observations of enhancing the crystallization by using seed crystals, we used 10 wt% (optimized quantity) of the seed crystals in synthesis experiments with TEBA-Br. Table 12-3 indicates that using 36.6 moles TEBA-Br a fully crystalline ZSM-5 was obtained after 21 days of crystallization. It is revealed from the data that as the TEBA-Br concentration was decreased from 36.6 to 6.6 moles the crystallization of ZSM-5 was favoured within 12 days with almost the same percent solid yield of the product (78 %). However, when TEBA-Br was further decreased to 3.3 moles the crystallinity decreased drastically dropped down to 10 % in 12 days. In the case of TPA-Br with lower (but not higher) than the optimized

Table 12-4

Synthesis of ZSM-5 (SiO_2/Al_2O_3 = 100) at 369 K in fluoride medium using EDA as template
KF = 18.0 moles; H_2O = 6608 moles; 10 wt% seed

EDA (moles)	Crystal- lization time(days)	Crystal- linity (%)	Solid yield	H_2SO_4 (moles)
80	15	10	-	11.0
40	10	100	80	5.5
24	10	100	81	nil
16	10	100	79	nil
16	17*	100	78	nil
8	10	50	20	-
8	25*	20	-	-

* Experiments in the absence of seed.

concentration of the template, the crystallization was much slower. But there exists an optimum concentration of TEBA-Br to yield a fully crystalline ZSM-5, below and above which the crystallization becomes slower. Ethylenediamine (EDA) was also found (Schwieger et al. 1989) to act as a template producing fully crystalline ZSM-5 at elevated temperatures up to 473 K. Table 12-4 summarizes the experiments carried out with EDA as templating agent. In this case, most of the experiments were carried out with 10 wt% seed in the gel. Therefore, when EDA concentration was lowered from 80 to 16 moles, the period needed to get fully crystalline product was lowered from 15 to 10 days. At this lower concentration of EDA (16 moles) a fully crystalline product was obtained after 17 days even in the absence of seed crystals. With the decrease in EDA concentration, the quantity of sulfuric acid was also simultaneously reduced. Further decrease in EDA up to 8 moles was found to be compensated by the increase in crystallization time up to 14 days to get fully crystalline product. The percent solid product yield was around 80% in all those experiments which yielded fully crystalline product. EDA concentration lower than 16 moles, i.e.; 8 moles failed to yield fully crystalline product both in the absence and in the presence of seed crystals.

In the above discussion synthesis experiments have often been carried out in the presence of seed crystals to enhance the rate of crystallization and the concentrations of different templates (TPA-

Table 12-5

Comparision of different templates (TPA-Br, TEBA-Br and EDA) in the synthesis of ZSM-5 at 369 K in fluoride medium
$KF = 18.0$ moles; $H_2O = 6608$ moles; 10 wt% seed

Template	Quantity (moles)	Crystal-lization time(days)	Solid yield (%)	H_2SO_4 (moles)
TPA-Br	4	5	79	11
TEBA-Br	6	12	78	21
EDA	16	10	80	5
{EDA+ TEBA-Br}	{16+ 6}	4	85	11

Br, TEBA-Br and EDA), KF and water have been optimized from a commercial viewpoint. Table 12-5 summarizes the synthesis experiments carried out with optimized quantities of templates, KF (18 moles), and water (6608 moles) with 10 wt% of seed crystals. Four moles of TPA-Br produced fully crystalline product after 5 days of crystallization, whereas, it took almost 12 days using 6 moles of TEBA-Br. Comparatively, a higher concentration of EDA (16 moles) was found to be necessary to give fully crystalline zeolite after 10 days of crystallization. It was, however, found interesting that a combination of EDA (16 moles) and of TEBA-Br (6 moles) could give a crystalline product in as many as 4 days only. The percent solid yield (85%) also was marginally higher in case of combinations of template than that (80%) obtained when single templating species was used. These experiments were also subjected to the optimization of sulfuric acid and the optimized quantity of sulfuric acid varied from system to system.

GENERAL REMARKS

The average silica to alumina ratio from all these experiments producing fully crystalline product was always around 80 and was sometimes lower depending upon the percent solid yield. XRD and I.R. data showed that samples were fully crystalline without the occlusion of any amorphous impurities except occasional occlusion of KF when the product washing was not adequate. The morphology and crystalline size was found to be spheroidal in shape having an average crystallite size ranging between 0.5 and 1.5 μm. SEM microphotographs also indicated an absence of amorphous or crystalline impurities. The sorption studies showed that all the crystalline samples sorbed about 8.2 wt% water, 12.6 wt% n-hexane and 6.5 wt% cyclohexane. These values are in close agreement with earlier reported data (Kulkarni et al. 1982; Kotasthane et al. 1986). The choice of synthesis of ZSM-5 with the product $SiO_2/Al_2O_3 = 80$ was not purely arbitrary and was commensurate with the commercial applications disclosed (Babu et al. 1986; Derewinski et al. 1984) so far. Synthesis experiments indicated that a product of lower SiO_2/Al_2O_3 ratio ($\simeq 35$) could be synthesized with these optimized quantities of templates, KF and water, but seed crystals were found to be helpful in getting a product of better quality. However, experiments aimed at getting product with higher SiO_2/Al_2O_3 300 failed to yield fully crystalline ZSM-5 and the product was found to contain an impurity contribution of alpha-quartz, hydrous silicon dioxide (Narita et al. 1985). To achieve fully crystalline ZSM-5 of higher $SiO_2/Al_2O_3 \approx 300$ the template concentration more than the optimized quantity was found to be essential.

REFERENCES

Argauer, J. and Landolt, G.R. 1972. Crystalline zeolite ZSM-5 and the method of preparing the same. U.S. Pat. 3,702,886.

Babu, G.P., Santra, M., Shiralkar, V.P. and Ratnasamy, P. 1986. Catalytic transformation of C_8 aromatics over ZSM-5 zeolites. J. Catal. 100: 458.

Derewinski, M., Haber, J., Ptaszynski, J., Shiralkar, V.P. and Dzwigaj, S. 1984. Influence of Ni, Mg and P on selectivity of ZSM-5 class zeolite catalysts in toluene-methanol alkylation and methanol conversion. Proc.

Intl. Conf. on Str. and Reactivity, modified zeol., p. 209. Eds. Jacobs, P.A. et al., Elsevier, Amsterdam.

Dwyer, F.G. 1976. Crystalline silicates and method of preparing the same. U.S. Pat. 3941871.

Erdem, A. and Sand, L.B., 1979. Crystallization and metastable phase transformation of zeolite ZSM-5 in the $(TPA)_2O - Na_2O - K_2O - Al_2O_3 - SiO_2 - SiO_2 - H_2O$. J. Catal. 60: 241.

Gubitosa, G. and Gherardi, P. 1983. Eur. Pat. Appl. EP 129239.

Guth, J.L., Kessler, H., Burgogue, M., Wey, R. and Szobe, G. 1986. Compagnic, Francaise de Raffikage, S.A. Delande FR 2567818.

Kotasthane, A.N. and Shiralkar, V.P. 1986. Thermoanalytical studies of high silica ZSM-5 zeolites containing organic templates. Thermo chimica Acta 102: 37-45.

Kotasthane, A.N., Shiralkar, V.P., Hegde, S.G. and Kulkarni, S.B. 1986. Synthesis and characterization of alumino- and ferrisilicate pentasil zeolites. Zeolites 6: 253-260.

Kouwenhoven, H.W. and Stork, W.A.J. 1980. U.S. Pat. 4,208,305 assigned to Shell Oil Corporation U.S.A.

Kulkarni, S.B., Shiralkar, V.P., Kotasthane, A.N., Borade, R.B. and Ratnasamy, P. 1982. Studies in the synthesis of ZSM-5 Zeolites 2:   313-318.

Marosi, L., Stabenow, J. and Schwarzmann, M. 1980. Ger. Pats. 2,831,630 and 2,909,929   assigned to BASF.

Narita, E., Sato, K., Yatabe, N. and Okabe, T. 1985 Synthesis and crystal growth zeolite ZSM-5 from sodium aluminosilicate systems free of organic templates. Ind. Eng. Chem. Prod. Dev. 24: 507.

Plank, C.J., Rosinski, E.J. and Rubin, M.K. 1979. Method for producing zeolites. U.S. Pat. 4175114.

Rao,G.N., Joshi, P.N., Shiralkar, V.P., Rao, B.S. and Kotasthane, A.N. 1990. Influence of synthesis variables on the formation of ZSM-5 zeolites. Ind. J. Tech. 28: 697-703.

Ratnasamy, P., Borade, R.B., Kulkarni, S.B., Kotasthane, A.N. and Shiralkar, V.P. 1983. A process for the preparation of crystalline catalyst composite material designated Encilite. Ind. Pat. 160212.

Rubin, M.K., Plank, C.J., Rosinski, E.J. 1979. Eur. Pat. 3144 assigned to Mobil Oil Corpn., U.S.A.

Schwieger, W., Bergk, K.H., Freude, D., Hun   ger, M. and Pfeifer, H. 1989. ACS Symp. Ser.   398. Amer. Chem. Soc., Washington, D.C., p. 274. Eds. Occelli, M.L. and Robson, H.E.

Shiralkar, V.P., Joshi, P.N., Eapen, M.J. and Rao, B.S. 1991. Synthesis of ZSM-5 with variable crystalline size and influence on physico-chemical properties. Zeolites 11: 511-516.

13

Gallosilicate Derivative of the EUO Framework Zeolites

G.N. Rao, V.P. Shiralkar, A.N. Kotasthane, and P. Ratnasamy *Catalysis Division, National Chemical Laboratory, Pune 411 008, India*

A gallium containing derivative of the EUO framework zeolite has been synthesized from Al-free hydrogel systems in the presence of dibenzyldimethylammonium or hexamethonium cations. Evidence for the presence of Ga^{3+} in the lattice framework has been obtained utilizing XRD, framework -IR, ^{29}Si and ^{71}Ga MASNMR techniques. With increasing Ga content the framework (T-O) asymmetric stretching frequencies shift to lower wavenumbers. The insertion of the larger gallium atoms in the EUO framework increases the unit cell volume. The material has a significant ion exchange capacity. The MASNMR spectrum of ^{71}Ga also suggests the presence of tetrahedrally-coordinated Ga^{3+} in the EUO framework zeolites.

INTRODUCTION

Breck has reviewed the early literature (Breck 1974) where Ga^{3+}, Ge^{4+} and P^{5+} were incorporated into a few zeolite structures via a primary synthesis route. More recently number of different gallosilicate analogs with known zeolite structures have been briefly reviewed (Szostak 1989). The Al-analog of zeolite EU-1 was first described in EPA 0042226 (Casci, Lowe, and Whittam 1981) and its isostructural phase ZSM-50 has been disclosed in EPA 127 399 patent (Chu, Vartuei, and Herbst 1984). Both the materials have been classified under EUO framework topology (Meier and Olson 1987). The framework topology of the EU-1 material confirms (Briscoe et al. 1988) the space group to be Cmma with lattice parameters a = 13.695 Å, b = 22.236 Å and c = 20.178 Å, having unidimensional 10 T-ring apertures in the main channel bounded by 5.8 X 4.1 Å and 12 T-ring apertures leading to side pockets of 6.8 X 5.8 Å in cross section and 8.1 Å in depth. The systematic investigations in the synthesis and characterization of Al-analog of EU-1 zeolites using hexamethonium (Casci, Whittam and Lowe 1984) and with dibenzyldimethyl cations (Rao, Joshi, Kotasthane and Ratnasamy 1989) have been published previously. The sorption (Rao, Joshi, Kotasthane and Shiralkar 1990) as well as its specific catalytic properties have also been reported (Pradhan, Kotasthane and Rao 1991). As the detailed characterization of the gallium silicate analogs of EU-1 has not been addressed in the patent literature, this study was undertaken to better understand the behaviour of gallium during Ga-EU-1 synthesis and the nature of gallium species in the final crystalline product. In this paper, we describe the first example of an extended framework substitution of Ga via the direct synthesis of gallo-EU-1 zeolites.

EXPERIMENTAL

Synthesis Procedures

Reagents used for gallium analog preparations under the direct synthesis were microsil-silica II (95.15 wt% SiO_2, 4.85 wt% H_2O and Al_2O_3 < 50 ppm, Leuchtstoffwerk, India), silica-sol (Sycol 1230, India, 28.9% SiO_2, 0.4% Na_2O and 70.7 wt% H_2O), gallium (III) sulfate (99.99%, Aldrich), H_2SO_4 (98% AR, BDH), sodium hydroxide pellets (97%), benzyldimethylamine and benzyl chloride (98%, Fluka), hexamethonium bromide monohydrate (Aldrich). The gallium analog of EU-1 was synthesized according to the previously published method (Rao, Joshi, Kotasthane and Ratnasamy 1989) replacing the Al source with $Ga_2(SO_4)_3$. Two different slurry compositions defined by aR : bNa_2O : Ga_2O_3 : $cSiO_2$: dH_2O were adopted to yield a wide range of gallium substitution, where R represents either dimethyldibenzylamine (DBDM$^+$) or hexamethonium (HM^{++}) cation, values a, b, d are constants and c varied in the range 40-600. For more siliceous products (higher SiO_2/Ga_2O_3 > 200 ratio) DBDM$^+$ cation was used in the precursor gel under static autoclave conditions, whereas HM-Br$_2$ was utilized under stirred reaction conditions to yield higher gallium containing (lower SiO_2/Ga_2O_3 ratio) products. All the crystallization experiments were performed in Parr pressure autoclave (Series 4568, 300 mL) under autogenous pressure between the temperature range 398-443 K.

Typically, gallo-EU-1 (SiO_2/Ga_2O_3 = 200) was synthesized using dimethylbenzylamine and benzylchloride having the slurry composition 19.6 R: 12.2 Na_2O : Ga_2O_3 : 200 SiO_2 : 2170 H_2O, where R represents $DBDMN^+$-Cl^-. A slurry was made containing 28.5 g of microsil silica with an acidified gallium sulfate (0.95 g) in 60.0 g of deionized water; to this was then added NaOH solution (2.17 g dissolved in 25 g H_2O) under stirring. After the mixture was stirred for 1 h at room temperature, an equimolar mixture of 5.7 g of benzyldimethylamine and 5.5 g of benzyl chloride was finally added. The resulting gel was then transferred to a stainless steel Parr autoclave of 300 mL capacity and crystallized at 443 K for 10 days. The samples with SiO_2/Ga_2O_3 300 and 600 were prepared in a similar way.

A slurry of composition 2.0 R_2O : 5.4 Na_2O : Ga_2O_3 : 40 SiO_2 : 970 H_2O was prepared as follows, where R_2O represents oxide moles of HM-Br_2. A solution of 4.35 g of NaOH in 35.0 g of H_2O was mixed with 81.0 g of silica-sol diluted by adding 20.0 g of H_2O. An acidic solution containing 4.3 g of $Ga_2(SO_4)_3$ and 2.6 g H_2SO_4 in 30 g of H_2O was slowly added to the former solution under stirring. Finally 15.2 g of hexamethonium bromide dissolved in 30.0 g of H_2O were added under continuous stirring. The resulting gel was transferred into a Parr autoclave and crystallized at 423 K under constant agitation (250 RPM) for 6 days. The samples with SiO_2/Ga_2O_3 70 and 110 were prepared in a similar manner.

The solid products were filtered, washed and dried at 383 K. The organic material was removed by calcination in flowing air at 813 K.

Methods

The crystallinity and the phase identification of the gallo-EU-1 products was performed on Philips X-ray diffractometer PW 1730 (Cu K_α, λ= 1.54041 Å) in 2θ range 3-50. The percent crystallinity of gallo-EU-1 was also estimated using XRD powder patterns as previously reported (Casci, Whittam and Lowe 1984; Rao et al. 1989). Framework ir spectra of the samples (KBr pellet) were recorded on a Perkin-Elemer 221 spectrometer. The sorption properties of the samples for different probe molecules (sorbates) were measured gravimetrically in a conventional high-vacuum adsorption system using a McBain balance. The amount adsorbed at a relative pressure of $P/P_o = 0.8$ and 298 K was recorded. The morphology and the crystal size of the crystalline gallo-EU-1 framework systems were mounted on Au-Pd evaporated films and examined using scanning electron microscope, Stereoscan 150, UK. ^{29}Si and ^{71}Ga MASNMR spectra for the crystalline gallium analog samples were recorded at 59.6 MHz and 91.5 MHz respectively with a Bruker MSL- 300 NMR spectrometer at ambient temperature. The spinning rate of rotor was kept between 3.0 and 3.2 KHz. The chemical composition of the samples was analyzed by wet chemical methods using atomic absorption (Hitachi, Z-8000, Japan) and inductively coupled plasma (Jobin Yuon JY-38 VHR, France) spectrometers.Thermoanalytical measurements were made on a thermobalance (Netzsch STA-490), in the temperature range 298-1273 K. The heating rate was 10 K min^{-1}, the air flow was 3.4 dm^3 h^{-1} and the amount of sample was 0.05 g. After the air calcination of the as-synthesized gallium analogs (5g) for 10 h at 823 K to remove the organic material, Na-forms of these gallium analogs were ion exchanged under reflux (363 K) with an aqueous 5N, NH$_4$NO$_3$. The degree of exchange was estimated by conventional elemental analysis of the solid as well as the filtrate. The results are summarized in Table 13-1.

RESULTS AND DISCUSSION

A list of gallo-EU-1 samples investigated is reported in Table 13-1 together with data concerning the chemical composition. The gallium content in the reaction mixture was controlled by varying the hydrogel composition.

X-ray Powder Diffraction

The peak positions and the relative intensities of the XRD patterns (Fig. 13-1) for samples synthesized using HM^{++} cations were consistent with those reported for Al-analog (Casci, Whittam and Lowe 1984). On the other hand, samples obtained using DBDM$^+$ cations show a more resolved weak intensity low angle peaks at ca. 2θ 4.40°, 5.60° and 9.10° which are in good agreement with those reported previously for Al analog (Chu, Vartuei and Herbst 1984). However, in both the cases all the XRD peaks for gallium analog show a minor shift downward in 2θ values (higher d Å), as compared to that of Al analog having similar SiO_2/Al_2O_3 ratios. This effect is ascribable to the expansion of the unit-cell volume when the larger Ga^{+3} (0.62 Å) ions are incorporated into the framework

Table 13-1. Composition & physico-chemical data for crystalline gallo-EU-1 products.

Sr. No.	Hydrogel input SiO_2/Ga_2O_3	Anhydrous wt.% Bulk product SiO_2	Ga_2O_3	Na_2O	Analytical SiO_2/Ga_2O_3
1.	40	92.4	5.24	2.36	55.0
2.	70	94.2	3.90	1.50	78.5
3.	110	96.3	3.00	1.00	100.0
4.	200	97.9	1.76	0.59	173.0
5.	300	98.1	1.12	0.40	271.6
6.	600	98.9	0.52	0.29	580.0

Table 13-1 contd...

Ads. cap. g/100 g H_2O	C_6H_{14}	C_6H_{12}	Exch.cap. m mol/g	NH_4/Ga ratio	Unit-cell Vol.$(A)^3$
13.2	9.7	1.9	0.45	0.39	6180(6164)[*]
12.1	9.4	1.7	0.40	0.47	6173(6159)
9.7	9.1	1.7	0.20	0.30	6169(6155)
5.5	9.0	2.0	0.15	0.38	6164(6149)
4.2	9.0	1.9	0.09	0.37	6160(6145)
3.5	8.5	1.8	0.04	0.35	6151(6134)

[*] Unit cell volumes for Al analog of EU-1.

during the synthesis. The insertion of gallium ions in the faujasite framework causes unit-cell expansion has been reported by Kuhl in the early 1970's (Kuhl 1971). The computed values of the unit-cell volumes (Table 13-1) from the XRD data reveal significant expansion in unit-cell of gallo-EU-1 with decreasing SiO_2/Ga_2O_3 ratios, indicating successful insertion of gallium ions during the hydrothermal syntheses. The trivalent metal ion substitution causes similar changes in powder diffraction patterns in other zeolite frameworks including MFI (Ruren and Wenqin 1985) borosilicate (Meyers et al. 1985) and FAU (Ratnasamy et al. 1989). The degree of crystallinity of the gallium analog samples was estimated by measuring the total intensity of the characteristic diffraction peaks occurring at 2θ values between $18°$ and $24°$, 100% crystallinity being arbitrarily assigned to the most crystalline material obtained during the study. The X-ray peak intensities and n-hexane adsorption capacities (Table 13-1) of these samples show that the materials are highly crystalline. Over the range of SiO_2/Ga_2O_3 between 40 and 600 and in the temperature range of 398-443 K, the main crystalline phase of the solid product exhibited the XRD patterns of the EUO framework (Chu, Vartuei and

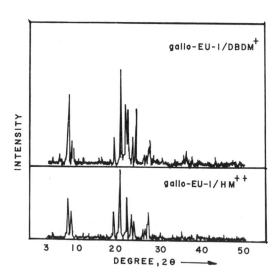

Fig. 13-1. X-ray diffraction powder patterns of as-synthesised gallo-EU-1 products.

Herbst 1984; Casci, Whittam and Lowe 1984; Rao et al. 1989). However, during the synthesis of gallo-EU-1 at $SiO_2/Ga_2O_3 < 40$ or > 600, either hydrated silica or an amorphous phase was obtained.

Framework IR

Fig. 13-2 shows the framework IR spectra of the crystalline Na-gallo-EU-1 samples with varying SiO_2/Ga_2O_3 ratios. The spectra are similar to those of the Al-analogs (Rao et al.1989). The band due to asymmetric (1100 cm^{-1}) stretching vibra tions shifts to lower wavenumbers with decreasing SiO_2/Ga_2O_3 ratios as would be expected if the heavier Ga atoms replace Si(Al) atoms in the lattice framework (Thomas and Lin 1986; Endoh et al. 1989).

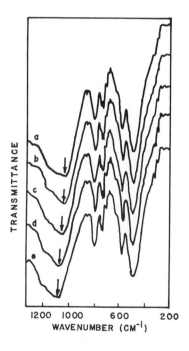

Fig.13-2. Lattice IR vibrations for various SiO_2/Ga_2O_3 ratio, (a) 55, (b) 72.5, (c) 100, (d) 173 and (e) 580.

Sorption Measurements

Water sorption decreases with a decrease in gallium in the zeolite (Table 13.1). This suggests the increased hydrophobicity with the increase in SiO_2/Ga_2O_3 ratio. A similar trend has already been reported for Al and Fe analogs of ZSM-5 (Kotasthane et al. 1986). The low cyclohexane sorption capacity for all the samples used in this study suggests that although the pore entry is characterized by 10 T-ring apertures in the main channel, its elliptical nature (5.8 X 4.1 Å) seems to make it inaccessible to the cyclohexane molecules. These results reconfirm earlier suggestions (Casci, Whittam and Lowe 1984) that the entry pore size is slightly less than 0.6 nm.

Ion Exchange Capacity

The presence of ion exchange sites, which would arise from framework gallium, was determined by the conventional technique of slurry exchange with NH_4^+ ions. The ion exchange capacity of the gallium analog samples ranges between 0.45 and 0.04 m mol g^{-1} with increasing SiO_2/Ga_2O_3 (Table 13-1), indicating that Ga-(III) present in the samples gives rise to ion-exchange sites. However, values for the NH_4/Ga ratio (Table 13-1) show a very different trend than that at the higher SiO_2/Ga_2O_3 samples, indicating loss of some framework gallium from the lower silica/gallia ratio samples upon calcination (Simmons et al. 1987).

Thermal Analysis

TG/DTA results for typical samples of gallium analog synthesized using both the HM^{++} (sample 3, Table 13-1) and $DBDM^+$ (sample 4, Table 13-1) cations in as-synthesized form are shown in Figure 13-3. The small steady weight loss observed at low temperatures is due to the dehydration of water on the external surface or located in pores not blocked by occluded templating ions. The major transformation and weight loss which begins at around 573 K is considered to be mainly due to organic material undergoing oxidative decomposition. This is well reflected in DTA curves exhibiting a characteristic two step exotherm between 573 and 873 K. The low temperature peaks between 573 and 773 K are believed to be due to the decomposition of loosely occluded $HM-Br_2$ and DBDM-chloride in gallo-EU-1, while the high temperature peaks between 773 & 873 K correspond to the decomposition of organic ions which are strongly bonded and associated with gallium acid sites in the channels (Kotasthane and Shiralkar 1986).

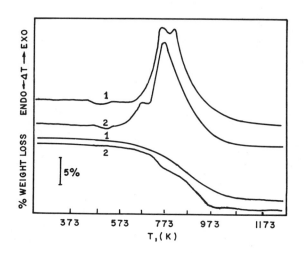

Fig.13-3. TG/DTA curves of gallo-EU-1 synthesized using (1) DBDM and (2) HM-Br$_2$ cations.

Scanning Electron Microscopy

Typical SEM micrographs of two EUO framework zeolite samples are compared in Fig. 13-4. The scanning electron micro photograph (sample 5, Table 13-1) in Fig. 13-4(a) consists of spheriodal 2-5μm sized crystal aggregates; this morphology stems from the reaction mixtures containing DBDM$^+$ cations under static autoclave conditions. However, the scanning electron micrograph (sample 2, Table 13-1) in Fig. 13-4(b) consists of homogeneously distributed and better outlined 2-3 μm crystallites obtained under stirred autoclave conditions while using reaction mixtures containing HM-Br$_2$.

13-4(a)

13-4(b)

2 μm

Fig. 13-4. SEM photographs of as-synthesized gallo-EU-1 zeolites.

High Resolution Solid State MASNMR

The ^{71}Ga MASNMR of Na-gallo-EU-1 (sample 3, Table 13-1) is shown in Fig. 13-5. For comparison, the ^{71}Ga MASNMR spectrum of gallium sulfate [in which gallium is known to have an octahedral coordination (Akitt 1972)] is also shown. As seen from Figure 13-5, the chemical shift δ= -87 ppm for gallium atoms in the solid $Ga_2(SO_4)_3$ is close to octahedral Ga sites in solution. The ^{71}Ga MASNMR spectrum of NA-gallo-EU-1 containing gallium showed a large chemical shift δ= 170 ppm for Ga^{3+}. Thus, the position of the line at 170 ppm obviously attributes to a gallium(III) in a tetrahedral environment with respect to oxygen in the EUO framework (Akitt 1972; Thomas et al. 1983). The ^{29}Si MASNMR spectrum of the sodium gallium analog samples 6 and 2 (Table 13-1) is also shown in Fig. 13-5. The spectrum shows two types of Si ordering with a chemical shift of about -111.0 ppm, which corresponds to Si(1Ga) indicating Si atoms having gallium atoms in their second coordination sphere (Mastikhin and Zamareav 1987). The intensity of this signal rises with increasing the Ga concentration in Na-gallo-Eu-1 (Fig. 13-5 A, R = 580. B, R = 78.5).

Fig. 13-5.

(1) ^{71}Ga spectra
of gallo-EU-1 zeolite
indicating tetrahedral environment of Ga δ= 170 ppm.
The octrahedral coordination
of Ga within gallium sulfate
δ= -87 ppm. The chemical
shifts are given relative
to $Ga(H_2O)_6^{3+}$ in
aqueous solution.

(2) ^{29}Si spectra of
gallo-EU-1. The shifts
are given relative to TMS.
SiO_2/Ga_2O_3
ratio A = 580, B = 78.5.

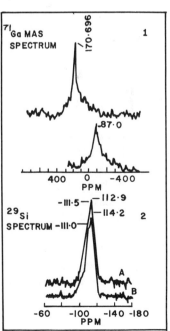

CONCLUSION

The isomorphous substitution of Ga^{3+} in EUO framework with varying SiO_2/Ga_2O_3 ratios has been demonstrated. Results from ^{71}Ga MASNMR, IR and XRD experiments support the conclusion that Ga^{3+} ions occupy tetrahedral positions in the EUO framework zeolites.

ACKNOWLEDGMENTS

G. N. Rao thanks the CSIR, New Delhi, India, for a fellowship. Fruitful discussions with Prof. C. Naccache, Sous Directeur, Inst. De Res. Sur La Catalyse, CNRS, Villeurbanne, France, are also gratefully acknowledged.

REFERENCES

Akitt, J.W. 1972. Nuclear magnetic resonance spectroscopy in liquids containing compounds of aluminium & gallium. Annu. Rep. NMR Spectroscopy Vol. 5A. Pg. 465.Ed. Mooney, E.F. London & New York : Academic Press.

Breck, D.W. 1974.Zeolite Molecular Sieves, Strcuture Chemistry & Use Pg.320, New York : John Wiley & Sons.

Briscoe, N.A., Johnson, D.W., Shannon, M.D., Kokotailo, G.T. and McCusker, L.B. 1988. The framework topology of zeolite EU-1. Zeolites 8, 74.

Casci, J.L., Lowe, B.M., and Whittam, T.V. 1981 Eur. Pat. Appl. 0042226 assigned to Imperial Chemical Industries.

Casci, J.L., Whittam, T.V. and Lowe, B.M. 1984. The synthesis and characterization of zeolite EU-1. Proceedings of the 6th International conference on zeolites, Pg. 894, Eds. Olson, D.H. and Bisio, A. Guildford, U.K: Reno, Butterworths.

Chu, P., Vartuei, J.C. and Herbst, A.J. 1984. Eur. Pat. Appl. 127 399 assigned to Mobil Oil Corp., USA.

Endoh, A., Nishimiva, K., Tsutsmi, K., and Takaishi T. 1989. Galliation and ^{18}O-exchange reactivities of ZSM-5 and ZSM-11. In Zeolites as catalysts, sorbents & detergent builders, applications & innovations. Proceedings of an international symposium, Wurzburg, F.R.G., Sept. 4-8, 1988. Stud. Surf. Sci. and Catalysis 46, 779. Eds. Karge, H.G. and Weitkamp, J. Amsterdam : Elsevier.

Kotasthane, A.N., Shiralkar, V.P., Hegde, S.G. and Kulkarni, S.B. 1986. Synthesis and characterization of alumino and ferrisilicate pentasil zeolites. Zeolites 6, 253.

Kotasthane, A.N. and Shiralkar, V.P. 1986. Thermoanalytical studies of high silica ZSM-5 zeolites containing organic templates. Thermochemica Acta 102, 37-42.

Kuhl, G.H. 1971. Preparation of gallosilicate faujasite in the presence of phosphate. J. Inorg. Nucl. Chem., 20, 3261-3268.

Mastikhin, V.M. and Zamareav, K.I. 1987. NMR studies of heterogeneous catalysis. Zeitschrift fur physikalischa Chemie Neue Folge, 152, 59-80.

Meier, W.M. and Olson, D,.H. 1987. Atlas of Zeolite Structure Types, 2nd edition. Pg. 60-61. New York: Butterworths.

Meyers, B.L., Ely, S.R., Kutz, N.A. and Kaduk, J.A. 1985. Determination of structural boron in borosilicate molecular sevies via X-ray diffraction. J. Catal. 91, 321-323.

Pradhan, A.R., Kotasthane, A.N. and Rao, B.S. 1991. Isopropylation of benzene over high silica zeolite EU-1. Applied Catalysis 72, 311-319.

Rao, G.N., Joshi, P.N., Kotasthane, A.N. and Ratnasamy, P. 1989. Synthesis and characterization of high silica EU-1. Zeolites 9, 483.

Rao, G.N., Joshi, P.N., Kotasthane, A.N. and Shiralkar, V.P. 1990 Sorption properties of EU-1 zeolites. J. Phys. Chem. 94, 8589-8593.

Ratnasamy, P., Kotasthane, A.N., Shiralkar V.P., Thangaraj, and Ganapathy, S. 1989. Isomorphous substitution of iron in the faujasite lattice. ACS Symposium   Series 398. Zeolite Synthesis. Chapter 28 Pg. 405.

Ruren, Xu and Wenqin, P. 1985. The synthesis, crystallisation and structure of heteroatom containing ZSM-5 type zeolite (M-ZSM-5). In Stud. Surf. Sci. and Catal. vol. 24. Pg. 27. Synthesis, structure, technology & application. Eds. Drzaj, B., Hocevar, S. and Pejovnik S. Amsterdam: Elsevier.

Simmons, D.K., Szostak, R., Agrawal, P.K. and Thomas, T.L. 1987. Gallosilicate molecular sieves: The role of framework and nonframework gallium on catalytic cracking activity. J. Catal. 106, 287-291.

Szostak, R., 1989. Molecular Sieves: Principles of synthesis & identification. Pg. 212-222. New York : Van Nostrand Reinhold.

Thomas, J.M., Klinowsky, J., Ramdas, S., Anderson, M.W., Fye, C.A. and Gobbi, G.C. 1983. New approaches to the structur al characterization of zeolites: Magic-angle Spinning NMR (MASNMR). Intrazeolite Chemistry, ACS Symposium Series 218 Pg. 159. Eds. Stucky, G.D. and Dwyer, F.G.

Thomas, J.M. and Lin, X.S. 1986. Gallozeolite catalysts: Preparation, characterization and performances, J. Phys. Chem. 90, 4843-4847.

14

On the Accessibility of the Zeolite MFI Structure for T Atom Introducing Reagents

R. de Ruiter, J. C. Jansen and H. van Bekkum
Laboratory of Organic Chemistry, Delft University of Technology, Julianalaan 136, 2628 BL Delft, The Netherlands

The effective dimensions and flexibility of the zeolite MFI pores were studied by adsorptive reaction of metal chlorides, metal alkyls and alkyl metal chlorides with silanol nests created by deboronation of [B]-MFI. The channel aperture appeared to be circular with a diameter of 6.1 Å at 300 °C. Consequently molecules like $SiCl_4$ and $TiCl_4$ cannot penetrate into the interior of the MFI structure under these conditions. The observed inaccessibility for $AlCl_3$ at 300 °C leads to the conclusion that the 10-ring zeolite pore opening adopts an ellipsoid form only with difficulty.

Upon (partly) substituting Cl by CH_3 in the reagents silanation and alumination can be achieved at 300 °C (e.g. by $SiMe_3Cl$ and $AlMe_2Cl$).

INTRODUCTION

Catalytically active sites in molecular sieve catalysts can be obtained by hydrothermal synthesis as well as by post-synthesis incorporation of T-atoms in the zeolite framework.

Hydrothermal synthesis of various metallosilicate catalysts is sometimes hampered by the specific chemical behavior of the metal, which is to be incorporated. Titanium and zirconium containing molecular sieves, for example, are hard to synthesize due to the extremely low solubility (Baes and Mesmer 1976) of both metals (as monomers) in aqueous zeolite synthesis mixtures. Boron (Kutz 1988) and aluminum (Jacobs and Martens 1987; Jansen 1991) zeolites, on the other hand, are obtained easily by direct synthesis.

Crystal morphology and size are determined amongst others by the synthesis mixture composition and thus are dependent upon the metal, which is to be incorporated hydrothermally. Boron and aluminum zeolite single crystals of various sizes and morphologies can be obtained (Jansen, Biron and van Bekkum 1988; Jansen et al. 1989).

The often described heterogeneous distribution (von Ballmoos and Meier 1981; Derouane et al. 1981) of active sites (e.g. [Al]-MFI) could be a disadvantage of the direct hydrothermal synthesis of aluminum zeolites. By contrast a homogeneous T atom distribution has been reported for boron zeolites of the MFI type (Jansen et al. 1989).

Post-synthesis substitution, starting with boron zeolites, could therefore be useful for synthesizing metallosilicate catalysts, which, for several reasons, are not easily synthesized by direct hydrothermal procedures.

Post-synthesis substitutions are often claimed to occur in so-called silanol nests (Breck 1971; Kerr 1968), which are believed to be present in as-synthesized frameworks (Rabo 1988; Yamagishi, Namba and Yashima 1990; Anderson, Klinowski and Xinsheng 1984; Chang et al. 1984) or can be generated during, for example, dealumination procedures (Kerr 1968; Kerr 1964).

The stability of silanol nests is questioned in the literature (Barrer and Trombe 1978; Thakur and Weller 1973). For example nests created by dealumination of high alumina zeolites are thought to recondense (Kerr 1968) as shown in

Figure 14-1.

Of course, this reaction requires framework flexibility caused by, for example, many lattice vacancies, which is not likely to occur in high silica zeolites (e.g. MFI). Removal of paired aluminum sites from 4-rings of mordenite (Bodart et al. 1986) also leaves a highly dynamic structure which might recondense easily (Barrer and Trombe 1978; Thakur and Weller 1973).

Obviously, too severe dealumination conditions can also stimulate the above mentioned recondensation as well as formation of a secondary pore system by rearrangement of framework T atoms (Reschetilowski et al. 1989).

In the present work the silanol nests are created by very mild *deboronation* of [B]-MFI zeolite frameworks containing up to 4 boron sites in the unit cell (Si/B > 23), which are homogeneously distributed (Jansen et al. 1989). These nests proved to be thermally stable up to 400 °C in flowing nitrogen and are therefore appropriate for gas phase T atom substitution.

MFI-type zeolites have often been subjected to post-synthesis reactions. For example alumination and dealumination have been reported to occur by reaction with aluminum trichloride (Yamagishi, Namba and Yashima 1990; Anderson, Klinowski and Xinsheng 1984; Chang et al. 1984; Jacobs et al. 1984; Voogd and van Bekkum 1991) and silicon tetrachloride (Jacobs et al. 1984) , respectively. Of course these reactions can only

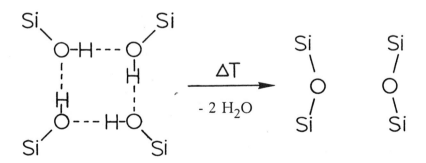

Figure14-1. Possible condensation mechanism for silanol nests at high temperature

take place when the MFI channels allow access to the substituting molecule, which is doubtful for $SiCl_4$ and dimeric $AlCl_3$ in view of their dimensions. Reduced diffusivity of silicon tetrachloride in mordenite (12-ring pores) at 600 °C, which leads to partial dealumination only, has been reported (B.Nagy and Derouane 1988).

Table14-1: Crystallographic pore dimensions (Å) of four different types of the MFI structure: As-synthesized MFI (containing tetrapropylammonium template) (van Koningsveld, van Bekkum and Jansen 1987), previously calcined MFI (<u>mono</u>clinic, <u>ortho</u>rhombic) (van Koningsveld 1990) and MFI loaded with 8 p-xylene molecules per unit cell (van Koningsveld et al. 1989)

	Channel type		d_{hc}[a]	
MFI type	Straight	Sinusoidal	Str.	Sin.
AS-SYNTH.	5.75x5.22	5.55x5.29	5.49	5.42
		5.60x5.28		5.44
CALCINED	5.78x5.18	5.89x5.35	5.49	5.63
(<u>mono</u>)	5.83x5.27	5.78x5.01	5.56	5.41
CALCINED	5.71x5.29	5.61x5.36	5.50	5.49
(<u>ortho</u>)		5.68x5.25		5.47
p-XYLENE	6.06x5.07	6.37x4.76	5.59	5.62
loaded	6.18x4.80	6.15x4.58	5.53	5.42

[a] d_{hc}= hypothetical circular diameter, see text

In Table 14-1 the crystallographic pore dimensions for MFI-type zeolites in the as-synthesized (van Koningsveld, van Bekkum and Jansen 1987) and calcined (van Koningsveld 1990) (orthorombic and monoclinic) forms as well as for a single crystal which contains eight p-xylene molecules in the unit cell (van Koningsveld et al. 1989), are given. Considering the pronounced elliptical pore dimensions of p-xylene loaded silicalite the lattice appears to be able to adapt itself to the adsorbate. The various elliptical channel openings can be mathematically transformed into circular apertures having the same circumference and a hypothetical circular pore diameter (d_{hc}). This hypothetical diameter for both channel openings and molecule cross-sections will be used throughout the paper to predict accessibility for adsorbates. For the straight and sinusoidal channels d_{hc} is 5.54 ± 0.05 and 5.52 ± 0.11 Å, respectively. Obviously these values indicate that the different features of the MFI pore opening are just caused by deformation of the silicon-oxygen rings.

Extensive adsorption studies (Wu, Landolt and Chester 1986) on MFI have been carried out using (almost) circular as well as ellipsoidal adsorbate molecules at temperatures below 100 °C. From this work the effective maximum dimensions of the MFI channel apertures can be estimated. For MFI crystals (>2 μm) cyclohexane (5.9 x 4.9 Å, d_{hc}=5.4 Å) was found to sorb easily whereas 2,2-dimethylbutane (6.0 x 5.4 Å, d_{hc}=5.7 Å) sorbed much slower with a much lower equilibrium adsorption capacity. As to aromatic adsorbates p-xylene (3.6 x 6.3 Å, d_{hc}=5.1 Å) adsorbed easily, o-xylene (3.8 x 6.9 Å, d_{hc}=5.6 Å) adsorbed slowly with a low equilibrium capacity, whereas mesitylene (1,3,5 - trimethylbenzene) (3.7 x 7.7 Å, d_{hc}=6.0 Å) did not adsorb. Assuming the hydrocarbon adsorbates to be able to deform the pore opening to an ellipse at these temperatures d_{hc} can be estimated to be 5.4 - 5.6 Å and the longest axis of the ellipse to be 6.3 - 6.9 Å.

In another study (Dessau 1980) neopentane (5.5 x 5.5 Å) was found to penetrate MFI whereas TMS (tetramethylsilane) (5.9 x 5.9 Å) did not adsorb at room temperature in the liquid state.

So the effective hypothetical circular diameter of the MFI pore seems to be between 5.5 and 5.8 Å at ambient temperature. This value is close to that determined with gas phase adsorption at 100 °C. In Table 14-2 critical dimensions for some hydrocarbons are given.

Table14-2: Critical dimensions[a] for some hydrocarbons

Compound	Critical dimensions (Å)
n-HEXANE	3.9x4.3
3 Me-PENTANE	4.7x5.1
BENZENE	3.4x6.3
p-XYLENE	3.6x6.3
CYCLOHEXANE	4.9x5.9
o-XYLENE	3.8x6.9
2,2 diMe-BUTANE	5.4x6.0
MESITYLENE	3.6x7.7

[a] Determined by measuring Courtauld models; these models use 1.0 Å for the van der Waals radius of hydrogen atoms. This comparatively low value was chosen because the energy required to reduce the van der Waals radius of hydrogen is less than in the case of other atoms.

In view of the diameter of $SiCl_4$ (6.5 Å) this molecule should not be able to penetrate the MFI pore opening. For aluminum chloride adsorption one should take into account the monomer-dimer equilibrium. At high temperatures a substantial amount of monomeric $AlCl_3$ is present. For Al_2Cl_6 and $AlCl_3$ $d_{hc} = 6.8$ and 5.4, respectively (*cf.* Table 14-3 and 14-4).

Table14-3: Vapor pressures of the reagents used at 20 °C and fraction of monomers of the aluminum compounds present at the reaction conditions

Compound	Vapor pressure (torr)	% Monomer	Ref.
$SiCl_4$	193	-	b
$SiCl_3Me$	134	-	b
$SiCl_2Me_2$	114	-	b
$SiClMe_3$	191	-	b
$SiMe_4$	589	-	c
$AlMe_3 \rightleftharpoons Al_2Me_6$	11	9.9	d,e
$AlMe_2Cl \rightleftharpoons Al_2Me_4Cl_2$	11	10.4[a]	d
$AlCl_3 \rightleftharpoons Al_2Cl_6$	1.2	27.4	f
$TiCl_4$	9	-	g

[a] No equilibrium data available; because the dimer Al_2Cl_6 is assumed to be more stable than $Al_2Me_4Cl_2$ (Rytter and Kvisle 1986) the lower limit of the amount of monomer is predicted by taking into account the dissociation constant for the $AlCl_3/Al_2Cl_6$ equilibrium.
[b] (Jenkins and Chambers 1954). [c] (Petrarch Systems 1989/90).
[d] (Texas Alkyls, Inc. 1988). [e] (Henrickson and Eyman 1967). [f] (Fischer and Rahlfs 1932). [g] (Weast and Tuve 1972/73).

Table14-4: Critical dimensions of $SiMe_xCl_{(4-x)}$ and $AlMe_yCl_{(3-y)}$ (and dimers) reagents[a]

Compound	Dimensions (Å)	Compound	Dimensions (Å)	Ref.
$SiMe_4$	5.9x5.9	$AlMe_3$	3.6x6.15	b
$SiMe_3Cl$	6.0x6.0	Al_2Me_6	8.36x6.71	c
$SiMe_2Cl_2$	6.1x6.1	$AlMe_2Cl$	3.6x6.3	d
$SiMeCl_3$	6.3x6.3	$Al_2Me_4Cl_2$	6.86x6.66	e
$SiCl_4$	6.5x6.5	$AlCl_3$	3.6x6.71	d,f
$TiCl_4$	6.7x6.7	Al_2Cl_6	6.45x7.15	f

[a] The critical dimensions of the Si compounds and $TiCl_4$ have been determined by measuring Courtauld models (see footnote Table 14-2), whereas the dimensions of the Al compounds were calculated from electron/neutron scattering data (see refs.).
[b] (Almenningen, Halvorsen and Haaland 1971). [c] (Vranka and Amma 1967).
[d] (Kata et al. 1965). [e] (Brendhaugen, Haaland and Novak 1974). [f] (Palmer and Elliott 1938).

From this dimensional point of view $SiCl_4$ and Al_2Cl_6 are not expected to penetrate the MFI pore system and alumination may only be possible at higher temperatures, depending on the deformation ability of the framework and/or the entering molecule. Furthermore it should be noted that the channel openings are more circular in the orthorhombic phase and deformation to an ellipse probably becomes more difficult at temperatures above the monoclinic-orthorhombic transition point (Hay, Jaeger and West 1985).

Another aspect of adsorption in zeolites is the interaction between guest molecules and the channel wall. The water adsorption capacity of highly siliceous MFI materials is poor (Flanigen et al. 1978) and the Cl -- O repulsion in case of attempted MCl_x adsorption could be a parameter which is until now underestimated. On the other hand the attractive interaction between alkyl groups and framework oxygens is well known (Baerlocher and Meier 1969; McMullan, Mak and Jeffrey 1966) and might contribute to adsorption of hydrocarbons, which are boundary cases as to their dimensions. In the present paper the substitution of internal silanol nests is studied by reaction of deboronated [B]-MFI crystals with two series of reactive adsorbates, i.e. $SiMe_xCl_{(4-x)}$ ($x=1$-4) and $AlMe_yCl_{(3-y)}$ ($y=1$-3). Also $TiCl_4$ has been explored as a T atom introducing reactant in MFI systems.

EXPERIMENTAL

[B]-MFI crystals were prepared in 5 days by hydrothermal synthesis at 180 °C in stainless steel Teflon-lined autoclaves. In Table 14-5 the molar oxide ratio of the synthesis mixture and the crystals formed are given.

Table14-5: Molar oxide ratio of synthesis mixture and [B]-MFI crystals obtained

	Synthesis mixture	Crystals
SiO_2	4.5	93.5
B_2O_3	1.9	1.25
$(TPA)_2O$	7.6	2
Na_2O	7.9	0.25
H_2O	2000	n.d.[a]

[a] n.d. = not determined

Reagents used were Aerosil 200 (Degussa) as silicon source, boric acid (Merck, p.a.) as boron source, tetrapropylammonium bromide (TPABr) (CFZ Zaltbommel) as template and sodium hydroxide (J.T. Baker, reagent grade) for pH adjustment. After synthesis the crystals were calcined and deboronated in a mild way (aqueous 0.01 M HCl, 100 °C), resulting in so-called silanol nests in the framework, of which the characterization will be reported in detail elsewhere (de Ruiter et al. to be published). Boron removal amounted to more than 95%.

These deboronated vacancy-rich crystals were used for post-synthesis reactions with $SiMe_xCl_{(4-x)}$, $AlMe_yCl_{(3-y)}$ as well as with $TiCl_4$. The reactions were carried out in a horizontal quartz tube-reactor. Before reaction the crystals were dried by raising the temperature to 400 °C (rate: 1 °C/min) in flowing nitrogen. Subsequently the oven temperature was decreased to reaction temperature (generally 300 °C) and the reagent was introduced by bubbling nitrogen through the reagent solution at 20 °C. In the case of solid $AlCl_3$ saturation took place at 100 °C.

After 4 hours of reaction (10 hours in the case of aluminum compounds) the excess of reagent was purged by flowing nitrogen at reaction temperature for 4 hours. Then the reactor was cooled down slowly to room temperature.

The progress of the reaction of the reagent with the silanol groups was determined by FT-IR microscope spectroscopy using an IFS-66 Bruker spectrometer. Infrared spectra were recorded before and after reaction or heat treatment at the same position of an MFI crystal. Particularly the absorption band at 950 cm^{-1} served this purpose. The crystals were measured *in transmission* (Jansen et al. 1989) in the crystallographic b direction, along which the straight channels are situated. The IR beam spot diameter was 20.8 μm.

RESULTS

The synthesized [B]-MFI crystals have a lath type morphology (*cf.* Figure 14-2) and a homogeneous framework boron distribution (Jansen et al. 1989) . The as-synthesized crystals contain approximately 2.5 boron atoms per unit cell (Si/B = 37). The size of the crystals is 120 x 35 x 20 μm. After calcination and deboronation (de Ruiter et al. to be published) the Si/B ratio is 960.

Figure14-2.SEM picture of [B]-MFI crystals

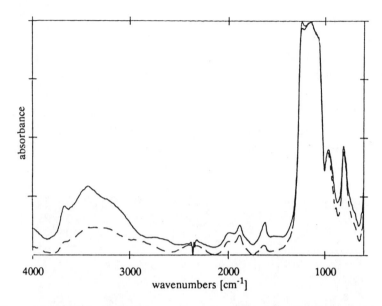

Figure14-3.FT-IR absorption spectra of a deboronated MFI crystal before
(—) and after (---) SiCl$_4$ exposure at 300 °C for 4 hours

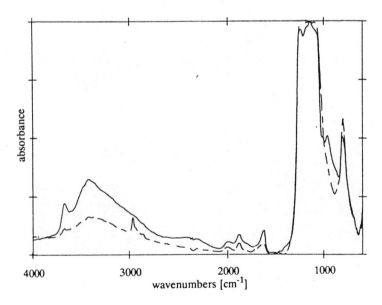

Figure14-4.FT-IR absorption spectra of a deboronated MFI crystal before
(—) and after (---) SiMe$_3$Cl at 300 °C for 4 hours

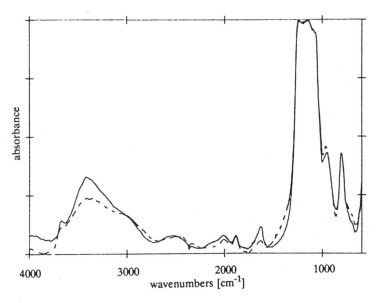

Figure14-5.FT-IR absorption spectra of a deboronated MFI crystal before
(—) and after (---) AlCl$_3$ exposure at 300 °C for 10 hours

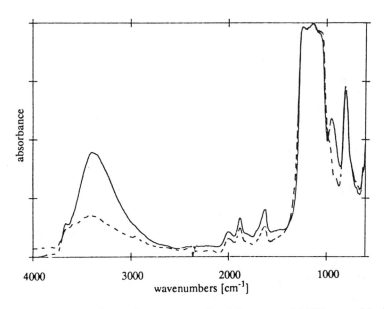

Figure14-6.FT-IR absorption spectra of a deboronated MFI crystal before
(—) and after (---) AlMe$_3$ exposure at 300 °C for 10 hours

The deboronated MFI crystals show an IR absorption band at 950 cm^{-1}, which is typical for silanol nests (de Ruiter et al. to be published). By measuring the integral intensity of this band at the same position on a crystal, before and after reaction with a T atom introducing reagent, the percentage of unreacted silanol groups was estimated. By using an IR microscope pure crystals could be measured *in transmission*, so outer surface effects are negligible, which contrasts with the KBr pellet- or self supported-wafer technique.

In Figures 14-3 and 14-4 microscope FT-IR spectra of a deboronated MFI crystal before and after exposure to $SiCl_4$ and $SiMe_3Cl$ vapor, respectively, are shown. In the case of $SiMe_3Cl$ treatment the 950 cm^{-1} band decreases strongly, whereas it is unaffected by exposure of a crystal to gaseous $SiCl_4$ at 300 °C.

From Figures 14-5 and 14-6 it can be concluded that $AlMe_3$ enters the zeolite pores of MFI and reacts with the silanol nests, whereas $AlCl_3$ does not influence the Si-OH group of the inner zeolite surface at 300 °C.

Table14-6: Percentage of unreacted SiOH groups in MFI crystals after exposure to volatile reagents at 300 °C for 4/10^a hours

Compound	Percentage of unreacted OH groups
$SiCl_4$	93
$SiCl_3Me$	95
$SiCl_2Me_2$	73
$SiClMe_3$	4
$SiMe_4$	35
$AlMe_3$	7.5
$AlMe_2Cl$	32
$AlCl_3$	93
$TiCl_4$	89

[a] 10 hours of exposure for the aluminium compounds; 4 hours for all other compounds.

In Table 14-3 the vapor pressures of the reactants at 20 °C (for $AlCl_3$ at 100 °C) and the amount of monomer of the used Al compounds at reaction temperature and pressure are given.

In Table 14-4 the molecular dimensions of the used reagents are given. $TiCl_4$ and the silicon compounds are spherical, whereas monomeric aluminum compounds have an elliptical cross-section.

In Table 14-6 the percentage of unreacted Si-OH functions after reaction is shown for the various reagents.

DISCUSSION

In this study the accessibility of MFI zeolite channels for commonly used post synthesis reagents is discussed. The most important parameter seems to be the effective radius of the channel openings in relation to the dimension of the reagent.

Alumination (Yamagishi, Namba and Yashima 1990; Anderson, Klinowski and Xinsheng 1984; Chang et al. 1984; Jacobs et al. 1984; Voogd and van Bekkum 1991) and dealumination (Jacobs et al. 1984; Kraushaar and van Hooff 1988) have been studied extensively. Also dealumination followed by titanization has been claimed (Kraushaar and van Hooff 1988). However, large crystals were used in relatively few studies and often the post-synthesis reactions were carried out under severe conditions. Thus Yamagishi, Namba and Yashima (1990) reported alumination using $AlCl_3$ vapor at 650 °C and concluded that aluminum was incorporated in framework imperfections. During this procedure, however, silicon was continuously released from the framework whereas the amount of tetrahedral aluminum (max. 1.25/unit cell) leveled off within one hour and the concentration of 6-coordinate aluminum increased slightly with increasing reaction time. In this case the formation of a secondary pore system, which has been described recently (Reschetilowski et al. 1989) cannot be excluded. Applying 500 °C (Voogd and van Bekkum 1991) incorporation of Al in silicalite upon treatment with $AlCl_3$ was reported.

Chang et al. (1984) used less severe alumination conditions (375 °C, dry He flow), but found substantial amounts of amorphous alumina or silica alumina species, probably at the external surface, which might explain the high alpha factors observed for aluminated samples. Moreover, reaction with $AlBr_3$, which has even larger critical dimensions than $AlCl_3$, proved to be a better alumination reagent. Jacobs et al. (1984) questioned the $AlCl_3$ post synthesis treatment, with respect to the alumination of the inner zeolite surface.

Therefore in this study the accessibility of the MFI channel system was studied using large crystals under relatively mild conditions (300 °C, 1 atm).

Silanation

From Table 14-6 it is concluded that $SiCl_4$ and $SiCl_3Me$ are not able to penetrate the MFI channel system. $SiCl_2Me_2$ obviously diffuses slowly, whereas $SiMe_3Cl$ reacts readily with the internal silanol groups. $SiMe_4$ exposure leaves 35% of the SiOH groups unreacted under the present conditions and reaction time. This is probably due to the relative inertness of this reagent. It should be noted that the silicon compounds are all (almost) circular and not readily deformed to other geometries. On the basis of these results the effective d_{hc} for the MFI channel opening at 300 °C is estimated to be 6.1 Å.

A literature comparison (Breck 1971) of the kinetic diameters and liquid phase densities of neopentane and carbon tetrachloride implies, in contrast to our measurements of Courtauld models that the four methyl groups occupy more space than the four chlorine groups. The fact that the critical dimensions of the molecule cross-section can be minimized by adjusting the conformation of the methyl groups, which is likely to happen during adsorption, may explain this discrepancy.

The remarkable difference in the adsorption/reaction behavior of $SiMe_2Cl_2$ and $SiMe_3Cl$ with almost equal dimensions could be due to the difference in affinity of both molecules to the zeolite wall. We assume Cl---O_{wall} repulsion and CH_3---O_{wall} attraction. In Figure 14-7 it is shown that $SiMe_3Cl$ can avoid Cl---O_{wall} repulsion whereas $SiMe_2Cl_2$ cannot while diffusing through the MFI channels. These repulsion/attraction phenomena are quantified (Hertle and Hair 1968) in Table 14-7. In this table the isosteric heats of adsorption for $SiMe_xCl_{(4-x)}$ on amorphous silica are given for $x = 0-3$. Clearly substitution of a Cl ligand by a methyl group increases the heat of adsorption, the affinity for the surface. These repulsion/attraction phenomena seem to play an important role when the guest molecule's dimensions are close to those of the MFI channel opening.

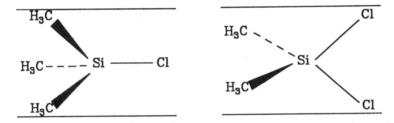

Figure14-7. Schematic drawing of the different adsorption behavior of $SiMe_3Cl$ and $SiMe_2Cl_2$ with respect to repulsive and attractive interactions with the zeolite wall

Table14-7: Isosteric heats of adsorption of $SiMe_xCl_{(4-x)}$ compounds ($x = 0-3$) on amorphous SiO_2 (Hertle and Hair 1968)

Compound	$-\Delta H_{ads}$ (isosteric) (kcal/mole)
$SiCl_4$	5.4
$SiCl_3Me$	6.5
$SiCl_2Me_2$	7.8
$SiClMe_3$	8.7

Titanization

The data for $TiCl_4$ (6.7 x 6.7 Å) adsorptive reaction clearly show that this compound cannot enter the channel system of zeolite MFI, which is in accordance with the dimensions of the guest molecule and the zeolite pores. The literature claims of titanization at 370 °C (Whittington and Anderson 1991) and of dealumination of H-[Al]-MFI followed by titanization at 400 or 500 °C (Kraushaar and van Hooff 1988) should perhaps be restudied as the observed catalytic activity may be due to outer surface titanium. Furthermore, the unit cell volume of titanized dealuminated MFI turned out to be smaller than that of the parent H-[Al]-MFI which contradicts the fact that the Ti-O exceeds the Al-O bond length.

Alumination

The large dimensions of all dimers of the aluminum compounds (see Table 14-4) justify the conclusion that they will not be able to enter the MFI channels. For the flat monomers with elliptical cross-sections the conclusion may be drawn that $AlCl_3$ does not adsorb, $AlMe_2Cl$ slowly adsorbs and $AlMe_3$ readily adsorbs into the inner pore system under the present conditions. If either the entering molecule or the channel opening could deform and adapt, these experimental results are unexpected on the basis of the above reported effective d_{hc} of the MFI pore opening, because all monomeric triangular molecules have d_{hc} values below 6.1 Å. Apparently the deformation of the 10-ring zeolite pore opening is limited and the conclusion is drawn that the longest axis of the ellipsoidal pore opening apparently has a maximum value of ~6.3 Å. The slow diffusion of $AlMe_2Cl$ can be explained by Cl -- O repulsion; by rotation of the flat molecule around an axis perpendicular to the zeolite channel direction the molecule cross-section dimensions can be minimized. The Cl---O_{wall} distance, however, decreases upon such a rotation which leads to Cl -- O repulsion.

Considering the adsorption of the trigonal aluminum monomers it may be noted that the adsorbing species may enlarge its coordination number by interaction with zeolite wall oxygens. In this way tetrahedral intermediates of $AlCl_3$ $AlMe_2Cl$ and $AlMe_3$ with (Me)Cl-Al-Cl(Me) angles changed from 120 to 109° (Churchill and Wasserman 1983) might play a role.

Courtauld models show that the smallest critical dimension of the base-plane of such adsorbed complexes (cf. Figure 14-8) is 6.6, 6.15 and 5.8 Å for O---$AlCl_3$, O---$AlMe_2Cl$ and O---$AlMe_3$, respectively. For the O---$AlMe_2Cl$ species, however, the distance from the wall-oxygen atom to the Cl-Me-Me plane is calculated to be 2.42 or 2.67 Å, assuming O-Al-Cl angles of 109° and an Al---O bond length of 1.75 Å (Churchill and Wasserman 1983) or 2.00 Å (Nelson, Kine and Shriver 1969), respectively. The channel width at these levels is 4.62 or 5.02 Å, which is much smaller than the critical dimension of the Cl-Me-Me plane (6.15 Å).

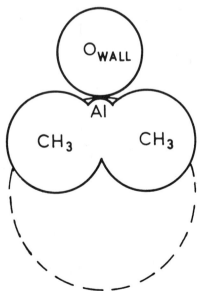

Figure14-8. Schematic projection of hypothetical tetrahedral adsorbing aluminum species coordinated with oxygen of the zeolite wall

In conclusion tetrahedral aluminum adsorbing species are improbable due to severe steric hindrance by the curved channel wall (*cf.* Figure 14-8).

Thus the effective hypothetical circular diameter of the MFI pore opening at 300 °C is approximately 6.1 Å. Upon adsorption of ellipsoidal reagents, e.g. $AlMe_2Cl$, the channel opening becomes ellipsoidal as well, but the longest axis of this ellipse (lea) is about 6.3 Å in length. If $[(lea-d_{hc})/d_{hc}] \times 100\%$ is taken as a measure of the adjustment of the channel opening to an ellipsoid form, this adjustment is at 100 and 300 °C, >15 and 3%, respectively, in agreement with crystallographic data. The monoclinic/orthorhombic phase transition (ellipsoidal/circular channel openings) is known to occur at about 70 °C for silicalite-1 (Hay, Jaeger and West 1985), which is in agreement with the adsorption studies of Wu, Landolt and Chester (1986) and Dessau (1980) and the present work.

Conclusions drawn from this study with respect to accessibility of the MFI channel system for guest molecules seem also valid for procedures such as dealumination by hydrochloric acid vapor, during which $AlCl_3$ vapor is thought to desorb from the MFI channel system. This is not likely to happen either.

CONCLUSIONS

From adsorptive reaction studies on large (120 x 35 x 20 μm), silanol nest containing crystals, the effective pore size, feature and flexibility of the MFI framework were determined.

For gas phase adsorption at 100 °C the hypothetical circular diameter is 5.4 - 5.6 Å, whereas elliptical molecules with 6.3 Å as largest critical dimension still adsorb easily (e.g. p-xylene).

At 300 °C the effective circular channel aperture diameter was estimated to be 6.1 Å, whereas the flexibility of the 10-ring opening turned out to be small. Apparently the pore opening can hardly become ellipsoidal; the longest axis of the channel opening is close to 6.3 Å. Close to the orthorhombic/monoclinic

Table14-8: Effective hypothetical circular dimension and longest ellips axis (lea) of the (adapted) MFI channel openings at 100 and 300 °C

	d_{hc} (Å)	lea (Å)
100 °C	5.4 - 5.6	> 6.3
300 °C	~6.1	~6.3

transition temperature the pore system seems to be more flexible (*cf.* Table 14-8).

Consequently $SiCl_4$ and $TiCl_4$ which have often been used for post-synthesis treatment of zeolite material *cannot enter or leave* the pore system of zeolite MFI at 300 °C due to the limited effective pore diameter whereas $AlCl_3$ reactive adsorption is *hampered* due to the relative inflexibility of the pore at this temperature.

By (partly) substitution of Cl by CH_3 in the reagents, molecular dimensions and properties (Cl -- O_{wall} repulsion vs. CH_3 -- O_{wall} attraction) are changed such that alumination and silanation of large MFI crystals can be achieved at 300 °C.

ACKNOWLEDGMENT

Financial support of the Koninklijke/Shell-Laboratorium Amsterdam is gratefully acknowledged. Prof. J. Reedijk is thanked for the stimulating discussions on aluminum coordination.

REFERENCES

Almenningen, A., S. Halvorsen, and A. Haaland. 1971. Acta Chem. Scand. 25: 1937.

Anderson, M.W., J. Klinowski, and Liu Xinsheng. 1984. J. Chem. Soc., Chem. Commun. 1596.

Baerlocher, Ch., and W.M. Meier. 1969. Helv. Chim. Acta 52: 1853.

Baes, C.F., and R.E. Mesmer. 1976. The Hydrolysis of Cations. New York: John Wiley & Sons.

Ballmoos, R. von, and W.M. Meier. 1981. Nature 289: 782.

Barrer, R.M., and J-C. Trombe. 1978. J. Chem. Soc., Farad. Trans. I 2786.

B.Nagy, J., and E.G. Derouane. 1988. In Perspectives in Molecular Sieve Science, eds. W.H. Flank and T.E. Whyte Jr., p. 2. Washington, DC: ACS Symp. Ser. 368.

Bodart, P., J. B.Nagy, G. Debras, Z. Gabelica, and P.A. Jacobs. 1986. J. Phys. Chem. 90: 5183.

Breck, D.W. 1971. Zeolite Molecular Sieves. New York: John Wiley & Sons.

Brendhaugen, K., A. Haaland, and D.P. Novak. 1974. Acta Chem. Scand. A28: 4.

Chang, C.D., C.T.W. Chu, J.N. Miale, R.F. Bridger, and R.B. Calvert. 1984. J. Am. Chem. Soc. 106: 8143.

Churchill, M.R., and H.J. Wasserman. 1983. Inorg. Chem. 22: 41.

Derouane, E.G., J.P. Gilson, Z. Gabelica, C. Mousty-Desbuquoit, and J. Verbist. 1981. J. Catal. 71: 449.

Dessau, R.M. 1980. In Adsorption and Ion Exchange with Synthetic Zeolites, ed. W.H. Flank, p. 123. Washington, DC: ACS Symp. Ser. 135.

Fischer W., and O. Rahlfs. 1932. Z. Anorg. Allg. Chem. 205: 1.

Flanigen, E.M., J.M. Bennett, R.W. Grose, J.P. Cohen, R.L. Patton, R.M. Kirchner, and J.V. Smith. 1978. Nature 271: 512.

Hay, D.G., H. Jaeger, and G.W. West. 1985. J. Phys. Chem. 89: 1070.

Henrickson C.H., and D.P. Eyman. 1967. Inorg. Chem. 6: 1461.

Hertle, W., and M.L. Hair. 1968. J. Phys. Chem. 72: 4647.

Jacobs, P.A., and J.A. Martens. 1987. Synthesis of High-Silica Aluminosilicate Zeolites. Amsterdam: Elsevier.

Jacobs, P.A., M. Tielen, J. B.Nagy, G. Debras, E.G. Derouane, and Z. Gabelica. 1984. In Proc. 6th Int. Zeolite Conf., eds. D.H. Olson and A . Bisio, p. 783. Guildford: Butterworths.

Jansen, J.C. 1991. In Introduction to Zeolite Science and Practice, eds. H. van Bekkum, E.M. Flanigen, and J.C. Jansen, p. 77. Amsterdam: Elsevier.

Jansen, J.C., E. Biron, and H. van Bekkum. 1988. Stud. Surf. Sci. Catal. 37: 133.

Jansen, J.C., R. de Ruiter, E. Biron, and H. van Bekkum. 1989. Stud. Surf. Sci. Catal. 49A: 679.

Jenkins A.C., and G.F. Chambers. 1954. Ind. Eng. Chem. 46: 2367.

Kata, H., K. Yamaguchi, T. Yonezawa, and K. Fukui. 1965. Bull. Chem. Soc. Japan 38: 2144.

Kerr, G.T. 1964. J. Catal. 15: 200.

Kerr, G.T. 1968. J. Phys. Chem. 72: 2594.

Koningsveld, H. van. 1990. Acta Cryst. B46: 731.

Koningsveld, H. van, H. van Bekkum, and J.C. Jansen. 1987. Acta Cryst. B43: 127.

Koningsveld, H. van, F. Tuinstra, H. van Bekkum, and J.C. Jansen. 1989. Acta Cryst. B45: 423.

Kraushaar, B., and J.H.C. van Hooff. 1988. Catal. Letters 1: 41.

Kutz, N.A. 1988. In Perspectives in Molecular Sieve Science, eds. W.H. Flank and T.E. White Jr, p. 532. Washington, DC: ACS Symp. Ser. 368.

McMullan, R.K., T.C.W. Mak, and G.A. Jeffrey. 1966. J. Chem. Phys. 44: 2338.

Nelson, N.J., N.E. Kine, and D.F. Shriver. 1969. J. Am. Chem. Soc. 91: 5173.

Palmer K.J., and N. Elliott. 1938. J. Am. Chem. Soc. 60: 1852.

Petrarch Systems Silanes and Silicones. 1989/90. Catalogue Silicon Compounds - Register & Review, p. 200.

Rabo, J.A. 1988. In Surface Organometallic Chemistry: Molecular Approaches to Surface Catalysis, eds. J-M. Basset and B.C. Gates, p. 245. Dordrecht: Kluwer (NATO ASI Symp. Ser., Ser. C 231).

Reschetilowski, W., W-D. Einicke, M. Jusek, R. Schoellner, D. Freude, M. Hunger, and J. Klinowski. 1989. Applied Catal. 56: L15.

Ruiter, R. de, A.P.G. Kentgens, J. Grootendorst, J.C. Jansen, and H. van Bekkum, to be published.

Rytter, E., and S. Kvisle. 1986. Inorg. Chem. 25: 3796.

Texas Alkyls, Inc. 1988. Catalogue Organometallics, p. 8, 9.

Thakur D.K., and S.L. Weller. 1973. In Molecular Sieves, eds. W.M. Meier and J.B. Uytterhoeven, p. 596. Washington, DC: ACS, Adv. Chem. Ser. 121.

Voogd, P., and H. van Bekkum. 1991. Ind. Eng. Chem. Res. 30: 2123.

Vranka R.G., and E.L. Amma. 1967. J. Am. Chem. Soc. 89: 3112.

Weast, R.C., and G.L. Tuve, eds. 1972/73. Handbook of Chemistry and Physics, p. D-176. Cleveland: CRC Press.

Whittington, B.I., and J.R. Anderson. 1991. J. Phys. Chem. 95: 3306.

Wu, E.L., G.R. Landolt, and A.W. Chester. 1986. In New Developments in Zeolite Science and Technology, eds. Y. Murakami, A. Lijima and J.W. Ward, p. 547. Tokyo, Amsterdam: Kodansha, Elsevier.

Yamagishi, K., S. Namba, and T. Yashima. 1990. J. Catal. 121: 47.

15

Quantitative Evaluation of Framework Gallium in As-Synthesized and Post-Synthesis Modified Gallo- and Gallo-Alumino MFI-Zeolites

Z. Gabelica[1], C. Mayenez[1], R. Monque[2], R. Galiasso[2] And G. Giannetto[3]

[1]Facultes Universitaires N.D. de la Paix, Namur, Dept. Chemistry, Lab. Catalysis, 61, Rue de Bruxelles, B-5000 Namur, Belgium
[2] INTEVEP S.A., Apdo. 76343, Los Teques, Caracas 1070A, Venezuela
[3]Universidad Central, Fac. Ingen., Los Chaguaramos, Caracas, Venezuela

Various synthesis methods leading to pure MFI-type gallosilicates or gallo-aluminosilicates have been tested in terms of the efficiency of Ga or Al incorporation. In most of the cases, a 100% synthesis efficiency was determined for both ions. The amount of tetrahedrally coordinated framework Ga or Al atoms could be directly and quantitatively measured in the as-synthesized zeolites by [71]Ga- and ^{27}Al-MAS NMR respectively, provided adequate calibration curves of NMR line intensities versus the trivalent element concentration are established. Each type of synthesis method, involving various counterions to Ga^{3+} in the final framework, needs a specific NMR versus Ga content correlation. Extra framework Ga species that contribute to bulk catalytic properties of gallo(alumino)silicate zeolites, are generated by various post-synthesis (hydro)thermal treatments. Neither the structural state of these species, nor their concentration, could be evaluated by [71]Ga-NMR, because of important line broadenings that occur in calcined materials. Nevertheless, their quantitative computing could be achieved directly, by [71]Ga-NMR probing the water equilibrated pre-calcined samples, or indirectly, in calcined (dry) samples, by combining ^{27}Al-NMR and TPD of ammonia generated upon heating the NH_4^+-exchanged zeolites. This latter technique also accounts for a quasi-homogeneous Ga^{3+} and Al^{3+} framework distribution, when syntheses are achieved in specifically defined alkali-free media.

INTRODUCTION

A decade ago, Barrer summarized the pre-1981 literature dealing with zeolites in which Ga was isomorphously replacing Al [1]. More recent studies regularly report synthesis and structural characterization of a variety of gallosilicate zeolites. These can be systematic and include a series of structures, such as the work of Newsam and Vaughan on Ga-bearing ABW, FAU, LTL, MAZ and SOD frameworks [2], or may concern selected and specific structure types, for exemple, the recent studies conducted on Ga-sodalite [3,4], (Ga,TMA)-sodalite [5], Ga-faujasite [6,7], (Ga,Al)-erionite [8], (Ga,Al)-mordenite [9], Ga-analcime [10], Ga-Nu-1 [11], Ga-ZSM-12 [12] or gallophosphate-type microporous materials of various structures [13], and many others, such as those cited in a recent compilation by Szostak [14].

Extended studies on synthesis and characterization on pentasil-type (MFI or MEL) gallo(alumino)silicates have been published [15-43], since such materials proved to be very active and selective bifunctional catalysts (involving both acid and Ga_xO_y sites), in the transformation of lower alkanes to aromatics (BTX) [20,24,27,31,33,37,39,45-52].
In fact, highly dispersed gallium oxidic species were recognized to catalyze the initial dehydrogenation of alkanes to corresponding olefins [20,27,33,37,38,45,46,48,51,52], but also to show a high selectivity to stable aromatics [20,21,33,45,46,48,51] and to increase their yield [33,34].

Such highly dispersed Ga_2O_3 species can be prepared either by calcining in an appropriate way Ga-exchanged [20,24,33,45,50], Ga-impregnated [21,30,33,37,40] or Ga_2O_3-loaded (mechanical admixture) [24,27,38,39,40,44] H-ZSM-5 zeolites. However, the highest dispersed state of these gallium oxidic species can be obtained by heating, under various conditions and atmospheres, the as-synthesized MFI-gallo- or gallo-alumino silicates. Indeed, such materials can be currently prepared very easily by using conventional synthesis methods [19,20,22,24,30-

32,35-37,41,48] or other more "exotic" recipes, such as galliation of a defected silicalite by $NaGaO_2$ solutions [17,18,29], or by the recent atom planting method, consisting in inserting Ga ions on defected sites of a silicalite, at high temperature [42].

Numerous techniques were used solely or in combination to prove the unambiguous siting of the Ga^{3+} ions in the T positions of the MFI framework: IR [15,18,22,37], ion exchange capacity [22,44], acid site determination by using TPD of ammonia [15,19,24,27,44] or of amines [30], framework expansion (XRD) [44], ^{29}Si-NMR showing Si(1Ga) configurations [12,16,18,37,41], or direct methods of "visualization" of framework Ga, like ^{71}Ga-NMR [8,11,12,16,25,35,41,53] or EXAFS [35].

Less is known about the thermal and hydrothermal stability of the gallium framework species. While low silica Ga-bearing zeolites such as faujasite (X or Y) undergo a rapid structural collapse on heating [54], the crystal integrity of high silica ZSM-5 zeolites was not really affected by (hydro)thermal treatments, because their framework contains relatively few structural Ga^{3+} ions per unit cell [14,41,55].

However, except for a recent work dealing with high silica ZSM-12 [12], appreciable amounts of structural gallium are released from the lattice sites when the gallosilicates are submitted to hydrothermal calcination under various atmospheres [11,25,32,33,37,41,43,44,52]. Various parameters, such as the temperature or the duration of the treatment, or the partial water vapor pressure, were shown to markedly influence the framework stability of the Ga species [14]. In fact, a careful control of these parameters could result in the final monitoring of the amount of non framework Ga component eventually released, and desired for the final bulk catalytic properties of the material [44].

The structural state of these "highly divided Ga_xO_y oxidic species" is still subject to debate in the literature. Upon calcination, Ga was supposed to migrate from the MFI-lattice and ultimately to agglomerate as "amorphous Ga oxide dispersed species" [33,41,44]. It was also suggested that such species can stay located as tiny aggregates within the zeolite channels [52,55] or as more bulky isolated Ga_2O_3 entities onto the surface of the crystallites [52]. Khodakov et al. [37] recently suggested that two types of non framework Ga species can be generated on calcining (Ga)-MFI materials; oxidic species partly located on the crystallite surface, in agreement with another recent work by Kucherov et al. [32], and cationic Ga species (Ga^{3+} or GaO^+), that compensate the residual framework charges created after the partial Ga release.

[71]Ga-NMR was adequately used as an accurate tool to fully characterize coordination sites of Ga zeolites [53]. While lattice tetrahedrally coordinated Ga^{3+} ions are adequately defined by a NMR signal often located near 160 ppm, [8,11,12,16,25,35,41,53], this technique proved completely inefficient to even detect the presence of extra framework Ga species in freshly calcined (Ga)-MFI zeolites. Only very broad, ill defined [71]Ga-NMR "bumps" are observed at about 50 ppm [25,26,41]. Their broadness, probably caused by strong quadrupolar interactions [25], possibly in combination with other secondary effects [26] , does not allow to precise the coordination state of these Ga species. In some cases, octahedral Ga species could be detected in steam-treated gallosilicates equilibrated with acetylacetone [25], with the permanent question whether such a coordination was actually generated during such a treatment.

Very recently, the quantitative aspects of [71]Ga-NMR investigation of MFI-type gallosilicates have been questioned [26]. Indeed, in normal conditions, not only second order quadrupolar broadenings but also a distribution of chemical shifts, were shown to affect the [71]Ga-NMR linewidths and, hence, the corresponding line intensities. One of the conclusions was that limitations should arise from detectable variations of the quadrupolar interactions experienced by the framework Ga, due to the nature of the different counterions (in the zeolite pores), and their moving, or even elimination, during treatments.

The aim of this study is to determine quantitatively the framework Ga and/or Al concentrations in a series of (Al)-, (Ga)- and (Al,Ga)-MFI zeolites prepared under different conditions, and to further follow the structural changes of these species as a function of various post-synthesis (hydro)thermal treatments. For that purpose, three different techniques were used in combination: ^{27}Al- and [71]Ga-NMR, the wet chemical analysis of the total Ga contents and ammonia TPD investigations of the corresponding NH_4^+-exchanged samples. In particular, we will delineate the possibilities and limitations of the potential quantitative aspects of [71]Ga-NMR, applied to the various cases.

EXPERIMENTAL

Synthesis of various (Ga)-, (Al)- and (Ga,Al)-MFI zeolites

Twenty-eight different MFI samples containing Ga, Al or both elements have been synthesized by using a series of original preparation recipes. They were intentionally classified into 3 main categories (Table 15-1). Samples from the first category were prepared in the presence of Na^+ ions (along with TPA^+), either by dissolving the Ga^{3+} salts in NaOH at high pH [56], or by using commercial soluble Na^+ silicates, according to methods described earlier for aluminosilicates [57]. Samples of category 2 were obtained by using original recipes, in the presence of TPA^+ ions, from alkali-free media [57] . Methylamine was added for adequate pH adjustments to basic values, as already described in the case of germanosilicate-type MFI zeolites [58]. The third category comprises samples synthesized in the presence of fluoride ions as mineralizing agents [59]. In each category, the framework Ga^{3+} was supposed to interact preferentially with neighbors that might have a direct influence on the ^{71}Ga-NMR properties, namely Na^+ ions, TPA^+ ions and F^- ions respectively in each of the three series of syntheses. Ga^{3+} was indeed recently shown to be partly linked to F^- ions in the final framework [41]. The detailed gel preparations and crystallization conditions are described elsewhere [56]. All the synthesis data are schematically summarized in Table 15-1. The order of mixing of the ingredients, another important parameter that may influence the NMR properties, is indicated in each case, as to facilitate the discussion. All the as-synthesized materials were filtered, thoroughly washed with cold water and dried at 90°C for 24 h, prior to characterization.

Other characterization techniques

The nature and purity of the products was determined by XRD using a Philips P.W. 1349/30 diffractometer (Cu Kα radiation).

The water and organic contents were determined by heating about 0.05 g of sample under N_2 flow from 20 to 600°C, then maintaining it isothermally at 600°C in air flow for 4 h so as to burn off the coke formed upon thermal degradation of the organic molecules in an inert atmosphere.

TABLE 15-1
Synthesis Conditions and Preliminary Characterization of a Series of (Ga), (Al), and (Ga,Al)-MFI Zeolites

Sample (code)	Synth type (1)	M³⁺ conc in gel (2)		Gel prep.(order of mixing) (1)	Crystallization conditions			Final zeolite
		Ga/u.c.	Al/u.c.		T°C	Time(d)	cond(1)	(% cryst)(3)
ZN-11	Na(OH)	2.45	-	Ga in NaOH + TPA + SiO2	150	5	ST	MFI(93)
ZN-12	"	1.22	(2Cr3+)	Ga + Cr in NaOH + TPA + SiO2	150	5	ST	MFI(84)
ZN-16	Na(Na,Si)	2.45	-	Ga + NaCl + TPA in (Na,Si)	170	0.75	R	MFI(95)
ZN-17b	"	1.22	(2Cr3+)	(Ga+Cr) + NaCl + TPA in (Na,Si)	170	1.75	R	MFI(89)
ZN-19	"	2.45	-	(Ga+TPA) in (NaCl + (Na,Si))	170	0.75	R	MFI(94)
ZN-39	TPA+(MA)	2.45	-	(NH4F + TPA) + Ga + 4MA + SiO2	185	10	R	MFI(97)
ZN-40	"	2.45	-	Ga in (4 MA + TPA) + SiO2	185	5	R	MFI(100)
ZN-42	"	2.45	-	Ga in (4 MA + TPA) + SiO2	185	12	R	MFI(100)
ZN-61	"	2.45	-	Ga in (MA + TPA) + SiO2	185	3	R	MFI(100)
ZN-58	"	1.84	1	Ga in (MA+TPA)+Al+SiO2	185	3	R	MFI(98)
ZN-64	"	1.84	1	Ga in (MA+TPA)+Al+SiO2	185	3	ST	MFI(97)
ZN-59	"	1.22	2	Ga in (MA+TPA)+2Al+SiO2	185	3	R	MFI(103)
ZN-65	"	1.22	2	Ga in (MA+TPA)+2Al+SiO2	185	3	ST	MFI(-)
ZN-67	"	1.22	2	2Al in (MA+TPA)+Ga+SiO2	185	3	ST	MFI(-)
ZN-68	"	1.22	2	(Ga+NH4F+TPA)+2Al+MA+SiO2	185	3	ST	MFI(98)
ZN-69	"	1.22	2	(2Al+NH4F+TPA)+Ga+MA+SiO2	185	3	ST	MFI(98)
ZN-60	"	0.61	3	Ga in (MA+TPA)+3Al+SiO2	185	3	R	MFI(101)
ZN-66	"	0.61	3	Ga in (MA+TPA)+3Al+SiO2	185	3	ST	MFI(90)
ZN-62	"	-	4	4Al in (MA+TPA)+SiO2	185	3	R	MFI(95)
ZN-63	"		-	SiO2 in (MA+TPA)	185	3	R	MFI(100)
ZN-47	F-(TPA+)	2.45	-	(Ga in HF)+TPA+SiO2	185	8	(Parr)	MFI(-)
ZF-3	"	2.42	-	(Ga in NH4F) + TPA + SiO2	170	6	ST	MFI(95)
ZF-4	"	0.49	-	(Ga in HF) + Ga + TPA	170	10	ST	MFI(99)
AS-205(4)	"	1.88	-	(SiO2 + NH4F) + Ga + TPA	200	23	ST	MFI(100)
AS-206(4)	"	1.88	1.02	(SiO2 + NH4F) + Ga + TPA	200	23	ST	MFI(100)
AS-207(4)	"	0.95	-	(SiO2 + NH4F) + Ga + TPA	200	16	ST	MFI(100)
ZN-7	Na(NaOH)exc.	2.45	-	Ga in NaOH + TPA + (Na, Si)	150	5	ST	ANA (>95)
ZN-53	MA	2.45	-	Ga in (MA + n-propanol)	185	14	R	amorph.(0)

Footnotes to Table 15-1

(1):   Abreviations : Ga, Al, Cr = respectively Ga^{3+}, Al^{3+} and Cr^{3+} ions added in gel as nitrates; TPA = tetrapropylammonium ions added as bromide; (Na,Si) = commercial sodium silicate ("Waterglass" or "Glasven"); SiO_2=silica Aerosil 200 (from Aldrich); MA = methylamine (40% aq.); exc. = excess of NaOH; crystallization conditions :   St and R = respectively static and rotating Teflon-coated stainless steel 150 ml autoclaves; (Parr) = commercial "Parr" 300 ml. autoclaves, with incorporated mechanical stirrer.
(2):   Ratio of the number of M atoms (M=Ga or Al) per 96 T atoms (T=Si+M) involved in a "theoretical" MFI unit cell (u.c.), calculated from the initial gel composition.
(3):   Crystallinities evaluated from n-hexane sorption capacities (see Experimental).
(4):   Reference samples kindly provided by Dr. J.L. Guth (Mulhouse University, France). The detailed synthesis conditions are given in ref. (41).

A Stanton Redcroft, ST-780 thermobalance (simultaneous TG - DTA - DTG) was used for that purpose.

The crystallinities of the various samples were evaluated from n-hexane sorption capacities, measured in the same thermobalance, at 90 °C, on freshly pre-evacuated samples. The reference value was the sorption of 7.9 mol of n-hexane per unit cell of (Si)-MFI, the value averaged by measuring n-hexane sorptions on different silicalites synthesized under similar experimental conditions as the corresponding metallosilicates, and arbitrarily considered to be 100% crystalline.

TPD of ammonia was performed by heating the NH_4^+-exchanged samples in a Setaram TG - DTG - DSC 111 thermoanalyzer, in a helium flow, at a rate of 10°C/min . The desorbed ammonia was automatically titrated by a sulfamic acid solution, using a Metrohm-type titrimeter assembly coupled to the thermobalance, a procedure similar to the one proposed by Kerr and Chester [60].

The total gallium contents of the various samples were determined selectively in the presence of Al^{3+} ions, by complexometric titration with standard EDTA, using a copper-PAN indicator, after pre-dissolution of the samples in hydrofluoric acid [61].

NMR

All ^{27}Al- and ^{71}Ga-NMR spectra were recorded at 20°C on a pulsed Fourier transform Bruker MSL-400 spectrometer operating at a field of 9.4 T. The spectra were respectively recorded at frequencies of 104.3 and 121.9 MHz, with a MAS spinning rate of 10 kHz. Chemical shifts were measured relative to Al and Ga nitrate aqueous solutions, used as external references.

The sample holder was made of zirconia, capped with a fluorocarbon "Kelef" cover, containing about 100 mg (accurately weighed) of sample. A probehead equipped with a 4 mm double-bearing MAS assembly from Brucker was used. Typical spectra were obtained by accumulating 2,000 to 3,000 (Al) and 20,000 to 50,000 (Ga) free induction decays per sample, produced by 1 µs excitation pulses, with delay times of 1 µs for Al and 2 µs for Ga. Short pulse lengths (very small flip angles, typically 10°), were used so as to follow the quantitative conditions (within less than 5% accuracy), defined in detail for ^{27}Al-NMR in ref. [62]. The same conditions have been adopted for ^{71}Ga-NMR.

Analyses were achieved on MAS rotating samples, despite the danger of slight apparent intensity losses due to intensity rejections in the spinning sidebands [62,63]. To minimize this contribution, intensities were always computed in the same way, from the integrals drawn between 60 and 40 ppm for tetrahedrally coordinated aluminum ^{27}Al-NMR lines and between about 220 and 80 ppm for the tetrahedrally coordinated gallium ^{71}Ga-NMR line. The broader lines in certain cases may cause errors between 10 and 15% for ^{71}Ga-NMR.

RESULTS AND DISCUSSION

As synthesized Ga- and (Ga, Al)-MFI zeolites

Quantitative ^{71}Ga-NMR investigation

Data from Table 15-2 show that, for each compound, the total gallium concentration, as determined by the quantitative chemical analysis, is the same as the one introduced in the corresponding gel precursors, thereby suggesting that the synthesis efficiency for gallium, at least in the 0 to 3 Ga/u.c. concentration range, is close to 100%. However, the chemical

TABLE 15-2
71 Ga-NMR Determination of Ga Concentrations in Various As-Synthesized (Ga)-Zeolites, Prepared in the Presence of Na^+, TPA^+ or F^- Ions.

Sample(code)	Synth.type[1]	Ga/u.c. [2]		I(a.u.)[4]	^{71} Ga NMR Ga / u.c.[2]		
		"Theor."[3]	Chemical analysis		Curve(A)	Curve(B)	Curve(C)[5]
ZN-11	Na(OH)	2.45	2.47	91	2.92	2.43	2.08
ZN-12	"	1.22	1.25	46	1.48	1.24	1.05
ZN-16	Na(Na,Si)	2.45	2.42	87	2.81	2.36	2.00
ZN-17b	"	1.22	1.27	47	1.51	1.27	1.06
ZN-19	"	2.45	2.41	91	2.92	2.43	2.08
ZN-39	TPA⁺(MA)	2.45	2.39	106	3.41	2.85	**2.44**
ZN-40	"	2.45	2.44	110	3.56	2.96	**2.52**
ZN-42	"	2.45	2.47	106	3.41	2.85	**2.44**
ZN-61	"	2.45	2.38	102	3.30	2.75	**2.39**
ZN-58	"	1.84	1.75	75	2.42	2.02	**1.74**
ZN-59	"	1.22	1.18	50	1.62	1.36	**1.18**
ZN-60	"	0.61	0.66	27	0.88	0.74	**0.60**
ZN-47	F⁻(TPA⁺)	2.45	2.40	73	**2.38**	**1.96**	1.64
ZF-3	"	2.82	2.90	93	**2.98**	2.50	2.13
ZF-4	"	0.49	0.52	15	**0.49**	0.39	0.31
AS-205(6)	"	1.88	1.92	57	**1.87**	1.52	1.30
AS-206(6)	"	1.88	1.88	59	**1.90**	1.60	1.35
AS-207(6)	"	0.95	0.96	30	**0.96**	0.80	0.68
ZN-53	MA	2.45	(2.93)	(35)	-	-	-

(1) : Abreviations: see footnote (1), Table 15-1
(2) : The accurate molecular weight for each unit cell was evaluated from known framework composition (recurrent calculation of M^{3+} contents) and from the actual amount of H_2O and TPA, as determined by thermal analysis.
(3) : Calculated from gel composition, in number of M atoms (M=Ga or Al) per 96 T atoms (T=Si+M) (see footnote (2), Table 15-1).
(4) : Normalized intensity of the ^{71}Ga-NMR line located at about 150 ppm.
(5) : Respectivivy correlations A, B and C, as indicated in Figure 15-7.
(6) : Reference samples of known composition, determined by atomic adsorption in Mulhouse University (see also footnote (4), Table 15-1).

analysis does not indicate whether all the Ga is actually incorporated in the zeolitic framework. In fact, an accurate physicochemical method allowing one to determine quantitatively the framework Ga^{3+} ions in as-synthesized zeolites does not seem to have been tested so far. Methods like ^{29}Si-NMR, IR or framework expansion measurements by XRD present several limitations, especially in the case of high silica gallosilicates. Ion exchange capacity measurements or acid site determination are methods currently used for calcined materials. Unfortunately, these latter do not necessarily keep the same Ga concentration as in the precursor, because of the well known framework instability of gallium, upon thermal treatments (see introduction).

The recent in-depth study by Kentgens et al. [26] on the potentialities of ^{69}Ga- and ^{71}Ga-MAS NMR spectroscopies to be used as quantitative tools mentions several physical limitations, such as, the possible presence of NMR invisible extra lattice (tetrahedral) gallium species, or the changes in resonance positions and linewidths, due to a change in the quadrupolar interaction experienced by the framework Ga^{3+} ions, upon changing the counterions.

We therefore decided to derive empirical correlations, by plotting the normalized ^{71}Ga-NMR intensities of the strong line found in the 150-160 ppm range and currently attributed to tetrahedral framework Ga^{3+} species in zeolites (references cited in the introduction). Figure 15-1A shows the typical shape of such a resonance, definitely different from that characterizing Ga in bulky oxides that undergo strong magnetic perturbation (spectrum B), or octahedrally coordinated Ga^{3+} ions (spectrum C).

The results are shown in Figure 15-2, for 24 Ga- or (Ga, Al)- MFI zeolites, synthesized under various conditions. At first sight, the proportionality between the total Ga concentrations, as determined by chemical analysis, and the corresponding NMR intensities, is not linear. However, the different data tend to fill three different linear correlations, noted A, B and C in Figure 15-2. Interestingly, each correlation involves samples prepared under specific experimental conditions. More precisely, curves B, C and A are respectively related to samples synthesized in the presence of different ions , namely Na^+, TPA^+ (in absence of alkali) and TPA^+ together with F^-, that sooner or later are incorporated in the framework as close neigbors to the Ga^{3+} species, and that ultimately exert on them some specific "magnetic" influence [26].

After the drawing of the three curves, the three corresponding values of Ga concentrations for each intensity value have been derived

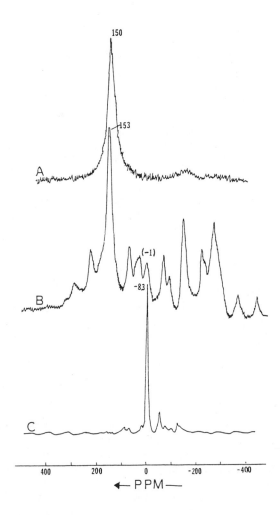

FIGURE 15-1 : ^{71}Ga-MAS NMR spectra of: (A): an as-synthesized (Ga)-MFI zeolite (ref.AS-206, Table 15-1); (B): commercial Ga$_2$O$_3$ b (Alpha Ventron 99.99%); (C): commercial Ga(NO$_3$)$_3$ .9H$_2$O (Aldrich, 99.999%)

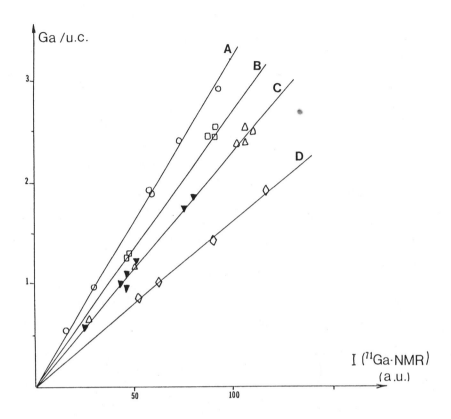

FIGURE 15-2 : Correlation between the Ga concentrations (in Ga per unit cell of zeolite) in various as-synthesized (Ga)-MFI zeolites, as determined by chemical analysis and the corresponding normalized ^{71}Ga-NMR line intensities (in arbitrary units). Correlations (A), (B) and (C) are respectively derived for samples synthesized in the presence of F⁻ ions (O), Na⁺ ions (□) and TPA⁺ ions (white triangles (△) for gallosilicates and black triangles (▼) for gallo-aluminosilicates). Samples as described in Tables 15-1, 15-2 and 15-4. Correlation (D) is derived for calcined, NH₄⁺-exchanged and water-equilibrated samples (◊), as described in Table 15-6.

(Table 15-2). Only one of the three values (printed in bold characters) actually appears to fit quasi-perfectly the "theoretical" or analyzed (total) Ga concentration found in the precursors. The other two values are different by a percentage larger than the experimental error. Aluminum does not seem to influence the Ga^{3+} magnetic resonance characteristics (black triangles on correlation B, Figure 15-2), obviously because of the probably large dilution of Al^{3+} and Ga^{3+} ions within the essentially silica-bearing framework of the corresponding gallo-aluminosilicates.

Our correlations seem to be valid only for well crystalline samples. Indeed, sample ZN-53 (amorphous gel), which exhibits a broad ^{71}Ga-NMR signal, also located near 150 ppm, does not fit any of the three correlations.

A more rigorous interpretation of the relation between the NMR line intensities and the nature of the Ga neighbor in the framework is beyond the scope of this work.

Quantitative ^{27}Al-NMR investigation

^{27}Al-NMR was shown to be of accurate use, under defined conditions, for a quantitative determination of framework·tetrahedrally coordinated Al atoms in zeolites [62,63] . Indeed, by using various well characterized zeolites with very different framework Al contents (Table 15-3), a perfect linear correlation between the Al framework concentration and the corresponding normalized ^{27}Al-NMR intensities is obtained (Figure 15-3). Data from Table 15-3 confirm that the different aluminosilicate ZSM-5 zeolites can be used as reference in their variously as-synthesized forms (with the different alkali cations as counterions), or as calcined and/or further NH_4^+-exchanged materials. No extra framework Al seems to have been produced in this latter case, confirming the good thermostability of aluminosilicates, with respect to gallosilicates, when treated under similar (hydro)thermal conditions [24].

The correlation from Figure 15-3 was used to calculate the actual number of Al atoms per unitcell in the various galloaluminosilicates prepared in the presence of TPA^+ and methylamine (MA), and in the absence of alkali cations (Table 15-4).

Surprisingly, the as-synthesized samples involve a higher amount of tetrahedrally coordinated Al species per unit cell (supposed to involve 96 T atoms) than the total amount (per 96 T atoms) calculated from the hydrogel composition. Such a phenomenon was never observed so far in

TABLE 15-3
Correlation Between the ^{27}Al-NMR Normalized Intensities of the NMR Line Located near 52 ppm and the Al Content in Variously Synthesized or Modified ZSM-5 Zeolites of Known Composition, used as Reference Materials

Sample (code)	Treatment (1)	ref.	Chemical analysis			approx (2) mol. wt.	mol Al/g (x 10^{-4})	I(^{27}Al-NMR) (a.u.)
			method(1)	Si/Al	Al/uc			
(Na,TPA) NTZ-16	as synthesized	(57)	PIGE	220	0.43	5470	0.78	80
(Na,TPA) NTZ-16 F$_1$	u.s. treated	(57)	PIGE	230	0.42	5470	0.76	83
[NH$_4$,TPA(Ga)]AS-206	as synthesized	(41)	A.A.	93	1.02	6550	1.56	151
(NH$_4$) MS-400	calc., ech.(NH$_4^+$)	(62)	PIGE	47	1.99	5800	3.43	327
(NH$_4$,TPA) NTZ-8	as synthesized	(57)	PIGE	41	2.28	6500	3.51	335
(H) NTZ-8	calcined	(57)	PIGE	39	2.40	5760	4.16	370
(NH$_4$) MS-200	calc., ech.(NH$_4^+$)	(62)	PIGE	15.9	5.68	5850	9.73	870

(1): Abreviations : calc = calcined (550°C, N$_2$, then air flow) ; ech.(NH$_4^+$) = calcined form, exchanged 3 times with NH$_4$NO$_3$ 0.5 M, 20°C; u.s. = ultrasonic cleaning treatment to remove traces of amorphous phase intermixed with ZSM-5 crystals (ref. 65); PIGE = proton induced gamma-ray emission (ref. 64); AA = atomic absorption (ref. 41).
(2): Molecular weight calculated from known framework composition and from the known amounts of H$_2$O and TPA, determined by thermal analysis.

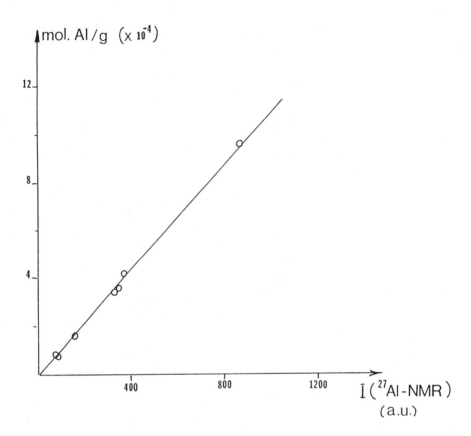

FIGURE 15-3 :Correlation between the Al concentrations (in mol. Al per g) in various
as-synthesized and modified reference (Al)-MFI zeolites, as determined by chemical or
physicochemical analyses (Table 15-3), and the corresponding ^{27}Al-NMR intensities (in
arbitrary units)

TABLE 15-4
Ga and Al Concentrations in Various As-Synthesized (Al), (Ga) and (Ga,Al)-MFI Materials, Prepared in the Presence of Methylamine (MA) or (MA+F-), as Mobilizing Agents.

Sample (code)	Ga/u.c. (1)		71 Ga NMR		Synth. eff. (4) = (c/a)	Al/u.c.(1)			Synth eff.(4) =(f/d)	Al + Ga / uc		Synth. Eff. (4) = (h/g)
	theor.(2) =(a)	chem. anal.= (b)	I (a.u.) (3)	Ga/u.c. =(c)		theor (2) =(d)	NMR(5) =(e)	NMR cal =(f) (6)		theor(2) (g)	NMR (h)=c+f	
ZN-61	2.45	2.38	102	2.39	96	-	-	-	-	2.45	2.39	96
ZN-58	1.84	1.75	75	1.74	97	1	1.36	0.97	97	2.84	2.71	95
ZN-64	1.84	1.83	80	1.85	100	1	1.52	1.05	(105)	2.84	2.90	(102)
ZN-59	1.22	1.18	50	1.18	97	2	2.41	1.98	99	3.22	3.16	98
ZN-65	1.22	1.20	46	1.10	90	2	2.40	1.96	98	3.22	3.16	98
ZN-67	1.22	1.00	43	1.00	82	2	2.42	2.07	(103)	3.22	3.07	95
ZN-68	1.22	1.23	51	1.19	98	2	2.39	1.95	98	3.22	3.15	98
ZN-69	1.22	0.94	46	1.02	84	2	2.96	2.07	(103)	3.22	3.09	98
ZN-60	0.61	0.66	27	0.60	98	3	3.75	2.90	97	3.61	3.50	97
ZN-66	0.61	0.59	24	0.56	92	3	3.85	2.80	93	3.61	3.36	93
ZN-62	-	-	-	-	-	4	4.92	3.98	100	4	3.98	100

(1), (2) and (3): See respectively footnotes (2) (3) and (4), Table 15-2
(4): "Synthesis efficiency", i.e., the percentage of trivalent element incorporated from the gel to the zeolite framework.
(5): Normalized intensity of the ^{27}Al-NMR line located at about 52 ppm and attributed to tetrahedrally coordinated Al^{3+} in the MFI- framework.
(6): The real framework Al content was determined from ^{27}Al-NMR line intensities of calcined materials (see text).

the synthesis of zeolites [66,67]. Indeed, in the presence of TPA+, ZSM-5 was always synthesized with an efficiency equal to or slighly lower than 100%, based on the alumina involved [66].

On the other hand, when the whole series of samples were calcined under usual conditions that do not generally lead to even a partial dealumination of (Al)ZSM-5, the intensities of the NMR lines located near 52 ppm decreased, while a smaller resonance was found near 0 ppm , a spectrum region where lines characterizing octahedrally coordinated Al^{3+} species usually occur (Figure 15-4, spectrum B). The new Al framework concentrations, as recalculated from the NMR intensities of the lines located at 52 ppm for the calcined samples, now perfectly correspond to those introduced in the initial hydrogel, thus showing now a 100% "synthesis efficiency" (Table 15-4 and Figure 15-5).

On calculating the actual amount of TPA+ ions per unit cell of as-synthesized zeolite by thermal analysis [56], we noticed that this amount was perfectly equivalent to the amount of framework Al determined after calcination, and obviously lower than the total amount of tetrahedral Al atoms in the corresponding as-synthesized materials. This clearly suggests that, in this case, NMR detects two types of tetrahedrally coordinated Al species. One type is obviously the true framework Al that is completely neutralized by TPA+, the only counterions supposed to be present in this system. These Al atoms integrally remain in their framework positions after calcination, as expected. It is concluded that the other kind of tetrahedrally coordinated Al atoms present in the as-synthesized phases are probably not in the zeolite lattice positions, although these species also contribute to the intensity of the 52 ppm ^{27}Al-NMR line.

A parallel study of that phenomenon [56] has shown that these extra species are in fact interacting with the residual methylamine so as to form fairly strong complexes or associations. These latter were found admixed with the as-synthesized MFI crystallites that they probably overcoat as thin films. These complexes also proved resistant to water washings, but they are less resistant to thermal treatments. Actually some methylamine was found released when the precursors were heated between 325 and 385°C (thys below the temperature at which TPA+ starts to decompose), as suggested by the combined TG-DSC-TPD data [56].

The higher the total Al content, the higher the decomposition temperature of these species. The amount of base released in these conditions, as measured by TPD [56], roughly corresponds (on a molar basis) to the amount of the extra lattice tetrahedral Al, indirectly determined by the present ^{27}Al-NMR data. This further suggests that we may deal

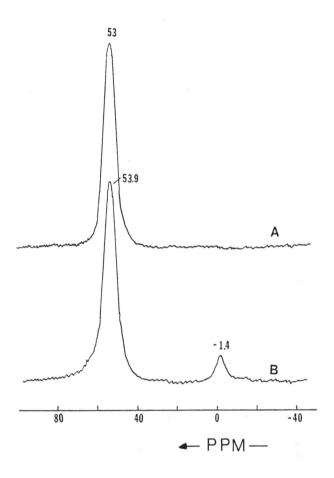

FIGURE 15-4: ^{27}Al-MAS NMR spectra of the gallo-aluminosilicate ZN-62; (A): as-synthesized material; (B): calcined (550°C, N$_2$, then air) material, equilibrated in ambient air

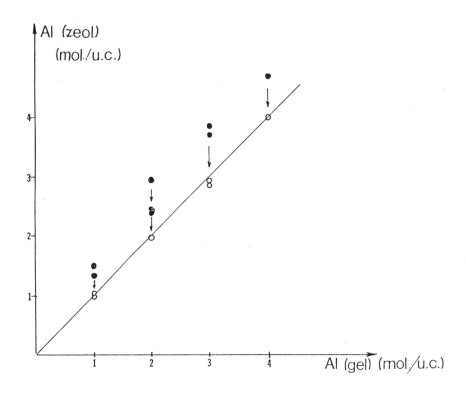

FIGURE 15-5: Correlation between the Al concentration in the precursor gel (from ingredient concentrations) and that determined in the final (Al) and (Al, Ga)-MFI zeolites, synthesized in the presence of TPA$^+$ (Table 15-4) : (●) = as-synthesized zeolites; (o) = calcined (550°C, N$_2$, then air) materials.

with methylamine-Al type associations (possibly also involving some residual silica species), having a stoichiometry close to 1:1.

The successive use of [71]Ga- and ^{27}Al-NMR spectroscopies to investigate the framework concentrations of the corresponding trivalent elements is straightforward for all other types of (Ga, Al)-MFI zeolites, synthesized in the absence of methylamine [56].

Table 15-4 finally shows that the series of materials prepared in alkali-free media involving TPA$^+$ and methylamine present a nearly 100% synthesis efficiency, thus leading to a total incorporation of both Al^{3+} and Ga^{3+} ions from the gel. Note that in all compounds, Ga^{3+} was admixed with the mobilizing species (MA+TPA$^+$), before Al^{3+} ions were added in the gel (see order of mixing of all the ingredients, Table 1). Apparently, Al^{3+} does not need a strong mobilizing action in the gel to be quantitatively incorporated. However, when Al^{3+} ions are first admixed with a mobilizing species (either F$^-$ ions, such as in ZN-69, or MA+TPA$^+$, such as in ZN-67), they are probably more rapidly incorporated in the growing zeolite during crystallization, possibly at the expense of Ga^{3+} ions, that are introduced later in the mixture. This is substantiated indirectly by [71]Ga-NMR (Table 15-4). Indeed, in these two samples, while the framework Ga contents determined by both chemical analysis and [71]Ga-NMR give reproducible values, these are lower than the initial amounts of Ga$^+$ introduced in the gel. This suggests that Ga^{3+} ions need to be sufficiently mobilized, for a 100% efficient incorporation.

Calcined (Ga)- and (Ga,Al)-MFI zeolites

Figure 15-6 shows the [71]Ga-NMR spectra of an as-synthesized (spectrum A) and of the corresponding calcined (spectrum B) (Ga)-MFI zeolite. Spectrum B exhibits a broader resonance, still located at about 150 ppm and probably still characterizaing (some of) the Ga^{3+} framework species. A very broad hump is also present upfield, centered at about 80 ppm. It probably corresponds to non-framework Ga species [26] that were more precisely identified as "amorphous Ga$_2$O$_3$ species" [41], involving undefined coordination states.

The intensity of the main [71]Ga-NMR resonance can still be evaluated from the integral curve drawn on the spectrum, which excludes the intensity partion due to the broad peak at 80 ppm. Firstly, such an estimation now becomes very inaccurate. In addition, as can be seen in

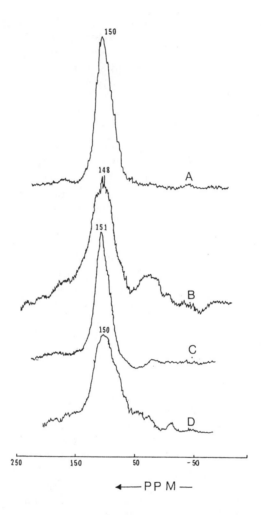

FIGURE 15-6: [71]Ga-MAS NMR spectra of sample ZN-19 [(Ga)- MFI/Na,Si]; (A) : as-synthesized; (B) : calcined (550°C, N₂, then air); (C): NH₄⁺-exchanged and water equilibrated; (D): sample (C), recalcined in air, not equilibrated in water

Table 15-5, the intensities so calculated are very random and actually drastically depend on the calcination conditions. Some samples, calcined under comparable conditions, may even exhibit different NMR intensities "characterizing" Ga^{3+} ions that remain in the zeolitic framework after calcination. In some other cases, no signal could be recorded (Figure 15-7, and Table 15-5, samples ZN-64 and ZN-65). We would agree with Kentgens et al. [25,26] in assuming that the remaining framework Ga^{3+} species, if any present, would give rise to resonance lines that are broadened beyond detection, due to the large quadrupolar interactions occurring in dry, not re-equilibrated samples. The quantitative use of the intensities so recorded for a calculation of the remaining framework Ga concentration is meaningless.

The total amount of trivalent ions (Al^{3+} and Ga^{3+}) in the calcined samples can be accurately determined by computing the total amount of acid sites generated by these ions in the zeolite framework. A currently used method is the TPD of ammonia evolved upon heating the corresponding NH_4^+-exchanged samples, as it was successfully applied in the case of gallium- [15, 19, 24, 27, 42, 44] and aluminium- [15, 19, 27, 44] bearing MFI zeolites.

The total amount of acid sites, as determined by NH_3-TPD, is given in Table 15-5. The knowledge of the total amount of trivalent (Al^{3+} + Ga^{3+}) ions by TPD and of the amount of framework Al^{3+} ions by ^{27}Al-NMR directly yields the residual amount of Ga^{3+} ions that remain in the zeolite framework after tha various calcination experiments (Table 15-5). As expected, this amount (column 6, Table 15-5) can not at all be correlated to the ^{71}Ga-NMR intensities really measured at least on the non-water equilibrated samples (column 3, Table 15-5).

Other interesting data can be derived from these determinations. We can, for example, evaluate the amount of framework Ga^{3+} ions that still (partly) generate the acidity of the calcined samples (column 7, Table 15-5), directly related to the percentage of Ga extracted from the lattice upon calcination (column 8). Undoubtedly, the amount of Ga^{3+} ions that remain in the lattice after calcination, or, in other words, the general stability of the framework Ga species, is noteworthily influenced by the actual Al^{3+} framework content. The richer the (Ga,Al)-MFI framework in Al^{3+}, the easier the Ga^{3+} extraction from it, upon calcination. Finally all Ga^{3+} ions are released from samples that initially involve 3 Al^{3+} and 0.6 Ga^{3+} in the framework, while only about 4% of the 1.8 framework Ga^{3+} ions are

TABLE 15-5

Ga and Al Contents of Various Calcined (1) MFI Materials (Atoms of M per Unit Cell of Zeolite)(2)

Sample (code)	Al(NMR) =(a)	Ga(NMR) =(b)	(Al+Ga)(3) (TPD)=(c)	Ac. strength (T°C max. TPD)	Ga/u.c. (d)=(c)-(a)	% Ga in acid sites=(d/c)x100	% Ga(4) extracted
ZN-61	-	0	1.91	386	1.91	100	20
ZN-58	0.97	1.23	2.68	398	1.71	64	4
ZN-64	1.05	0.73 and 0(5)	2.80	404	1.75	63	4
ZN-57	1.98	0.14	2.87	416	0.89	31	25
ZN-65	1.96	0.97 and 0(5)	2.80	420	0.84	30	24
ZN-67	2.07	0.12	2.74	418	0.67	24	33
ZN-68	1.95	0.77	2.60	420	0.65	25	47
ZN-69	2.07	0.73	2.65	422	0.58	22	38
ZN-60	2.90	0	2.86	438	0	0	100
ZN-66	2.80	0.37	2.68	436	0	0	100
ZN-62	3.98	-	3.84	450	-	-	-

(1) : As-synthesized materials, calcined under dry N_2 flow, then in air, at 550°C (see Experimental).

(2) : See footnote (2), Table 15-2.

(3) : From the total number of acid sites, as determined by NH_3-TPD measurements on pre-
calcined and NH_4^+-exchanged samples.

(4) : $\dfrac{Ga(as\ synth.) - Ga(calc.)}{Ga(as\ synth.)} \times 100$

(5) : Same precursor, heated under different conditions.

released when only 1 Al^{3+} completes the framework composition (Table 15-4, column 5).

In fact, more gallium is lost from the pure gallosilicate precursor ZN-61 (about 20%), possibly suggesting that small amounts of Al^{3+} in the lattice play a kind of stabilizing role towards framework Ga^{3+}. Such a preliminary observation undoubtely needs confirmation and generalization and is now subject of further investigations (56).

Another remarkable feature is the regular variation of the temperature of the TPD peaks corresponding to the NH_3 release from the acidic sites, generated by both Ga^{3+} and Al^{3+} framework ions (Table 15-5, column 5). This temperature smoothly increases from 360°C (pure Ga^{3+}-sample), to 450°C for the pure Al^{3+} counterpart. This is a good indication for a relative acid site homogeneity, thus suggesting that Al^{3+} and Ga^{3+} ions are quasi-simultaneously incorporated in the growing zeolite framework during the crystallization process, leading to a rather homogeneous final Al-Ga distribution within the lattice. This observation could be further confirmed by monitoring the TPA^+ decomposition in these samples by DSC (56). The relatively high TPD temperature, with respect to that reported for various MFI gallo- or gallo-aluminosilicates synthesized from Na^+-bearing gels (15,56), also suggests that the corresponding acid sites are substantially stronger in our series of samples than in zeolites currently prepared by using more conventional methods.

Post-synthesis treated (Ga)- and (Ga,Al)-MFI zeolites

The exchange treatments of the calcined materials with aqueous NH_4NO_3 solutions, followed by a further drying at moderate temperatures (e.g.100°C), actually result in a final re-equilibration of the dry samples in water. As predicted by Kentgens et al. (25,26), such a treatment can markedly reduce the quadrupolar broadenings and result in the possible restoration of the corresponding ^{71}Ga-NMR line intensities. Indeed, the resonances recorded for various NH_4^+-exchanged samples appear narrower, better resolved and increase in intensity (Figure 15-6, spectrum C). In some cases, when the resonance has completly disappeared due to an extensive line broadening (e.g. Figure 15-7, spectrum B), the NMR line reappears again upon NH_4^+ or water treatment of the sample.

Upon a second thermal tratment of the water-equilibrated samples, the lines disappear again, or at least decrease in intensity and become

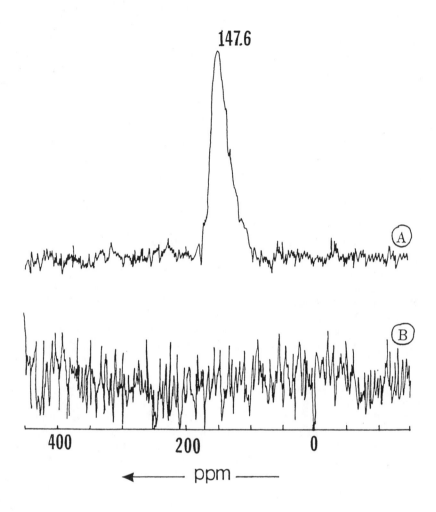

FIGURE 15-7: [71]Ga-MAS NMR spectra of sample ZN-61 [(Ga)-MFI/TPA+ +MA];
(A): as-synthesized; (B): calcined (550°C, N2, then air), not equilibrated in water

broader (Figure 15-6, spectrum D), thereby confirming that ^{71}Ga-NMR is of little use in such cases for quantitative evaluations.

In an attempt to check whether ^{71}Ga-NMR can be quantitatively used to compute the tetrahedral Ga^{3+} species present in equilibrated samples, we have examined by this technique a series of NH_4^+-exchanged samples that were further contacted with water vapor at ambient temperature, for several days, so as to achieve a more complete equilibration of the Ga^{3+} ions by creating a less strained framework.

The ^{71}Ga-NMR line intensities recorded for 4 of the equilibrated samples (Table 15-6, column 5) can be compared neither to those corresponding to the as-synthesized counterparts (column 2) nor to the pre-calcined (dry) precursors (column 4). However, if we plot them versus the Ga^{3+} content in the NH_4^+-exchanged samples, measured by ammonia TPD (column 7), we again obtain a good linear correlation (Figure 15-2, correlation D). The line intensity decreases again upon a subsequent thermal treatment of the NH_4^+-form (column 8).

More data are needed to conclude that ^{71}Ga-NMR could allow the accurate determination of framework Ga^{3+} species in variously NH_4^+-exchanged samples but our results appear promising. The fact that the NMR intensities versus the Ga^{3+} concentrations follow another correlation than those derived for the various as-synthesized materials is not surprising, as Ga^{3+} ions undergo a similar influence (from the NH_4^+ counterions) in all samples, definitely different from other ionic neighbors present in the precursors.

CONCLUSION

Our work shows for the first time that the ^{71}Ga-NMR technique can be very accurately used for a quantitative determination of framework Ga^{3+} ions present in a wide series of as-synthesized (Ga)- or (Ga,Al)-MFI zeolites. Calibration curves of the NMR intensities versus the Ga^{3+} concentrations accurately determined by a separate method, must be first established. Moreover, such correlations appear to be valid for materials synthesized by following similar recipes and involving gels that contain at least the same potential counterions supposed to be further incorporated in the zeolite framework and to interact with the trivalent ions.

^{71}Ga-NMR cannot be used to determine the trivalent concentration in the post-synthesis treated samples, unless they are further re-

TABLE 15-6
Variation of the ^{71}Ga-NMR Line Intensities, as a Function of Selected Post-Synthesis Treatments of Some (Ga)-MFI Zeolites

Sample (code)	As synthesized		Calcined (2) 550°C(dry)	NH4+ - exchanged (water equilibrated)			H+ form (dry)(4)
	I(NMR) (a.u.) (1)	Ga/u.c.(5) (NMR)	I(NMR)(1) (a.u.)	I(NMR)(1) (a.u.)	Ga/u.c(5) (NMR)	Ga/u.c. (TPD) (3)	I(NMR) (1)
ZN-11(NaOH)	91	2.43(B)	31	53	0.84(D)	0.86	–
ZN-12(NaOH)	46	1.24(B)	23	–	–	–	–
ZN-16(Na,Si)	87	2.36(B)	26 and 0(6)	90	1.44(D)	1.41	–
ZN-19(Na,Si)	91	2.43(B)	25	62	1.00(D)	1.03	42
ZN-58(TPA)	75	1.74(C)	60	116	1.88	1.91	–

(1) : Normalized ^{71}Ga-NMR intensities for the line located at about 150 ppm.

(2) : H+ or Na+ forms, obtained by calcining the as-synthesized materials at 550°C in air, not rehydrated in ambient (humid) air.

(3) : Equilibrated in water during NH4+ exchanges.

(4) : Obtained by calcining the NH4+ forms at 550°C, in air and not equilibrated in (humid) air.

(5) : In parentheses: correlation from which the calculation was derived; see Figure 15-2

(6) : Same precursor, heated under different. conditions.

equilibrated with water vapor at moderate temperatures. Indeed, strong quadrupolar interactions can occur in the framework of the dry samples and cause important NMR line broadenings, rendering the technique totally unsuitable for quantitative evaluations.

The determination of residual amounts of Ga^{3+} ions present in the framework of gallosilicate samples calcined under various atmospheres can be quantitatively achieved by using TPD of ammonia released from NH_4^+ exchanged samples. In the case of gallo-aluminosilicates, as ^{27}Al-NMR always remains a powerful means to determine the framework tetrahedrally coordinated Al^{3+} ions, in both as-synthesized and calcined materials, this technique can be used in combination with NH_3-TPD for the quantitative evaluation of both trivalent ($Al^{3+} + Ga^{3+}$) ionic species present in the post-synthesis treated MFI-frameorks.

None of the above techniques gives information on the structural state of the extra framework gallium oxidic species that are released upon (hydro)thermal treatments of the precursors. At best, these species are sometimes qualitatively characterized by a small and broad ^{71}Ga-NMR resonance occurring at about 80 ppm. Their quantitative computing can only be achieved indirectly, by evaluating the extent of Ga extraction upon each treatment on combining NMR, TPD and chemical analysis.

1. REFERENCES

[1]   R.M. Barrer, Hydrothermal Chemistry of Zeolites (Academic Press, London, 1982), p. 282.
[2]   J.M. Newsam and D.E.W. Vaughan, in New Developments in Zeolite Science and Technology, edited by Y. Murakami et al. (Proc. 7th Int. Zeolite Conf., Kodansha, Tokyo, 1986) pp. 457-464.
[3]   K. Suzuki, Y. Kiyozumi, S. Shin and S. Ueda, Zeolites, 5, 11 (1985).
[4]   J.M. Newsam and J.D. Jorgensen, Zeolites, 7, 769 (1987).
[5]   W. Lortz and G. Schön, J. Chem. Soc. Dalton Trans., 1987, 623.
[6]   J.M. Newsam, A.J. Jacobson and D.E.W. Vaughan, J. Phys. Chem., 90, 6858 (1986).
[7]   M.L. Occelli, U.S. Patent N° 4 803 060 (7 February 1989).
[8]   V.M. Mastikhin, N.V. Klyueva, K.G. Ione and K.I. Zamaraev, React. Kinet. Catal. Lett., 35, 465 (1987).
[9]   S. Ueda and H. Yamada, Nippon Kagaku Kaishi, 29, 159 (1989).
[10] W.B. Yelon, D. Xie, J.M. Newsam and J. Dunn, Zeolites, 10, 553 (1990).
[11] G. Bellussi, R. Millini, A. Carati, G. Maddinelli and A. Gervasini, Zeolites, 10, 642 (1990).
[12] Z. Yin-Xing, A. Tuel, Y. Ben Taarit and C. Naccache, J. Mol. Catal., 1991 (in press).
[13] See for example : J.B. Parise, Inorg. Chem. 24, 4312 (1985); J. Chem. Soc. Chem. Commun., 1985, 606; S. Hirano and P. Kim, Bull. Chem. Soc. Jpn. 62, 275 (1989); J.L. Guth in Zeolite Microporous Solids : Synthesis, Structure and Reactivity, edited by E.G. Derouane et al., (Kluwer Academic Publishers Dordrecht, 1992), pp ; M. Estermann, L.B. McCusker, C. Baerlocher, A. Merrouche and H. Kessler, Nature, 352, 320 (1991).
[14] R. Szostak Molecular Sieves, Principles of Synthesis and Identification (Van Nostrand Reinhold, New York, 1984), p. 212.
[15] C.T.W. Chu and C.D. Chang, J. Phys. Chem., 89, 1569 (1985).
[16] S. Hayashi, K. Suzuki, S. Shin, K. Hayamizu and O. Yamamoto, Bull. Chem. Soc. Jpn., 58, 52 (1985).
[17] X.S. Liu and J.M. Thomas, J. Chem. Soc. Chem. Commun., 1985, 1545.
[18] J.M. Thomas and X.S. Liu, J. Phys. Chem., 90, 4843 (1986).
[19] H.K. Beyer and G. Borbely, ref. 2, pp. 867-874.

[20] T. Inui, Y. Makino, F. Okazumi, S. Nagano and A. Miyamoto, Ind. & Eng.Chem. Res., 26, 647 (1987).

[21] E.A. Hyde and K.T. McNiff, Eur. Patent. Appl. 244 162 (23 April 1987).

[22] G.P. Handreck and T.D. Smith, J. Chem. Soc. Faraday Trans. 1, 85, 3215 (1989).

[23] V.N. Romannikov, L.S. Chumachenko, V.M. Mastikhin and K.G. Ione, React. Kinet. Catal. Lett., 29, 85 (1985).

[24] L. Petit, J.P. Bournonville and F. Raatz, in Zeolites: Facts, Figures, Future, edited by P.A. Jacobs and R.A. Van Santen (Proc. 8th Int. Zeolite Conf., Elsevier Sci. Publ. Amsterdam, The Netherlands, 1989), pp. 1163-1171.

[25] C.R. Bayense, J.H.C. Van Hooff, A.P.M. Kentgens, J.W. De Haan and L.J.M. Van de Ven, J. Chem. Soc. Chem. Commun., 1989, 1292.

[26] A.P.M. Kentgens, C.R. Bayense, J.H.C. Van Hooff, J.W. De Haan and L.J.M. Van de Ven, Chem. Phys. Lett., 176, 399 (1991).

[27] N.S. Gnep, H.Y. Doyemet and M. Guisnet, J. Mol. Catal., 45, 281 (1988); in Zeolites as Catalysts, Sorbents and Detergent Builders, edited by M. Kaye and J. Weitkamp, (Elsevier, Amsterdam, The Netherlands), 1989, p. 153.

[28] Y. Yang, X. Guo, M. Deng, L. Wang and Z. Fu, ibid., p. 98.

[29] A. Endoh, K. Nishimiya, K. Tsutsumi and T. Takaishi, ibid., p. 138.

[30] T.J. Gricus Kofke, R.J. Gorte and G.T. Kokotailo, Appl. Catal., 54, 177 (1989).

[31] T. Inui, Y. Ishikawa, K. Kamachi and H. Matsuda, ref. 24, pp. 1183-1192.

[32] A.V. Kucherov, A.A. Slinkin, G.K. Beyer and G. Borbely, J. Chem. Soc. Faraday Trans.1, 85, 2737 (1989).

[33] J. Kanai and N. Kawata, Appl. Catal., 55, 115 (1989).

[34] J. Kanai and N. Kawata in Scientific Bases for the Preparations of Heterogeneous Catalysts (Proc. 5th Int. Symp.- Preprints, Louvain-la-Neuve, Belgium, 1990), pp. 211-220.

[35] G. Bellussi, A. Carati, M.G. Clerici and A. Esposito, ibid., pp. 201-209.

[36] S.A. Axon, K. Huddersman and J. Klinowski, Chem. Phys. Letters, 172, 398 (1990).

[37] A.Yu. Khodakov, L.M. Kustov, T.N. Bondarenko, A.A. Dergachev, V.B. Kazansky, Kh.M. Minachev, G. Borbely and H.K. Beyer, Zeolites, 10, 603 (1990).

[38] V. Kamazirev, G.L. Price and K.M. Dooley, J. Chem. Soc. Chem. Commun., 1990, 712.

[39] G.L. Price and V. Kanazirev, J. Catal., 126, 267 (1990).

[40] G.L. Price and V. Kanazirev, J. Molec. Catal., 66, 115 (1991).

[41] A. Seive, Ph.D. Thesis, University of Mulhouse, France, 1990.

[42] T. Yashima, K. Yamagushi and S. Namba in Chemistry of Microporous Crystals (Stud. Surf. Sci. Catal., vol. 60, Kodansha, Tokyo, Japan, 1990), pp. 171-178.

[43] M.D. Dompas, W.J. Mortier, D.C.H. Kenter, M.J.G. Jansen and J.P. Verduijn, J. Catal., 129, 19 (1991).

[44] D.K. Simmons, R. Szostak, P.K. Agrawal and T.L. Thomas, J. Catal., 106, 287 (1987).

[45] H. Kitagawa, Y. Sendoda and Y. Ono, J. Catal., 101, 12 (1986).

[46] G. Sirokman, Y. Sendoda and Y. Ono, Zeolites, 6, 299 (1986).

[47] D.C. Martindale, J.A. Kocal and T.H. Chao, U.S. Patent 4 795 845 (3 January 1989).

[48] P. Meriaudeau, G. Sapaly and C. Naccache, ref. 24, p. 1423.

[49] G. Centi and G. Golinelli, J. Catal., 115, 452 (1989).

[50] K. Fujimoto, I. Nakamura and K. Yokota, Zeolites, 9, 120 (1989).

[51] J. Baumgartner and E. Iglesia, presented at the 12th North American Meeting on Catalysis, Lexington, Kentucky, May 1991 (in press).

[52] P. Meriaudeau and C. Naccache, ref.51 (in press); in Catalyst Deactivation 1991, edited by C.H. Bartholomew and J.B. Butt, Elsevier, Amsterdam, The Netherlands, 1991, pp 767-772.

[53] H.K.C. Timken and E. Oldfield, J. Am. Chem. Soc., 109, 7669 (1986)

[54] Z.G. Zulfugarov, A.S. Suleimanov and Ch.R. Samedov, Stud. Surf.Sci. Catal., 18, 167 (1984).

[55] A.A. Slinkin, personal communication.

[56] Z. Gabelica, G. Giannetto, F. Dos Santos, R. Monque and R. Galliasso, in Proc. 9th Intern. Zeolite Conf., Montreal, July 1992, (in press).

[57] Z. Gabelica, N. Blom and E.G. Derouane, Appl. Catal., 5, 227 (1983).

[58] Z. Gabelica and J.L. Guth, Angew. Chem. Int. Ed. Engl., 28, 81 (1989); Stud. Surf. Sci. Catal., 49A, 421 (1989).

[59] J.L. Guth, H. Kessler and R. Wey, Stud. Surf. Sci. Catal., 28, 121 (1986).

[60] G.T. Kerr and A.W. Chester, Thermochim. Acta, 3, 113 (1971).

[61] W. Wagner and C.J. Hull, Inorganic Titrimetric Analysis (Marcel Dekker, Inc., New York, 1971).

[62] C. Fernandez, F. Lefebvre, J. B.Nagy and E.G. Derouane, Stud. Surf. Sci. Catal., 37, 223 (1988).

[63] A. Samoson, E. Lippmaa, G. Engelhardt, U. Lohse and H.G. Jerschkewitz, Chem. Phys. Lett., 134, 589 (1987).

[64] A. Nastro, Z. Gabelica, P. Bodart and J. B.Nagy, Stud. Surf. Sci. Catal., 19, 131 (1984).

[65] G. Debras, E.G. Derouane, J.P. Gilson, Z. Gabelica and G. Demortier, Zeolites, 3, 97 (1983).

[66]   P.A. Jacobs and J.A. Martens, <u>Synthesis of High Silica Aluminosilicate Zeolites</u> (Stud. Surf. Sci. Catal. vol 33), Elsevier, Amsterdam, 1987, p 192.

[67] C. Pellegrino, R. Aiello and Z. Gabelica in <u>Zeolite Synthesis</u>, edited by M.L. Occelli and H.E. Robson (ACS Symp. Series 398, American Chemical Society, Washington, D.C., 1989), pp 161-175.

16

From Clathrasils to Large-Pore Zeolites: The Effects of Systematically Altering a Template Structure

Yumi Nakagawa and Stacey I. Zones

Chevron Research and Technology Company

Organic templates are believed to play an important role in the process of high-silica zeolite crystallization. We have prepared a series of related compounds using a versatile synthetic method in an effort to study how the template structure affects the final product obtained in hydrothermal zeolite preparations. By making small, systematic modifications on a parent structure, we have been able to suppress formation of a clathrasil (Nonasil) structure and favor the crystallization of large-pore zeolitic materials. The effects of varying substituting framework atoms will also be discussed.

Organic amines and quaternary ammonium cations were first used by Barrer in the synthesis of zeolites in the early 1960s.[1] This approach led to a significant increase in the number of new zeolitic structures discovered as well as an expansion in the boundaries of composition of the resultant crystalline products. Previously, products with low silica to alumina ratios ($SiO_2/Al_2O_3 \leq 10$) had been obtained, but upon using the organocations as components in the starting gels, zeolites with increasingly high SiO_2/Al_2O_3 were realized.[2]

Templating effects in zeolite synthesis have been observed for alkali as well as organic cations. In both cases, it has been postulated that the positive charge of the templating species (and its sphere of hydration) interacts favorably with negatively charged silicate subunits, resulting in the crystallization of the product observed. A classic example of templating involves the crystallization of sodalite in the presence of tetramethylammonium (TMA) cation.[3] The TMA cations are found within the cavities of the sodalite cages, yet the 6.9 Å diameter of the cation (7.3 Å when hydrated) precludes it from entering the cavity via the 6-membered ring portals *after* formation of the structure; therefore the sodalite cage must result from growth *around* the cation. Unfortunately, the relationship between structure of the organocation and the resultant zeolite is far from straightforward, as evidenced from the multitude of products which can be obtained using a single quaternary ammonium salt,[4] or the multitude of organocations which can produce a single zeolitic product.[5]

It is clear that the organic cation exerts its influence on the zeolite crystallization process in many ways. Aside from acting in a templating role, its presence also greatly affects the characteristics of the gel. These effects can range from modifying the gel pH to altering the interactions of the various components via changes in hydration (and thus solubilities of reagents) and other physical properties of the gel. It would be desirable to investigate how the presence of a particular quaternary ammonium salt influences many of these gel characteristics in order to determine more rigorously how such salts exert their templating effects.

Many of the organocations which have been used as templates for zeolite synthesis are conformationally flexible. These molecules can adopt many conformations in aqueous solution, therefore it is not surprising that several templates can give rise to a particular crystalline product (Table 1). Studies which involved alterations on such conformationally flexible organic amines and cations have been published. In one study, Rollmann and Valyocsik described how varying the chain length for a series of α,ω-linear diamines resulted in different intermediate-pore products.[6] In addition, it has been recently reported that three different products (Nonasil, EU-1, NU-87) which have related framework topologies, can be formed from three linear bis-quaternary ammonium templates of varying chain lengths.[7]

Altering the structure of a conformationally rigid organic molecule can also lead to a change in the zeolite obtained, presumably due to the differing steric demands of each template. Some preliminary investigations have addressed the correlation of template structure and zeolitic product. It has been demonstrated that in switching from 1,3-dimethylimidazolium hydroxide to 1,3-diisopropylimidazolium hydroxide as template, using the same starting gel ($SiO_2/Al_2O_3 = 100$), the former directs toward formation of ZSM-22 whereas the latter affords ZSM-23.[8] We have demonstrated the utility of using other large, rigid organic cations as templates in the discovery of new zeolitic materials.[9]

Table 16-1. Flexible organocations capable of producing high-silica zeolites.

Template	Zeolitic Product
$Et_4N^+X^-$	ZSM-5, ZSM-12, Beta
$n\text{-}Pr_4N^+X^-$	ZSM-5
$n\text{-}Bu_4N^+X^-$	ZSM-5, ZSM-11
$n\text{-}BuN^+Me_3X^-$	ZSM-5
$EtMe_2N^+(CH_2)_nN^+EtMe_2 2X^-$	ZSM-12
$(PhCH_2)Me_3N^+X^-$	ZSM-12
$Et_2Me_2N^+X^-$	ZSM-12

Our work has been directed toward exploring the role of conformationally restricted organocations in zeolite synthesis. In this paper, we present a general methodology to synthesize organic templates in a deliberate fashion using readily available, inexpensive starting materials. The results of a systematic study of the template effects in various zeolite synthesis reactions will also be presented.

SYNTHESIS OF ORGANIC TEMPLATES

The Diels-Alder reaction is one of the most useful transformations in synthetic organic chemistry. Two new bonds and a six-membered ring are formed in this reaction, formally a [4+2]cycloaddition of a 1,4-conjugated diene with a double bond (dienophile) (Figure 1). The versatility of this reaction is in part responsible for its usefulness:[10] the dienophile can consist of a carbon-carbon, carbon-heteroatom, or heteroatom-heteroatom double (or triple) bond, leading to a diverse pool of potential products. Electron-withdrawing groups on the dienophile greatly increase its reactivity, whereas electron-donating groups on the diene have the same effect. The diene must exist in the *cisoid* conformation in order for reaction to occur. The Diels-Alder reaction has been generally regarded as a symmetry-allowed one-step (concerted) reaction, although the synchronicity of the reaction has been the source of continual debate.[11] Frontier molecular orbital theory has been used to explain both the reactivity and selectivity of Diels-Alder reactions.[12] In a "normal" electron demand cycloaddition,

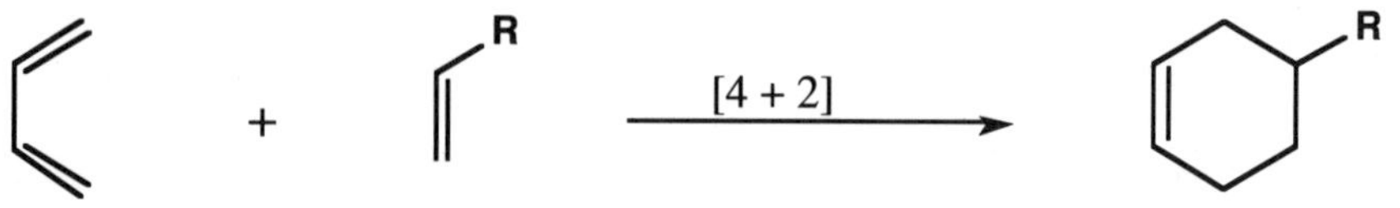

Figure 16-1. The Diels-Alder reaction.

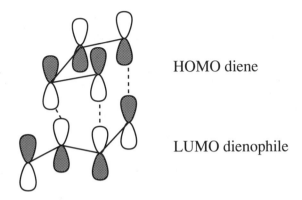

HOMO diene

LUMO dienophile

Figure 16-2. HOMO-LUMO orbital arrangement in an *endo* Diels -Alder transition state.

the dominant orbital interactions result from the highest-occupied molecular orbital (HOMO) of the diene and the lowest-unoccupied molecular orbital (LUMO) of the dienophile.

Another attractive feature of the reaction is its stereospecificity. Due to the concerted nature of the reaction, addition with respect to the dienophile is almost always *syn*, and the relative orientation of the substituents on the dienophile is therefore preserved in the product. In addition, a conjugating substituent on the dienophile tends to orient so that it interacts with the diene in the transition state, resulting in an *endo* product (Figure 2).[11] Thus, one often is able to predict which isomer will be favored (sometimes exclusively) in a given Diels-Alder reaction.

The reaction of cyclopentadiene with maleic acid derivatives results in formation of *endo* products due to favorable secondary orbital overlap interactions between the acid derivatives and the diene.[13] Thus, the reaction of cyclopentadiene with N-methylmaleimide affords the 4-aza-tricyclo[5.2.1.0^{2,6}]decane derivative shown in Figure 3. Reduction of the imide to the corresponding tertiary amine is achieved using lithium aluminum hydride, and the amine can be quaternized in high yield with methyl iodide.

By varying either the diene or the dienophile, we have been able to make small but specific modifications on the parent system, **A**. We have used the Diels-Alder methodology to generate the key intermediates necessary to synthesize the series of templates shown in Figure 4. These templates have been used to probe how a given change in structure affects the products obtained in a number of zeolite screening reactions. The rigid structures also allow us to determine the spatial characteristics of the cations, which we hope to explore in the future using molecular modelling.

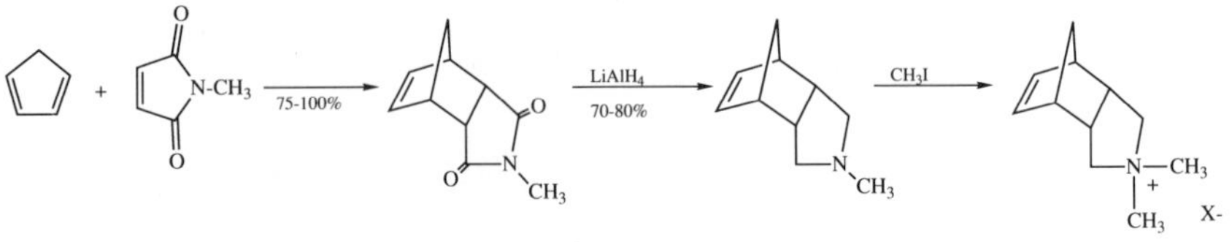

Figure 16-3. Synthesis of Template A.

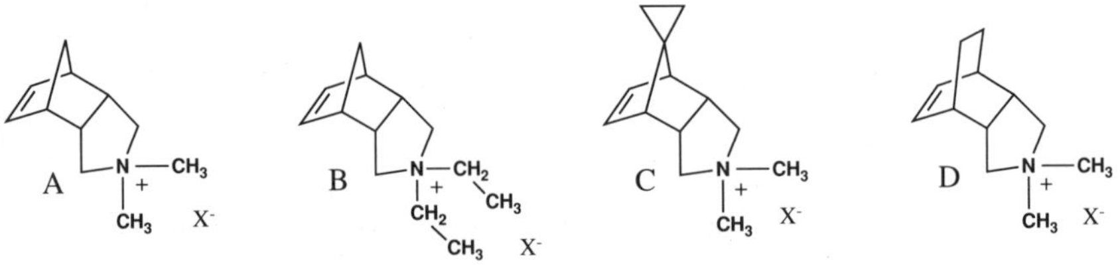

Figure 16-4. Templates synthesized via Diels-Alder methodology.

Template **A** is the smallest of the series; in **B** we have increased the size of the alkyl groups surrounding the positively charged nitrogen from methyl to ethyl, whereas in **C**, substitution around the nitrogen remains the same, but we have added a cyclopropyl group to the bridging carbon (C-10). In Template **D**, the bridging unit has been expanded from one to two carbons.

EXPERIMENTAL

A. Organic Templates

The procedure for the synthesis of Template **A** is representative of the synthetic reactions used to prepare the four templates discussed in this paper. Variations of the synthesis for the three other templates are listed below:

   Template **B**: Maleimide was used as the dienophile and CH_3CH_2I was used instead of CH_3I in the quaternization step.
   Template **C**: Spiro[2.4]hepta-4,6-diene used as the diene; $AlCl_3$ used as a Lewis acid.
   Template **D**: 1,3-cyclohexadiene used as the diene.

Preparation of Template **A**:

Diels-Alder adduct:

   Cyclopentadiene was obtained by cracking dicyclopentadiene:[14] dicyclopentadiene (bp 170°C) was heated in a 1-L round-bottomed flask fitted with a 30-cm Vigreux column to distill cyclopentadiene (bp 41°C). A 2-L, 3-necked flask was equipped with a magnetic stir bar, reflux condenser, and thermometer. The flask was charged with cyclopentadiene (295 g, 4.46 mol) and benzene (1.4 L). N-Methylmaleimide (30.1 g, 0.45 mol) was added at room temperature (exotherm noted), and the homogeneous yellow solution was heated to reflux for 24 hours. Thin layer chromatography (silica, 40% ethyl acetate/hexane) was used to monitor the disappearance of maleimide. The reaction mixture was concentrated by rotary evaporation to yield a mixture of oil and solid products, which was taken up in 200 mL of CH_2Cl_2 and transferred to a separatory funnel. Water (200 mL) was added and the pH of the aqueous layer adjusted to ≤ 1 using conc. HCl. The phases were separated and the organic phase was washed once more with H_2O (200 mL). After drying over $MgSO_4$, the organic phase was filtered and concentrated to yield an oil and solid mixture which was recrystallized from 500 mL of hot Et_2O. The ethereal solution was placed in the refrigerator overnight and the resulting white crystals of the Diels-Alder adduct were collected by vacuum filtration and washed with a small amount of cold ether (65.43 g, 82% yield, mp 103-105 °C).

Reduction of Diels-Alder imide:[15]

A 3-L, 3-necked flask was fitted with a mechanical stirrer, addition funnel, and reflux condenser. The imide from the Diels-Alder reaction (61.5 g, 0.35 mol) was dissolved in 495 mL of CH_2Cl_2 in the addition funnel. The flask was charged with $LiAlH_4$ (41.6 g, 1.04 mol) and anhydrous Et_2O (990 mL) and the system was placed under N_2. The imide solution was added slowly to the $LiAlH_4$ suspension. Gas evolution and an exotherm were noted. Addition of the imide solution was complete after approx. 1 h and the gray heterogenous solution was allowed to stir under N_2 overnight. Thin layer chromatography (silica plates, 5% MeOH/95% CH_3Cl) indicated the absence of starting material. The reaction was *carefully* worked up in the following manner:[16] 38.5 mL of H_2O was added slowly to the reaction. Vigorous gas evolution was noted as well as an exotherm. This step was followed by the cautious addition of 38.5 mL of 15% aq. NaOH solution. Another 115 mL of H_2O was added and the mixture, which turned from gray to white, was stirred for 1 h at room temperature. The solids were removed by filtration and washed with CH_2Cl_2. The aqueous layer was acidified with conc. HCl to pH ≤ 1 and the non-basic organic impurities were removed in the organic phase. The aqueous layer was then made basic (pH ≥ 12) with 50% NaOH and the crude tertiary amine was isolated by extracting twice with CH_2Cl_2. The organic layers were combined and dried over $MgSO_4$. Following filtration, the solution was concentrated to yield 41.4 g (52%) of the amine, which was taken directly to the next step. IR and ^{13}C NMR spectroscopy could be used to monitor the disappearance of the imide functionality (1700 cm^{-1} and 177.5 ppm, respectively).

Quaternization of the 4-Methyl-4-aza-tricyclo[5.2.1.02,6]dec-8-ene (Template A):

The amine (15.0 g, 0.10 mol) was dissolved in 100 mL of $CHCl_3$ in a 250 mL round-bottomed flask which was equipped with an addition funnel and magnetic stirrer. The reaction flask was immersed in an ice bath and the addition funnel charged with CH_3I (28.7 g, 0.20 mol). The CH_3I was added to the amine over a 10-minute period (exothermic reaction) and the homogeneous solution was stirred at room temperature for 3 days. Diethyl ether (100 mL) was added to the reaction mixture and the yellow solids were collected by filtration and washed with more ether. These solids were recrystallized from hot acetone/Et_2O (a small amount of MeOH was added to aid in dissolution of solid) to afford 21.2 g of Template **A** (mp 255-258 °C, 73% yield). Anal. Calc'd for $C_{11}H_{18}NI$: C, 45.38; H, 6.23; N, 4.81. Found: C, 45.28; H, 6.23; N, 4.85.

Ion-exchange of Template A to hydroxide form:

Bio-Rad AG1-X8 anion exchange resin was used to convert the iodide salt of **A** to the corresponding hydroxide form in 90.5% yield. The yield of the conversion was based upon titration of the resultant solution using phenolphthalein as the indicator.

B. Zeolite Synthesis

Reactions were carried out using Parr 4745 reactors (23 mL) and heating in Blue M ovens as described previously (Zones and Van Nordstrand 1988).[17] Silica was provided in the form of Cabosil M-5, alumina as Reheis F-2000 hydrated alumina (unless noted otherwise in text), and boron in the form of sodium borate decahydrate. Reagent grade 1.0 N NaOH solution was used as an additional source of hydroxide.

Typical reaction ratios are shown below:

R/SiO_2	$NaOH/SiO_2$	H_2O/SiO_2	$OH-/SiO_2$
0.15	0.15	44	0.25-0.30

R denotes the quaternary ammonium template in the hydroxide form. The hydroxide content of the reaction is the sum of the contribution from both the template and alkali hydroxide. Ratios of aluminum and boron to silica are given in the text, as well as temperatures, duration of reaction, and revolutions per minute. The products from these reactions were typically filtered, washed, dried (to 100°C), and analyzed by X-ray diffraction as described in reference 17.

A typical reaction preparation is described below:
2.29 grams of a 0.655 mmol OH-/g solution of Template B (1.5 mmol R, OH-; 12.7 mmol H_2O), 4.63 grams H_2O (25.7 mmol), and 1.0 gram 1 N NaOH

(Baker) were added to a teflon cup of a Parr 4745 reactor. Reheis F2000[18] (0.0195 g; 0.10 mmol Al_2O_3) was added to the reaction solution which was stirred until the Reheis completely dissolved. Cabosil M-5 (0.62 g; 10 mmol SiO_2) was next added to the reaction and the mixture was stirred until it was homogeneous. The reaction was heated to 160°C and tumbled at a rate of 43 rpm. After 9 days, the reaction was filtered and washed with water to yield 0.60 g of product, determined by XRD to be ZSM-12. (Elemental analysis: 45.5 wt % Si; 0.877 wt % Al; 13.0% loss on ignition)

C. Analysis

All organic templates were fully characterized using ^1H and ^{13}C NMR, melting point, and C, H, N analysis. ^{13}C CP-MAS NMR spectra were obtained using a Bruker CPX-300 NMR spectrometer (75.48 MHz), 3.92 KHz spinning rate, 1 msec contact time, 5 sec repetition rate and 16 Gauss ^1H decoupling field. The chemical shifts for these spectra were set using adamantane as a reference (upfield peak at 29.5 ppm).

Scanning electron micrographs were obtained on a Hitachi 570-1 instrument.

RESULTS

Screening Reactions

All-silica reactions

Under all-silica conditions (see Experimental section), the four templates give rise to three different products (Table 2). The silica reaction is of interest because use of an adamantyl cation with this system produced the unexpected all-silica AFI structure.[19] In the absence of any organocation, the reaction yielded quartz. In contrast, the presence of the parent template **A** results in the formation of Nonasil, a clathrate structure composed of 5- and 6-ring secondary building units, but with no connecting channels. By increasing the size of the alkyl substitutents on the nitrogen (Template **B**), the one-dimensional, large-pore zeolite ZSM-12 is formed. If the increased steric bulk is located in the area of the bridging C-10 carbon (in the form of the cyclopropyl ring in template **C**, or a 2-carbon bridge as in template **D**), yet another large-pore all-silica zeolite is produced: SSZ-31.[20] Scanning electron micrographs of the three different products are shown in Figure 5.

The presence of the organic templates within the zeolitic channels was confirmed by ^{13}C CP-MAS NMR (75.48 MHz). Peak assignments were made based on a comparison of the TOSS and TOSSOP (Total Suppression of

Table 16-2. Results of all-silica zeolite screening reactions.

Template	Product
none	QUARTZ
A	NONASIL
B	ZSM-12
C	SSZ-31
D	SSZ-31

All reactions run at 150°C, 0 rpm

Figure 16-5. Scanning electron micrographs of all-silica molecular sieves. Clockwise from upper left: Nonasil (Template A); ZSM-12 (Template B); SSZ-31 (Template C); SSZ-31 (Template D).

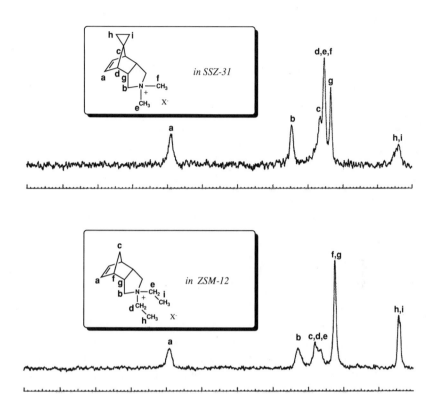

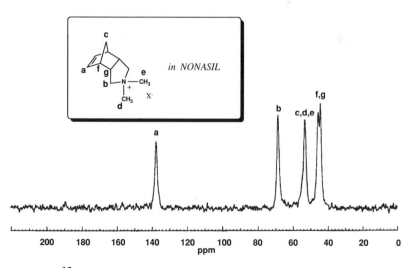

Figure 16-6. ^{13}C CPMAS NMR of all-silica molecular sieves containing templates.

Spinning Sidebands (Opella)) spectra with the solution ^{13}C NMR spectra for each template.[21] The CP-MAS spectra are shown in Figure 6. These spectra are important in confirming that our template organocations are being used in their intact forms, since we have observed in other systems that molecular sieve products can be formed from the decomposition products of the original organic templates.[4]

Elemental analyses on the organozeolites show that 10-12% by weight of organic matter (C,H,N content) is contained in each sample. This indicates the use of most of the existing pore volumes of these materials. As mentioned above, NMR experiments show that only one component (the template) is found to be filling the pores of the molecular sieve products obtained, therefore one can calculate the molar amount of template in the pores based on the organic content. Modelling studies indicate our templates are between approximately 6.25-7.65 Å in length,[22] therefore, the Template/TO_2 ratio will be in the range of 1/20 - 1/25, which is consistent with findings for a number of other systems.[23]

Boron-Containing Reactions

In zeolite screening reactions containing both silica and boron (starting gel $SiO_2/B_2O_3 = 50$) in the absence of any template, the layered product, Magadiite is formed.[24] Template **A** gives the Nonasil product and Template **B** once again forms ZSM-12. Template **C**, however, no longer affords the SSZ-31 product as it did in the all-silica case, but rather ZSM-12 (See Table 3). We believe this may be due to the inability of the SSZ-31 framework to accommodate such a high concentration of substituting element. As in the all-silica AFI (SSZ-24) case, crystallization may be favored by low substitution.[19] Even over a much higher SiO_2/B_2O_3 range, ZSM-12 is the only product obtained using Template **C** (Table 4). Use of Template **D** under the standard boron reaction

Table 16-3. Results of boron-containing zeolite screening reactions.

Template	Rxn Time	Zeolite Product
none	15 d	MAGADIITE
A	8 d	NONASIL
B	6 d	ZSM-12
C	10 d	ZSM-12
D	24 d	SSZ-33

Starting $SiO_2/B_2O_3 = 50$
All reactions run at 160°C, 43 rpm

Table 16-4. Zeolite reactions using Template C over a range of SiO$_2$/B$_2$O$_3$ ratios.

B/Si	M	M/Si	OH/Si	Rxn Time	Zeolite Product
0.01	Na	0.11	0.25	14 d	ZSM-12
0.02	Na	0.11	0.25	14 d	ZSM-12
0.03	Na	0.12	0.25	21 d	ZSM-12
0.04	Na	0.15	0.28	10 d	ZSM-12

All reactions run at 160°C, 43 rpm

Table 16-5. Zeolite reactions using Template A over a range of SiO$_2$/B$_2$O$_3$.

B/Si	M	M/Si	OH/Si	Rxn Time	Zeolite Product
0.02	Na	0.15	0.25	8 d[a]	NONASIL
0.04	Na	0.15	0.28	8 d[a]	NONASIL
0.06	Na	0.15	0.30	6 d[a]	NONASIL
0.07	Na	0.13	0.28	11 d[b]	CHABAZITE + NONASIL
0.10	Na	0.18	0.28	7 d[b]	CHABAZITE + NONASIL (minor)

(a) Reactions run at 160°C, 43 rpm
(b) Reactions run at 150°C, 0 rpm

conditions results in the formation of the multi-dimensional, large-pore zeolite SSZ-33.[25]

Exploration of a wide range of boron concentrations using Template **A** revealed that up to a SiO$_2$/B$_2$O$_3$ ratio of 33, Nonasil is the exclusive product formed. However, Chabazite becomes the favored product upon further increase of the boron content (Table 5).

Aluminum-containing reactions

In a series of reactions in which alumina is incorporated in the starting gel (SiO$_2$/Al$_2$O$_3$ = 100), Templates **A** and **B** yield the expected products (Nonasil and ZSM-12, respectively), whereas Templates **C** and **D** have failed to give any crystalline products after 30 days of reaction . Depending both on the concentration as well as the source of aluminum, a variety of crystalline products have been obtained using Template **A** (Table 6).

DISCUSSION

By choosing the appropriate reactants for our Diels-Alder reactions, we have been able to obtain a range of products which differ slightly in their spatial dimensions. Our screening reactions have demonstrated that members of this

Table 16-6. Aluminum-containing screening reactions using Template A.

Al/Si	M	M/Si	OH/Si	Temp	rpm	Time	Al source	Product
0.01	Na	0.10	0.25	160	43	15 d	F2000[a]	NONASIL
0.02	Na	0.10	0.25	160	43	11 d	F2000	NONASIL
0.03	Na	0.10	0.25	160	43	12 d	F2000	NONASIL
0.04	Na	0.10	0.25	160	43	20 d	F2000	GEL
0.04	Na	0.20	0.40	160	43	21 d	FAUJ[b]	CHABAZITE + trace NON-ZEOLITIC
0.06	Na	0.72	0.65	135	30	5 d	FAUJ	CHABAZITE
0.07	Na	0.07	0.41	150	0	23 d	NaAl[c]	AMORPHOUS + BETA

(a) Reheis F-2000 hydrated aluminum hydroxide
(b) Y-zeolite used as aluminum source[26]
(c) Sodium aluminate

Table 16-7. Large organocations found to produce the Nonasil structure (listed in order of increasing size).

Organic Structure	Reference
$\diagup\!\!\diagdown\!\!\diagup\!\!\diagdown N^+(CH_3)_3 X^-$	28
cyclohexyl-$N^+(CH_3)_2$ with CH_3 groups, X^-	29
cyclopentyl—$N^+(CH_3)_3 X^-$	27
cyclohexyl—$N^+(CH_3)_3 X^-$	27
norbornyl—$N^+(CH_3)_3 X^-$	30
$N^+(CH_3)_2$ ring structure, X^-	30
bicyclic —$N^+(CH_3)_2$, X^-	This work

series of rigid, polycyclic quaternary ammonium compounds are quite successful in promoting silicate crystallization to large-pore (12-ring) molecular sieves.
Only the smallest of the 4-aza-tricyclo[5.2.1.0]decanes (Template **A**) produces
the Nonasil structure. The Nonasil clathrate structure has been described previously in the patent literature (SSZ-15,[27] EU-4,[28] ZSM-51[29]) from syntheses
that use quaternary ammonium salts which are smaller than Template **A**. Table
7 lists a variety of organocations that are capable of producing this clathrate
structure.

Our strategy in using rigid polycyclic template molecules was designed to
avoid forming ZSM-5, a frequently encountered, stable product. Numerous
small organic molecules are effective pore-fillers (if not templates) in ZSM-5
syntheses.[31] We consider ZSM-5 to be a low energy (thermodynamially
favored) product in a variety of reactions for several reasons. First, it can form
at lower temperatures than most organic template-requiring zeolite syntheses.
ZSM-5 can also be synthesized at a lower pH (and therefore lower OH-/SiO_2)
than other aluminosilicates.[32] It is known that reduced OH- concentrations can
greatly retard crystallization rates, yet we fail to observe the formation of any
other product as the ZSM-5 crystallization rates decrease. In addition, we have
not observed any transformation of ZSM-5, once it has crystallized, to any other
phases, which indicates that the favorable formation of ZSM-5 is not simply a
kinetic phenomenom.

In our search for organic templates which would avoid formation of ZSM-
5, bicyclic systems derived from readily available precursors to [2.2.1] or
[2.2.2] systems (e.g. DABCO) were first examined. These templates were effective in preventing the formation of ZSM-5, at SiO_2/W_2O_3 (W = Al, B) values
above 50, and appeared to stabilize the Nonasil structure (see Table 7 for examples). Although we had been successful in creating large cavities, structures
with connecting channels eluded us. In this last regard, it is interesting to note
that the original discovery of ZSM-5 was based upon the use of tetrapropylammonium bromide as the template. A nice fit of the propyl chains extending
down the channels of ZSM-5 has been shown.[33] As other small, linear organics
(such as propyl- and butylamine) were found to be effective in synthesizing
ZSM-5, the emphasis continued to be placed upon stabilizing zeolitic channels.
Considerable discussion has addressed the unique and valuable 5.5 Å diameter
of these channels with respect to catalysis.[34]

The ZSM-5 structure can also be viewed as an array of larger cavities (with
~9 Å cross-sections) that have been connected. From the standpoint of synthesis, this has been recently discussed as a possible growth mechanism.[35] As template candidates become more spherical in nature (due to a cycloalkyl group or
perhaps the hydration sphere surrounding a trimethylammonio-moiety attached
to a small, fourth substituent), cavity formation becomes favorable, but no channel system develops for structures such as Nonasil.

Our smallest 4-aza-tricyclo[5.2.1.0]decane salt (Template **A**) appears to be close to the maximum cavity-stabilizing size for Nonasil. Substitution of ethyl groups at the ring nitrogen (Template **B**) gives the template additional length in its longest dimension (6.25 Å in **A** vs. 7.66 Å in **B**)[22] and now produces a molecular sieve which is *all* channel and *no* cavity: ZSM-12. Linear charged molecules have been shown to be good directors for ZSM-12 formation.[36]

By enlarging the spatial parameters at the bridging (C-10) carbon in the [5.2.1.0] structure, new large-pore zeolites are obtained. We accomplished this in two ways: first, by incorporating a cyclopropyl group at C-10 (Template **C**), and second, by enlarging the bridge to two carbons (Template **D**). In both SSZ-31 and SSZ-33, the internal spatial dimensions must be larger than that of ZSM-12, as demonstrated by the lower constraint index values obtained in catalytic cracking experiments.[20,25] The puckered ring observed in ZSM-12[37] must now be enlarged to at least a full 12-ring in these two new zeolites. In addition, use of Template **D** in boron-containing reactions has resulted in the formation of a multi-dimensional, large-pore zeolitic framework (SSZ-33).

The silicon-substituting, trivalent elements (Al, B) also play an important structure-directing role. From the data it is evident that structures such as Nonasil and SSZ-31 are disfavored as the concentration of substituting element is increased in the starting gel. This has been previously observed for the all-silica AFI structure, SSZ-24.[4] In contrast, SSZ-33 is favored by higher substitution.[25]

The Nonasil structure is composed of 5- and 6-rings, which is a common feature for a number of very high silica clathrate structures.[38] When the boron or aluminum content is increased for reactions with Template **A**, Chabazite (which is comprised of 4- and 6-rings) becomes the favored product. It has been suggested that aluminum may occupy the tetrahedral positions in 4-rings of zeolites preferentially, since the most aluminum-rich zeolites are also rich in 4-

Table 16-8. Conditions for ZSM-12 Synthesis Using Template B.

Al/Si	B/Si	M+/Si	RN/Si	OH/Si	Temp	Time
		0.10	0.15	0.25	150[a]	4 d
0.02		0.10	0.15	0.25	160	9 d
0.03		0.12	0.10	0.21	160	6 d
0.055		0.15	0.15	0.30	160	5 d
	0.04	0.15	0.15	0.28	160	9 d
	0.05	0.17	0.10	0.21	160	6 d

$M^+ = Na, K$
All reactions tumbled at 43 rpm unless noted otherwise
(*a*) 0 rpm

rings.[39] It is interesting to note that the slightly smaller templates that are based upon bicyclo[2.2.2] and -[2.2.1] systems and produce Nonasil at high silica concentrations also form Chabazite as the concentration of substituting elements is increased.[40]

ZSM-12 contains some 4-rings. It is possible that this provides the flexibility for the structure to be produced by Template **B** and **C** over a sizeable concentration range of various substituting elements (Tables 4 and 8). In some systems, if the amount of substituting element becomes great enough, Beta-zeolite will form preferentially over ZSM-12.[41] Beta-zeolite has 12-ring channels like ZSM-12, but they are connected in three-dimensions, resulting in a considerably more active catalyst. When the concentration of substituting element becomes too great, it may be impossible to accommodate the population of 4-rings in the ZSM-12 structure.

Because these readily accessible templates are so effective in directing formation of interesting, large-pore molecular sieves, we are currently studying the influence of trivalent element substitution on the formation of branched channels in molecular sieve structures.

CONCLUSIONS

In this study we have demonstrated the utility of using rigid, polycyclic ammonium cations as templates for molecular sieve synthesis. The four templates discussed were prepared in a straightforward fashion using Diels-Alder chemistry, and this method lends itself to expanding to an even larger number of related compounds. Although the templates possess a considerable amount of structural similarity, they exhibit high selectivities for the zeolites they form. Under all-silica conditions, each organocation exerts a unique enough effect to result in the formation of several different molecular sieve products.

Using high-silica starting gels, the smallest template consistently directs toward formation of the clathrate, Nonasil. Upon enlarging the spatial characteristics of this molecule, the Nonasil cavity can no longer accommodate the template, and zeolitic materials (with channels) result. We observed how the product selectivity for a given template can be profoundly influenced by the concentration of substituting elements in the synthesis gel; SSZ-31 prefers to have no or only a minimal amount of substitution by boron or aluminum in the framework, whereas the SSZ-33 structure can only be made in the presence of a significant amount of boron. In reactions containing a high amount of substituting element ($SiO_2/W_2O_3 \sim 25\text{-}35$), Chabazite often is the product observed using these rigid, polycyclic templates. Chabazite possesses a large internal cavity as well as a small-pore (8-ring) system by which to connect them, and we have observed this to be the favored product in reactions with low SiO_2/W_2O_3 using adamantyl cations as well.[42]

In the absence of any organocation, the synthesis gels afford vastly different, non-zeolitic products. It is therefore apparent that our templates are exerting some type of structure-directing effect on the gel mixtures. The Diels-Alder methodology described in this paper has provided us with a viable synthetic route for making a wide variety of related compounds which will enable us to investigate further the relationship between template structure, synthetic gel composition, and resultant zeolitic products.

ACKNOWLEDGMENTS

The authors wish to thank Gregory S. Lee for his laboratory prowess in synthesizing the four templates which were discussed in this study. We also acknowledge the contributions from R. A. Van Nordstrand (XRD and discussion), D. M. Wilson (^{13}CMAS NMR), D. Glazier (SEM photos), and Chevron Research and Technology Company for research support.

REFERENCES

1.   Barrer, R. M.; Denny, P. J. 1961. *J. Chem. Soc.* 971-982.
2.   Barrer, R. M. 1982. *Hydrothermal Chemistry of Zeolites.* New York: Academic Press, Inc.
3.   Baerlocher, Ch.; Meier, W. M. 1969. *Helv. Chimica Acta 52*, 1853.
4.   Zones, S. I.; Van Nordstrand, R. A.; Santilli, D. S.; Wilson, D. M.; Yuen, L.; Scampavia, L. D. 1989. In *Zeolites: Facts, Figures, Future*, ed. P. A. Jacobs and R. A. van Santen, pp. 299-309. Amsterdam: Elsevier Science Publishers.
5.   Barrer, R. M. 1989. In *Zeolite Synthesis*, ACS Symposium 398, ed. M. L. Occelli and H. E. Robson, pp. 11-27. American Chemical Society.
6.   Valyocsik, E. W.; Rollmann, L. D. 1985. *Zeolites 5*, 123.
7a. Shannon, M. D.; Casci, J. L.; Cox, P. A.; Andrews, S. J. 1991. *Nature 353*, 417-420.
7b. Casci, J. L. 1986. In *New Developments in Zeolite Science and Technology*, ed. Y. Murakami, A. Iijima, and J. W. Ward, pp. 215-222. Elsevier.
8.   Zones, S. I. 1989. *Zeolites 9*, 458-467.
9.   Zones, S. I. 1990. In *Synthesis and Properties of New Catalysts: Utilization of NovelMaterials Components and Synthetic Techniques (Extended Abstracts)*, ed. E. W. Corcoran, Jr. and M. J. Ledoux, pp. 7-9. Materials Research Society.
10. Fringuelli, F.; Taticchi, A. 1990. *Dienes in the Diels-Alder Reaction.* New York: John Wiley and Sons, Inc.
11. Sauer, J.; Sustmann, R. 1980. *Angew. Chem. Int'l. Ed. Engl. 19*, 779.
12a. Sustmann, R. 1971. *Tetrahedron Lett.* 2721.
12b. Houk, K. 1975. *Accts. Chem. Res. 8*, 361.
12c. Fukui, K. 1970. *Fortschr. Chem. Forsch. 15*, 1.
13. Hall, H. K., Jr.; Nogues, P.; Rhoades, J. W.; Sentman, R. C.; Detar, M. 1982. *J. Org. Chem. 47*, 1451.

14. Moffett, R. B. 1963. *Organic Syntheses Coll. Vol. IV*, ed. N. Rabjohn, pp. 238-241. New York: John Wiley and Sons, Inc.

15. Anderson, W. K.; Milowsky, A. S. 1985. *J. Org. Chem. 50*, 5423-5424.

16. Fieser, L. F.; Fieser, M. 1967. *Reagents for Organic Synthesis*, Vol. 1, pp. 581-594. New York: John Wiley and Sons, Inc.

17. Zones, S. I.; Van Nordstrand, R. A. 1988. *Zeolites 8*, 166.

18. Reheis F2000 from the Reheis Co. This was used as an aluminum source to eliminate the need for adding excess alkali hydroxide to neutralize protons generated from sources such as $Al_2(SO_4)_3$.

19. Van Nordstrand, R. A.; Santilli, D. S.; Zones, S. I. 1988. *Perspectives in Molecular Sieve Science*, ACS Symposium 368, ed. W. H. Flank and T. E. Whyte, Jr., pp. 236-245. American Chemical Society.

20. Zones, S. I.; Harris, T. V.; Rainis, A.; Santilli, D. S. 1990. Patent Appl. WO 90/04567.

21. The Opella spectra were obtained using a 100 μsec window of no ^{1}H decoupling before the beginning of signal acquisition. This results in supression of any carbon signals which are bound to protons, except methyl groups.

22. *Polygraf* software by Molecular Simulations Inc. was used for the minimizations and geometry analysis.

23. Valyocsik, E. W. 1987. U.S. Patent 4 698 217.

24. Lagaly, G. 1945. *Am. Miner. 60*, 642.

25. Zones, S. I. 1990. U.S. Patent 4 963 337.

26. Zones, S. I. 1991. *J. Chem. Soc. Faraday Trans. 87*, 3709-3716.

27. Zones, S. I. 1986. U.S. Patent 4 610 854.

28. Casci, J. L.; Lowe, B. M.; Whittam, T. V. 1982. E.P. Appl. 063,436.

29. Valyocsik, E. W. 1986. U.S. Patent 4 568 654.

30. Unpublished results.

31. Araya, A.; Lowe, B. M. 1986. *Zeolites 6*, 111.

32. Guth, J. L.; Kesser, H.; Wey, R. 1986. In *Zeolite Science and Technology*, Ed. Y. Murakami, A. Iijima, J. W. Ward, pp.121-128. Elsevier.

33. van Kroningsveld, H.; van Bekkum, H.; Jansen, J. C. 1987. *Acta Cryst. B43*, 127-132.

34. Csicsery, S. M. 1984. *Zeolites 4*, 202-213.

35. Chang, C. D.; Bell, A. T. 1991. *Catal. Lett. 8*, 305-316.

36. Davis, M. E.; Saldarriaga, D. 1988. *J. Chem. Soc., Chem. Comm.* 920.

37. La Pierre, R. B.; Rohrmann, A. C. Jr.; Schlenker, J. L.; Wood, J. D. 1985. *Zeolites 5*, 346.

38. Gies, H.; Liebau, F.; Gerke, H. 1982. *Angew. Chem. Int'l Ed. Engl. 21*, 206.

39. Ooms, G.; Van Santen, R. A.; Jackson, R. A.; Catlow, C. R. A. 1988. *Innovation in Zeolite Materials Science*, Vol. 37, ed. P. J. Grobet, W. J. Mortier, E. F. Vansant, G. Schulz-Ekloff, pp. 317-322. Amsterdam: Elsevier Science Publishers B.V.

40. Zones, S. I. 1985. U.S. Patent 4 544 538.

41. Zones, S. I.; Holtermann, D. L.; Jossens, L. W.; Santilli, D. S.; Rainis, A.; Ziemer, J. N. 1991. Patent Appl. WO 91/00777.

42. Zones, S. I.; Van Nordstrand, R. A. 1990. *Novel Materials in Heterogenous Catalysis*, ACS Symposium Series 437, ed. R. T. K. Baker and L. L. Murrell, pp. 14-24. American Chemical Society.

17

Clear Solution Synthesis of AlPO$_4$ Molecular Sieves

R. Szostak[1], B. Duncan[1], R. Aiello[2], A. Nastro[2], K. Vinje[3], K. Lillirud[3]
[1]Zeolite Research Program, Georgia Tech Research Institute, Atlanta, GA 30332 U.S.A.
[2]Department of Chemistry, University of Calabria, Rende, Italy
[3]Department of Chemistry, University of Oslo, Oslo, Norway

The synthesis of molecular sieves from clear solution is used to identify transient phases formed during the crystallization of AlPO$_4$ molecular sieves. AlPO$_4$-11 forms readily from clear solution with a batch composition: Al$_2$O$_3$: 1.26 P$_2$O$_5$: 1.5 HF : 2.4 DPA : 70 H$_2$O. Crystallization produces low yields of very thin, 1 mm size needles after 2 hours. After 18 hours, the yield increases 50% to 75% at the expense of the crystal size, with smaller 5 to 10 micron crystals resulting. Such results suggest that a secondary nucleation occurs, with the redistribution of nutrients between the larger, initially-formed crystals and the smaller growing crystals. The solution phase plays a critical role in AlPO$_4$ crystallization, while both transient amorphous and crystalline phases appear to act as a source of Al and P nutrients.

INTRODUCTION

The method by which zeolites crystallize has been under debate for the last thirty years (1,2). The critical role of the solution phase has been well documented for a variety of materials including mordenite (3), NaY (4,5), NaA (6), offretite (7), erionite (7), and silicalite (8). These zeolites were shown to grow from solutions void of a solid or gel phase. Both the nuclei necessary to initiate zeolite crystallization as well as the nutrients for zeolite growth are found in the solution phase. In general, clear solutions are realized under high-pH conditions and low aluminum concentrations for the zeolite molecular sieves.

The aluminophosphates (AlPO$_4$s) form structures similar to the zeolite molecular sieves. In the AlPO$_4$s, the aluminum is in strict alternation with phosphorous in the tetrahedral framework sites. Not only have zeolite topologies been prepared, but new framework structures with no known aluminosilicate counterpart have also been identified. A critical role was assigned to the solid phase in the preparation of these unique structures, though systematic study of these aluminophosphate gel systems has not yet been undertaken (9). To date, only one study of an AlPO$_4$-based molecular sieve, FAPO-5, has shown that these materials can be crystallized from clear solutions (10). Unlike the silicate-based molecular sieves, the aluminophosphates crystallize at low pH, thus mineral acids can be used to solubilize the alumina and aluminophosphate components of the reactive mixture.

The objective of this study was to explore the crystallization of the aluminophosphate molecular sieves, starting from clear homogeneous solutions. The advantage of beginning with a clear solution is that transient phases, either crystalline or amorphous, can be identified. The nature of these phases and their role in the generation of the final product can be seen more easily. In the gel-based systems, the presence of undissolved alumina starting material or initially precipitated gel can obscure the observation of transient amorphous or low crystalline phases, which may or may not be critical in the subsequent crystallization process.

Clear Aluminophosphate Solutions

The acidic mineralizing agent, HF, was best suited for the generation of a clear aluminophosphate solution at low pH when organic amines are present in the crystallization mixture. In a wholly inorganic system, the addition of HCl also provides solutions free of the solid/gel phase. The batch composition used in this study which provided a transparent clear solution* is:

$$Al_2O_3 : 1.26\ P_2O_5 : 1.5\ HF : 2.4\ DPA : 70\ H_2O \quad (i)$$

* solution lacking turbidity and which does not contain solid matter that can be removed by centrifugation.

$AlPO_4$-11 crystallizes from this composition. In the inorganic system free of organic amine, the batch composition of:

$$Al_2O_3 : 0.86\ P_2O_5 : 1.5\ HCl : 50\ H_2O \quad (ii)$$

produced a clear or transparent solution phase, resulting in $AlPO_4$-H1 (or VPI-5) crystals forming. A clear solution could also be obtained from the batch composition:

$$Al_2O_3 : 1.26\ P_2O_5 : 1.5\ HCl : 70\ H_2O \quad (iii)$$

producing only crystobalite. The clear solutions used in this study were dependent on the type of mineral acid used. Replacing HF with HCl in equation (i) did not result in a clear solution phase. Replacing HCl with HF in equation (iii) did result in a clear solution; however, the products formed were substantially different. The differences in the crystalline phases produced is discussed below.

Inorganic Systems, No Added Amine (Figures 1 and 2) *
Using a batch composition of:

$$Al_2O_3 : 1.26\ P_2O_5 : 70\ H_2O \quad (iv)$$

did not result in a clear solution. When this gel was heated to 190°C, the dense phase $AlPO_4$-H4 crystallized after 4 hours, with quartz beginning to form after 18 hours. When 1.5 HCl was added to this basic recipe (eqn. (iii)), a transparent solution was obtained and crystobalite was the phase observed after 18 hours. When 1.5 HF was used:

$$Al_2O_3 : 1.26\ P_2O_5 : 1.5\ HF : 70\ H_2O \quad (v)$$

---

* All figures appear at end of chapter.

a new phase, GTRI-B, was observed. The threshold temperature for the crystallization of this pure phase appears to be 120°C. After 18 hours at this temperature, some amorphous precipitate is mixed with the crystalline GTRI-B phase. Completely crystalline material is formed between 150 and 190°C after 18 hours. The x-ray powder diffraction pattern of this phase is shown in figure 3. These long fibrous needles produced only one peak of high intensity. The infrared spectrum is shown in figure 4. Features found in the i.r. spectrum are similar but not identical to that of minyulite, $Al_2K(PO_4)_2(F,OH)*3H_2O$ (11). The x-ray diffraction pattern does not match that of minyulite. The data shown in figure 5 indicates two endotherms corresponding to water loss from the structure. When heated to 200°C, complete structure collapse is observed. The second endotherm, therefore, corresponds to loss of water from that amorphous phase.

When half the amount of HF was used, a different phase, GTRI-A, appeared. The x-ray powder diffraction pattern of this phase is shown in figure 6a. GTRI-A is sensitive to humidity, changing the nature of the diffraction pattern depending on the degree of hydration. The diffraction pattern found in figure 6a is that of the dry sample. Upon standing in high humidity, the pattern changes to that shown in figure 6b. Evacuation overnight causes a partial return to original dry material. The partially dehydrated sample is shown in figure 6c. The infrared spectrum shown in figure 7 possesses features similar to that of variscite and metavariscite, which also convert to other phases upon dehydration.

A two-step synthesis was examined. The solution with a batch composition shown in figure 2 was digested for 2 days, the autoclave quenched, and DPA added to this mixture of solid and liquid. The temperature was elevated to 190°C; and quenching took place for the second time, after 4 hours. The product was found to be a mixture of GTRI-B and AlPO$_4$-11. No other phases could be observed. In the presence of DPA, under the batch compositions and conditions chosen for these experiments, only AlPO$_4$-11 was found as a crystalline product. The presence of the solid phase does not redirect the type of structure form.

When HCl was used under identical conditions, only crystobalite was observed to form. Such product differences cannot be accounted for by the presence of the extra H+ alone, as the nature of the halide appears to contribute to the final product. In the HCl system shown in equation (ii), the crystalline phase which results is that of AlPO$_4$-H1, a material with properties previously shown to be similar to that of VPI-5. Replacing HCl with HF in this system did not lead to the AlPO$_4$-H1 phase. These results further reinforce the role the halide may have in the crystallization of the aluminophosphate molecular sieves. In all the inorganic systems, there were no intermediate amorphous phases observed to form from the clear solutions prior to the growth of the crystals at temperatures of 150°C and above.

Crystallization from DPA-containing solutions (Figure 8)

For the batch compositions chosen, the addition of the DPA does not cause precipitation to occur. All solutions were transparent prior to heating. At low temperatures (80 to 120°C), the clear solutions produced amorphous or low crystalline dense phases after 18 hours, which were not further identified. To ensure that the initial heating did not immediately result in the precipitation of the solid, the reaction was quenched after 4 hours at 120°C -- only the transparent solution was observed. No precipitate could be isolated when the solution was centrifuged. At 150°C, clear solutions were observed after 2 and 4 hours of heating. The viscosity of the solution increased during that time. After 18 hours, $AlPO_4$-11 crystals were collected from the vessel. At 190°C, a clear solution was observed to remain after 2 hours of heating. Three hours of crystallization produced $AlPO_4$-11 in very low yield. The crystals which were collected after 3 hours at 190°C were extremely long, thin (ca. 1 mm x 1 micron), and fragile. SEM images of fragments of these crystals are shown in figure 9. After 18 hours, 70% more $AlPO_4$-11 crystals were isolated. The size of the crystals diminished significantly, as shown in figure 10.

In order to illustrate the dynamic nature of the crystallizing solution, crystals of $AlPO_4$-H3 were added to the clear solution, and the mixture was heated overnight. The x-ray powder diffraction pattern of the resulting product was that of pure crystalline $AlPO_4$-11. The morphology of the crystals is similar to those shown in figure 10. All of the $AlPO_4$-H3 solid added converted into $AlPO_4$-11. The yield obtained reflected the conversion of $AlPO_4$-H3 to $AlPO_4$-11.

The role of the organic was examined in this study by replacing the DPA with i-propylamine. The results are shown in figure 11. Crystallization from this system differed somewhat from that of the DPA containing system. At 150°C, a stiff but transparent gel (much like gelatin or aspic) was obtained after 3 hours. Crystal growth was observed after raising the temperature of this gel to 190°C. $AlPO_4$-34 crystals were found imbedded in the stiff gel. The transparent stiff gel was not observed when the reaction vessel was heated to 190°C. After 3 hours, only large crystals of $AlPO_4$-34 in low yield were found. With time, the yield of $AlPO_4$-34 increased. After 18 hours, a mixture of small and large crystals were isolated. The SEM images of these crystals are shown in figure 12.

Metal-containing Aluminophosphates

The addition of cobalt acetate to the crystallization mixture with the batch composition:

$$0.8 \ Al_2O_3 : P_2O_5 \ : 0.2 \ CoO : 2.0 \ DPA : HF : 50 \ H_2O$$

produces a clear blue solution. Upon raising the temperature to 190°C, crystalline CoAPO-11 results. SEM micrographs of these materials at high magnification show the presence of two phases, the crystalline CoAPO-11 phase and a secondary phase associated with the crystals. Analysis of this second phase shows the presence of high concentrations of cobalt.

DISCUSSION

Crystallization of the aluminophosphates can occur from solutions containing aluminum and phosphorus sources. The addition of mineral acid acts as a solubilizing agent, much the same way that sodium hydroxide solubilizes the silica for the zeolite molecular sieves. Under the conditions employed in these experiments, no amorphous or other solid intermediate phase is observed in the temperature range normally used for molecular sieve aluminophosphate crystallization. At the onset of crystallization, a few very large crystals are observed to form. With time and the generation of more nuclei in the solution, more crystals which are smaller in size are observed at the expense of the large crystals. The addition of the organic to the crystallization mixture changes both the viscosity of the transparent solution as well as the crystalline product resulting. In an organic free system, crystalline product can be formed. With the preparation of AlPO$_4$-H1 from clear solutions in the presence of HCl, it is not evident that the organic plays a direct role in the production of this molecular sieve structure nor is this crystalline product formed from a direct solid-state conversion mechanism without the need for the solution phase, as previously suggested (9).

It was hoped that the generation of clear homogeneous metal-containing aluminosilicate solutions would encourage further incorporation of the metal in the framework structure. However, the results of these studies suggest that a coprecipitation can occur during the crystallization of the metal containing AlPO$_4$, with the second phase being richer in the metal component than the crystalline molecular sieve phase. There may be a limitation on the amount of metal that can be incorporated into these structures, when a pathway exists for precipitation of a non-molecular sieve phase containing that metal under these same conditions.

EXPERIMENTAL PROCEDURE

A typical procedure for the generation of the clear solutions used in this study took 1.18 g $Al(OH)_3$ and added to it 2.16 g of H_3PO_4. This was stirred until effervescence was completed. To this was added a solution containing 0.6 g HF (36%) and 10 grams of water. This mixture was stirred briefly. 1.8 grams of DPA were added prior to introduction of the clear solution to the teflon-lined/stainless steel autoclaves. Upon heating, the resulting solution produced $AlPO_4$-11. The identical procedure was followed for the crystallization of $AlPO_4$-34, substituting isopropylamine (1.05 g) for the DPA. Aluminum hydroxide (Pfalz and Bauer) was used as the source of aluminum, and phosphoric acid (85% Fisher) was used as the source of phosphorus. Fisher HCl and HF (36%) was used. After synthesis, the autoclaves were rapidly cooled and the solids were recovered immediately through filtration and dried in air at room temperature overnight.

Optical microscopy provides a rapid method of identifying unique phases and morphologies. The scanning electron microscope (Cambridge 150 Stereoscan with Tracor Northern TN5500 energy-dispersive x-ray spectrometer) provided further details on the morphology of the samples.

The x-ray powder diffraction data was collected on a Rigaku x-ray powder diffractometer with CuKa radiation. All spectra were recorded at 60-70% humidity. The infrared spectra were recorded as thin KBr wafers containing approximately 5 wt% sieve using a Perkin-Elmer Infrared Spectrometer Model 698, computer-assisted. DATA/TGA data were recorded on a Perkin-Elmer DATA/DATA, using a heating rate of 10°C/min in air.

REFERENCES

(1) R. Szostak, Molecular Sieves: Synthesis and Identification, Van Nostrand-Reinhold, New York, 1989.

(2) C. T. G. Knight, Zeolites, 10:140 (1990).

(3) S. Ueda, H. Murata, M. Koizumi and H. Nishimura, Am. Mineral., 65:1012 (1980).

(4) S. Ueda, N. Kageyama and M. Koizumi, Proc. of the 6th International Zeolite Conference, D. Olson, A. Bisio eds., Butterworths, UK, 1983, 905 (1984)

(5) S. Kasahara, K. Itabashi and K. Igawa, "New Developments in Zeolite Science and Technology," Studies in Surface Science and Catalysis 28:185 (1986), Y. Murakami, A. Iijima, J.W. Ward, eds., Elsevier Publishing Co., Amsterdam.

(6) P. Wenqin, S. Ueda, M. Koizumi, "New Developments in Zeolite Science and Technology" Studies Surface Science and Catalysis 28:177 (1986), Y. Murakami, A. Iijima, J. W. Ward, eds., Elsevier Publishing Co., Amsterdam.

(7) S. Ueda, M. Nishimura, M. Koizumi, "Zeolites (Synthesis, Structure, Technology and Applications)," Studies in Surface Science and Catalysis 24:105 (1985), eds. B. Drazaj et al., Elsevier Publishing Co., Amsterdam

(8) F. Testa, R. Szostak, A. Nastro, R. Aiello, manuscript in preparation

(9) M. E. Davis, C. Montes, P. E. Hathaway, J. M. Garces, "Zeolites: Facts, Figures, Future", Studies in Surface Science and Catalysis 49A:281 (1989), P. A. Jacobs, R. A. Van Santan, eds., Elsevier Publishing Co., Amsterdam

(10) P. Wenqin, Q. Shilun, K. Qiubin, W. Zhiyun, P. Shaoyi, "Zeolites: Facts, Figures, Future", Studies in Surface Science and Catalysis 49A:281 (1989), P. A. Jacobs, R. A. Van Santan, eds., Elsevier Publishing Co., Amsterdam

(11) J. R. Lehr, E. H. Brown, A. W. Frazier, J. P. Smith, R. D. Thrasher, Crystallographic Properties of Fertilizer Compounds, National Fertilizer Development Center, Muscle Shoals, Alabama

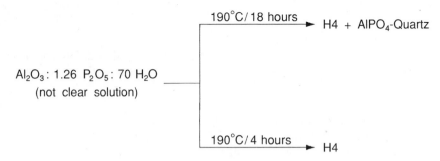

Figure 1. Aluminophosphate batch composition and conditions of synthesis.

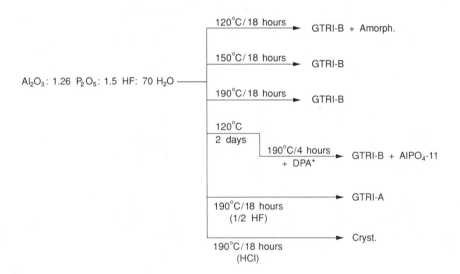

Figure 2. Aluminophosphate batch composition with HF added and conditions of synthesis

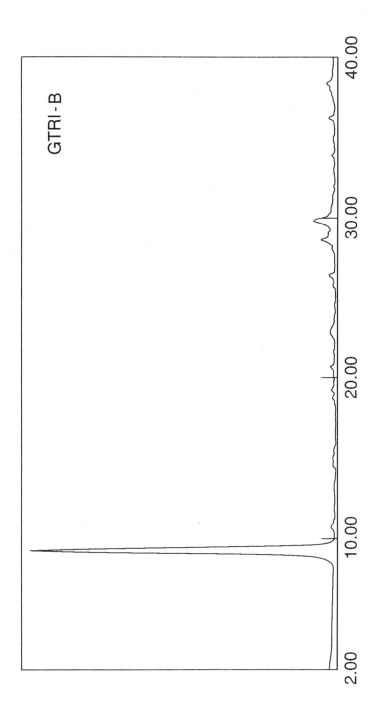

Figure 3. X-ray powder diffraction patter of GTRI-B.

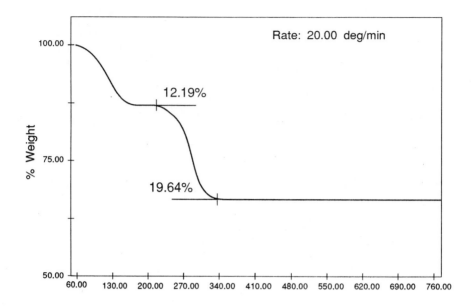

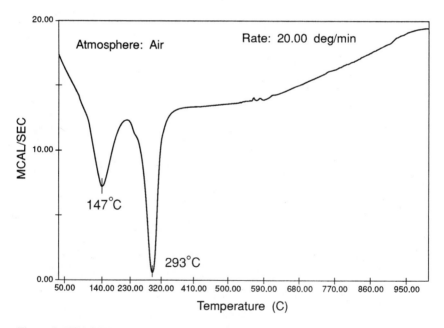

Figure 4. TGA/DTA of GTRI-B showing the two-step water loss. Crystallinity is lost upon heating to 200° C.

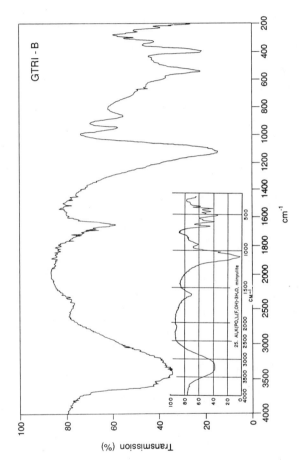

Figure 5. Infrared spectrum of GTRI-B. Insert is the infrared spectrum of minyulite to show the similarity between the two materials.

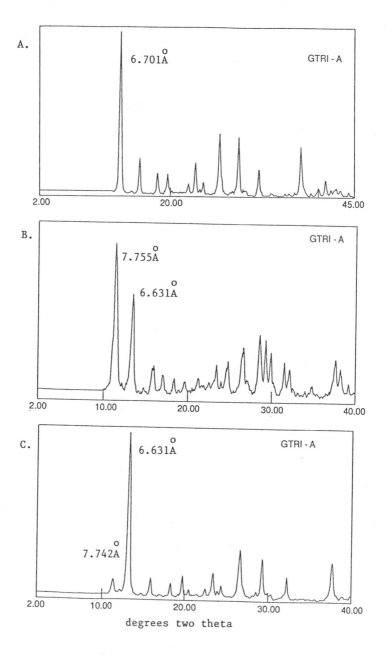

Figure 6. X-ray powder diffraction pattern of GTRI-A. A. Initially dried at 100° C (note that the scale between A, B and C differs). B. Identical sample after sitting in high humidity for an extended period of time C. The sample evacuated at room temperature for 24 hours. Removal of the excess humidity causes partial return to the hydrated material in B.

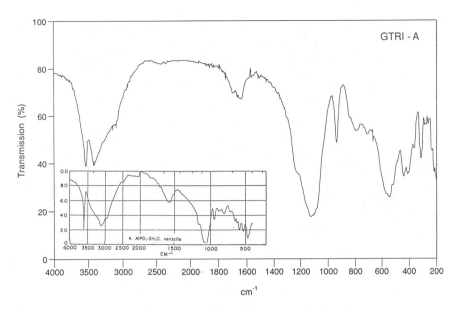

Figure 7. Infrared spectrum of GTRI-B. Insert is the infrared spectrum of variscite to show the similarity between the two materials.

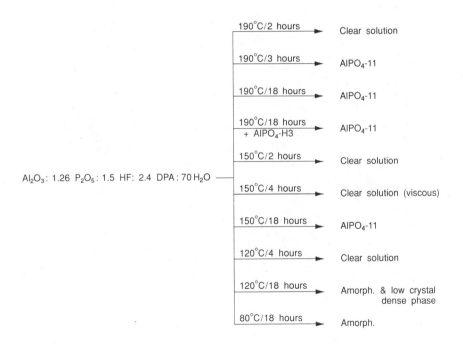

Figure 8. Aluminophosphate batch composition with HF and DPA and clear solution conditions of synthesis.

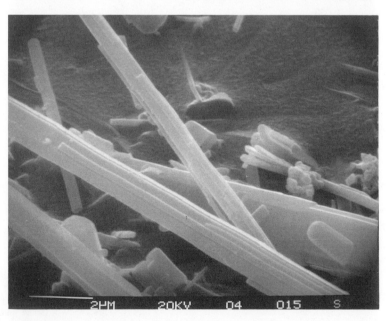

Figure 9. SEM images of AlPO$_4$-11 showing the formation of very long 1mm crystals after 3 hours of crystallization at 190° C.

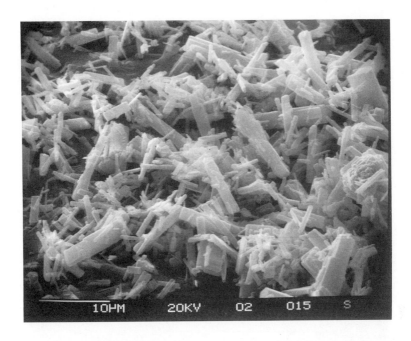

Figure 10. SEM images of AlPO$_4$-11 showing the presence of smaller size crystals after 18 hours of crystallization.

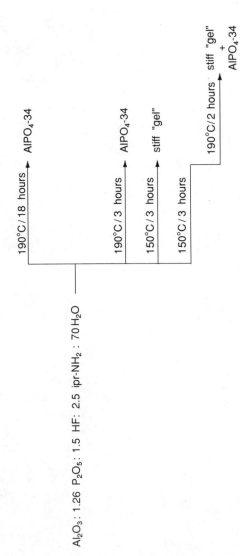

Figure 11. Aluminophosphate batch composition in the presence of isopropylamine and HF and conditions for clear solution synthesis.

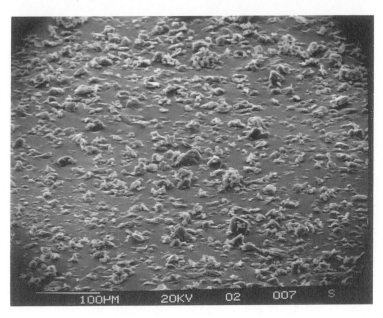

Figure 12. SEM micrographs of AlPO$_4$-34. (top) After 3 hours of synthesis, (bottom) After 18 hours of synthesis.

18

Investigation of the Crystallization of Very Large Pore Aluminophospate Phases and Their Physico-chemical Characterization

E. Jahn,[1] D. Mueller,[2] J. Richter-Mendau [3]

[1]*Ruhr-Universitaet Bochum, Institut fuer Mineralogie, Postfach 102148, Universitaetsstr. 150, 4630 Bochum 1, Germany*
[2]*Zentralinstitut fuer anorganische Chemie Berlin, Rudower Chaussee 5, 1199 Berlin, Germany*
[3]*Zentralinstitut fuer physikalische Chemie Berlin, Rudower Chaussee 5, 1199 Berlin, Germany*

On the basis of the investigation of various synthesis procedures it is concluded that the preparation of very large pore aluminophosphate molecular sieves involves techniques known for pillaring of layered solids. In the case of VPI-5 this means a) the partial transformation of the starting pseudo-boehmite into a layered aluminophosphate phase, b) the expansion of the latter material by substances containing amine and phosphate molecules and c) the crystallization of the expanded product. The use of compounds which contain P-O-P linkages instead of monophosphoric acid as P source improves the conditions for occuring of steps a) and b). In the present work the sequence of the different synthesis parts is characterized by n.m.r. spectroscopy and X-ray diffraction.

INTRODUCTION

During the course of our synthesis work on aluminophosphate based-molecular sieves, we remarked that some features of the crystallization process of VPI-5 are very distinct for VPI-5 on the one side, and for nearly all the other members of the ALPO family on the other.

First, the organic amine used to crystallize VPI-5 does not reside within the extra large-pores of this phase in a space-filling capacity. This is in line with observations presented by Davis et al. (Davis et al. 1989), although they employed different amines. In the case of other microporous aluminophosphate phases, the amount of the adsorbed organic additive is typically equivalent to pore filling.

Second, the presence of only a small amount of the organic species within the large channels of VPI-5 can be explained with the result of the Rietveld refinement of the VPI-5 structure (McCusker et al. 1991). There is a regular arrangement of water molecules within these channels.

Third, the successful synthesis of a pure VPI-5 phase has been found to be strongly dependent on the pretreatment of the pseudo-boehmite used as the aluminum source. The synthesis of VPI-5 fails when a rather well crystallized boehmite is used, whereas the formation of VPI-5 is favored when the pseudo-boehmite is wet milled before it is added to the synthesis mixture. It seems to be reasonable to correlate the influence of the form of the starting pseudo-boehmite on the VPI-5 synthesis with its structure. Pseudo-boehmite is essentially finely crystalline boehmite with the same or similar octahedral layers in the x, z plane, but lacking three-dimensional order because of a restricted number of unit cells along y. It consists of a significant number of crystallites that involve a single unit cell along y, or single octahedral layers, and contains more water than boehmite. This water is intercalated between octahedral layers (Tettenhorst and Hofmann 1980). Thus, pseudo-boehmite can be recognized as a layered material and should be capable of being permanently expanded under appropriate conditions.

Fourth, the most striking improvement of the VPI-5-synthesis, with respect to both the phase purity of the reaction product and the development of well-shaped crystals, is achieved by the use of polyphosphoric acid instead of monophosphoric acid as the source of phosphorus (McCusker et al 1991). This reveals that the VPI-5 crystallization is sensitive to the replacement of a small molecule with medium acidic character by a bulky molecule with polyelectrolytic properties.

The ensemble of the noted tendencies involved in the VPI-5 formation provides the background for the conclusion that this crystallization is not dominated by the templating effect of an organic additive. On the contrary, when pseudo-boehmite is used as the Al source, the generation of the very large pores of this material is probably due to a synthesis route that bears much resemblance to the pillaring of layered substances. Moreover, the possibility exists that pseudo-boehmite represents a precursor for the preparation of new sheet structures, which in their turn can be converted into large-pore materials. In order to substantiate this hypothesis, the focus of the present material is the detailed investigation of the crystallization of VPI-5, with special reference to the sheet-like structure of pseudo-boehmite. Intermediate and end products are examined by means of X-ray diffraction and solid-state MAS n.m.r. spectroscopy. The combined utilization of these two techniques offers the advantage of monitoring the changes in both short- and long-range ordering during the course of the synthesis.

EXPERIMENTAL PROCEDURE

Synthesis

The organic additive di-n-propylamine was used exclusively. The Al sources were pseudo-boehmite (25 wt% H_2O) and an aqueous suspension of aluminium hydroxide (2,3 wt% Al_2O_3), both commercially available. Monophosphoric acid (H_3PO_4, 85 wt%), diphosphoric acid ($H_4P_2O_7$), and polyphosphoric acid ($H_{10}P_8O_{25}$) were employed as various P starting components. The latter two materials were obtained by dissolving P_2O_5 in the necessary mole ratio. The average chain lenght of the polyphosphoric acid was determined by means of n.m.r. spectroscopy to be eigth P atoms. The method of preparing the synthesis gels was as follows: 1) a mixture of the Al and P components, the amine and water was milled for a period of 90 minutes by using a ball mill, 2) the gel obtained was further diluted to different preselected molar H_2O/P_2O_5 ratios and stirred at ambient temperature for 24 hours.

Deviations from this preparation procedure will be described for each individual case below. The batch compositions are listed in Table 1 - 1. The final aluminophosphate gel was divided into several portions and sealed in identical autoclaves. These were kept at 130 °C without rotating and progressively removed from heating in order to get materials with different degrees of crystallinity. After cooling, each sample was filtered, washed with cold distilled water, and dried at 50 °C for 8 hours.

Characterization

The crystalline phases of the samples were identified by X-ray powder diffraction, using Kα radiation. The ^{27}Al and ^{31}P MAS n.m.r. spectra were recorded on a Bruker MSL 400 spectrometer at 4.9 T. ^{31}P MAS n.m.r. cross - polarization experiments were carried out with a contact time of 1 ms. The chemical shifts are reported relative to AlCl$_3$ and 85 wt% H$_3$PO$_4$, respectively. Spinning frequencies of 4 - 5 kHz were used.

RESULTS AND DISCUSSION

P source: Monophosphoric acid

X - ray diffraction study

Figure 1 shows the X-ray powder diffraction patterns of the pseudo-boehmite used as the Al starting material, and of intermediate phases of the synthesis run 1 (Table 1) collected after 64, 80, 90, and 120 hours of hydrothermal heating. The broad diffraction peaks in the pattern of pseudo-boehmite result from the poorly ordered arrangement of the sheet-like units that essentially characterize the structure of pseudo-boehmite. The diffraction pattern of the phase isolated after 64 hours of hydrothermal treatment also indicates the presence of a layered material. Compared with the pattern of pseudo-boehmite, however the following differences have to be noted: the first peak of the pattern is shifted toward lower Θ values, and comprises additional reflections in the region between 9 and 11°Θ. The increased d-spacing of the basal reflection can be reasonably interpreted as the consequence of an intercalation process, in which the sheet-like units are regularly separated from each other by the accommodation of guest molecules in

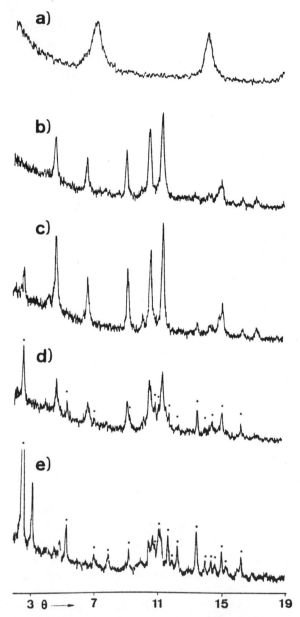

Figure 1. X-ray diffraction patterns of samples of run 1: a) Starting pseudo-boehmite; b) after 64 hours; c) after 80 hours; d) after 90 hours; e) after 120 hours. Asterisks in d and e indicate reflections due to VPI-5.

Table 1. Synthesis conditions and results of crystallization runs.

Batch composition (mole ratios) $Al_2O_3:P_2O_5:H_2O:Pr_2NH$	Time (hours)	Products
P source: H_3PO_4		
1) 1 : 3 :120: 2	64	L+unreacted AlO(OH)
	80	L+unreacted AlO(OH)
	90	L+VPI-5
	120	VPI-5>L
P source: $H_4P_2O_7$		
2) 1 : 1 : 40: 1 [1]	20	VPI-5>>AlPO$_4$-11
3) 1 : 1 : 40: 1	20	VPI-5+L
	72	VPI-5+L
4) 1 : 1 : 80: 1	15	VPI-5+H3
	48	H3>VPI-5
P source: $H_{10}P_8O_{25}$		
5) 1 : 1 : 80: 1	24	VPI-5>L
	48	VPI-5>L+H3
6) 1 : 1 : 80: 1 [2]	24	VPI-5+H3
7) 1 : 1 :120: 1	24	VPI-5>L
	48	VPI-5
8) 1 : 1 :300 [3]	20	VPI-5+L
	30	VPI-5>L
	44	VPI-5>>H3
9) 1 : 1 :300 [4]	12	L1
	24	H3

L: Layered phase with Θ values as follows (copper radiation):
4.6; 6.65; 9.2; 10.6; 11.4; 15.1.
L1: Layered phase with Θ value as follows: 2.6.
H3: Aluminophosphate hydrate $AlPO_4·1.5\ H_2O$.
[1] Pseudo-boehmite milled in presence of water
[2] Gel agitation for 48 hours
[3] Presence of VPI-5 seeds
[4] Absence of VPI-5 seeds

the space between these bodies. This parallels findings reported by Tettenhorst and Hofmann (Tettenhorst et al. 1980) which reveal the dependence of the basal reflection in the X-ray pattern of pseudo-boehmite on its water content. The occurrence of the peak group in the Θrange between 9 and 11° will be considered in more detail later.

With the prolongation of the hydrothermal heating, the shift of the first peak in the diffraction pattern continues (see Figure 1c) until, in the pattern of a sample isolated after 90 hours of reaction time, peaks representative of VPI-5 can be clearly distinguished from those of the sheet-like material (Figure 1d). Longer heating periods improve the yield of VPI-5, but a pure VPI-5 phase cannot be obtained by applying the synthesis conditions of run 1.

MAS n.m.r. investigations

As already mentioned, solid-state n.m.r. investigations are included in the present study in order to demonstrate the changes taking place on the atomic level during the course of the crystallization.

In the starting pseudo-boehmite, the Al atoms occur exclusively as AlO_6 structural units, which are responsible for the appearance of a signal at about 5 ppm in the ^{27}Al spectrum (not presented).

Figures 2a and 3a show the ^{27}Al and ^{31}P spectra recorded from the sample collected after 64 hours of hydrothermal treatment.

The ^{27}Al spectrum (Figure 2a) involves several resonance lines. The larger peaks at 38.4 and 41.2 ppm are both attributable to AlO_4 tetrahedra with neighboring PO_4 tetrahedra (Müller et al. 1983). The peak at 5.3 ppm is due to the existence of unchanged pseudo-boehmite in the sample, and the signal with a chemical shift of -11.2 ppm can be assigned to Al atoms in octahedral coordination with 4 PO_4 tetrahedra and 2 water molecules completing the first coordination sphere of the Al atoms (Müller et al. 1983).

The interpretation of the ^{31}P spectrum of this sample (Figure 3a) is more complicated. A broad dominant peak at -26.8 ppm and small additional peaks at -15.9 and -18.9 ppm are observed. The broadening of the peak at -26.8 ppm gives rise to the assumption that there are various kinds of PO_4 tedrahedra in the sample, with differences in their first neighbors. The position of this peak fits into the range of chemical shifts of P atoms that are surrounded by 4 AlO_4 tetrahedra. This assignment of the signal corresponds to the ^{27}Al spectrum. However, in view of the chemical composition

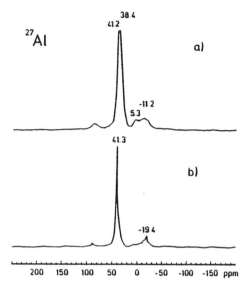

Figure 2. ²⁷Al MAS n.m.r. spectra of samples of run 1: a) 64-hour sample; b) 90-hour sample.

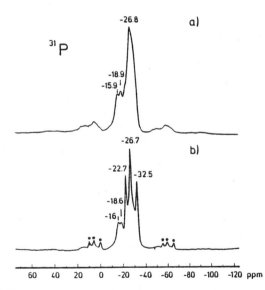

Figure 3. ³¹P MAS n.m.r. spectra of samples of run 1: a) 64-hour sample; b) 90-hour sample. Asterisks indicate spinning side bands.

of the reaction mixture, PO_4 tetrahedra connected with further PO_4 tetrahedra have to be taken into account. Systems containing both phosphoric acid and organic amine compounds are known to generate complexes in which PO_4 tetrahedra are bound via hydrogen bridging bonds to further PO_4 tetrahedra. In these species, the local environment of the P atoms is quite similar to that of P atoms in polyphosphate chains. Indeed, the ^{31}P spectrum taken from polyphosphoric acid in the presence of pseudo-boehmite contains an intense resonance line at -29.0 ppm. The shift of the peaks at -15.9 and -18.9 ppm toward the value of the free phosphoric acid suggests that these PO_4 tetrahedra are not fully linked to either AlO_4 or other PO_4 tetrahedra, but that some of the oxygen atoms are connected with protons, forming hydroxyl groups.

According to this n.m.r. investigation, the 64-hour intermediate product of synthesis run 1 represents an aluminophosphate phase in which Al is preferentially tetrahedrally coordinated exclusively by PO_4 groups. In contrast, the P atoms reside in various chemical environments. In addition to this aluminophosphate phase, there is residual pseudo-boehmite in the sample.

Figures 2 and 3 also show the [27]Al and [31]P n.m.r. spectra of an intermediate product after 90 hours reaction time. A comparison of the spectra clearly demonstrates the strong influence of the continued hydrothermal heating on the local environment of both the Al and P atoms.

The [27]Al spectrum of the 90-hour sample (Figure 2b) has a sharp signal at 41.3 ppm, which is attributable to Al(4P) tetrahedra. In contrast to the spectrum of the 64-hour sample, the peak at 38.4 ppm is only a shoulder on the peak at 41.3 ppm. Moreover, the maximum of the resonance line due to octahedral Al is shifted to -19.4 ppm. The signal at 5 ppm indicates that unreacted pseudo-boehmite is still present.

In the case of the ^{31}P spectra, the most obvious difference between the two spectra of Figure 3 is the occurrence of quite well resolved lines at -22.7, -26.7, and -32.5 ppm in the spectrum of the 90-hour sample. This phenomenon is consistent with the significant enhancement in the degree of crystallinity of the reaction product by prolonged heating. The peaks at -16.0 and -18.6 ppm, and the shoulder on the signal at -26.7 ppm, reflect the existence of residual amounts of the phase found after 64 hours.

The dominant lines in both the [27]Al and the [31]P n.m.r.spectra of the 90-hour sample correspond to those of VPI-5 (see Figure 5 g and h) (McCusker et al. 1991; Grobet et al. 1989)

Discussion

In view of the results of the X-ray diffraction and the MAS n.m.r. investigation of synthesis run 1, the following hypothesis can be advanced concerning the crystallization of VPI-5. Initially, most of the pseudo-boehmite is converted into an aluminophosphate phase. During this process, the layered character of the pseudo-boehmite, which is related to the existence of sheet-like units in this material, is transferred to the aluminophosphate phase (Figure 1, a and b). It can be assumed that, at the beginning, small aluminophosphate islands are generated within the sheet-like units, and these grow by consuming Al from the sheets and phosphoric acid from the liquid phase. This newly formed layered phase differs from the starting pseudo-boehmite mainly in two respects. First, its interlayer distance is increased due to the accommodation of the bulky polyphosphate chains. The precursors of the latter, phosphoric acid and the organic amine, probably penetrate into the interlayer space during the milling of the reaction mixture. The assumed occupation of the interlayer space by volatile substances is supported by the fact that, in the X-ray pattern, the "large spacing" peak disappears upon heating (Figure 4). Second, the alumino-

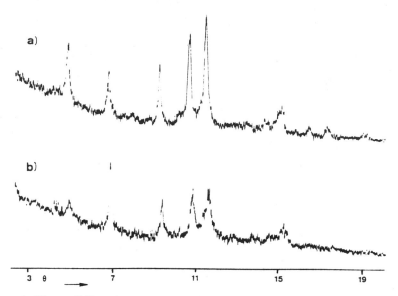

Figure 4. X-ray diffraction patterns of the layered phase L (a) and of the heated phase L (b).

phosphate phase is characterized by an unique structure. This is indicated by the occurrence of reflections in its X-ray pattern that are not present in that of pseudo-boehmite.

The water content of the sample increases from 9.0 wt% after 64 hours of hydrothermal treatment to 16.4 wt% after 90 hours. Most of the water is physically, adsorbed because it can be removed easily at 110°C. The water uptake is accompanied by a remarkable drop in the amount of adsorbed amine. The amount of carbon present diminishes from 4.5 wt% for the 64-hour sample to 2.6 wt% for the 90-hour one. Similarly, the value for nitrogen drops from 0.9 wt% to 0.5 wt%. According to the proposed layered structure of the 64-hour sample, the water uptake during the reaction can be recognized as a swelling process. The interpretation is consistent with the continuous decrease of the   value of the basal reflection in the corresponding X-ray patterns (figure 1, b, c and d).

The diffusion of the water into the interlayer space is consistent with the hydrophilic nature of the polyphosphate chains that are present in this space.

This hypothesis is corroborated by the behavior of Al(4P) tetrahedra, located within the sheets, upon the water uptake and the amine loss. In $AlPO_4$ molecular sieve synthesis batches of low amine/water ratios, Al(4P) tetrahedra have been found to be irreversibly converted to octahedra by the penetration of water molecules into the first coordination sphere of the Al. This leads to the formation of the aluminophosphate hydrate H3, or of the dense $AlPO_4$-phase tridymite (Jahn et al. 1989). The comparison of the [27]Al n.m.r. spectra of Figure 2 does not reveal an enhancement of the intensity of the peak at -11.2 ppm, which is attributable to octahedral Al, with increasing time of hydrothermal treatment. This confirms that, in the synthesis under study, the transformation of tetrahedral Al into octahedral Al does not take place, and parallels the results of the X-ray examination (Figure 1); peaks representative of H3 or $AlPO_4$-tridymite are not detectable in the X-ray patterns. Apparently, the hydrophilic nature of the polyphosphate-containing species inhibits the attack of water on the Al(4P) tetrahedra.

The information obtained regarding the composition and the properties of the layered phase suggests that the last step of the crystallization of VPI-5 (the transformation of the layered structure into a three-dimensional network) occurs via a mechanism involving the crosslinking of the aluminophosphate layers in such a manner that the very large pores of VPI-5 are formed. Support for this proposal is supplied by the two experimental results. First, the crystallization of VPI-5 is obviously induced, when the expansion of the layered phase attains a layer spacing equal to the diameter of the large pores of VPI-5. The similar value of the basal reflection in the X-ray patterns of the two phases substantiates this correlation (see Figure 1). Second, there is little delay between the end of the swelling process and the first appearance of VPI-5 in the reaction product.

P source: Di- or polyphosphoric acid

The proposed crosslinking mechanism raises questions regarding the origin of the species that are acting as pillars in this process. We have found that it is advantageous to use di- or polyphosphoric acid as the P source for these investigations. The results of this set of synthesis runs are shown in Table 1.

As can be seen from Table 1 a pure VPI-5 phase is not obtained at all by using diphosphoric acid as the P source. The use of polyphosphoric acid, however, provides the conditions for the production of well-crystallized VPI-5 samples. Furthermore, for the latter case, the dominant synthesis parameter is the choice of the appropriate H_2O/P_2O_5 ratio for the synthesis batch. For example, increasing the H_2O/P_2O_5 ratio improves the yield of VPI-5, whereas the prolongation of the gel agitation does not remarkably influence the composition of the reaction product. The results of this set of synthesis runs also indicate that VPI-5 forms preferentially at the expense of the layered phase. This parallels the information from the n.m.r. investigation of the products of these crystallization experiments (figure 5). The comparison of the ^{31}P spectra, especially, reveals that the dilution of the reaction mixture causes the disappearance of those peaks in the n.m.r. spectra that are assigned to the layered phase (see Figure 5b, f, and h). These observations can be explained as follows: On the one hand, a certain degree of condensation of the phosphoric acid is necessary to meet the requirements for the swelling of the layered intermediate phase, on the other hand, however, the polyphosphoric acid has to be depolymerized prior to its incorporation into the advancing three-dimensional network, because of the alternating occupation of the T positions by P and Al atoms in the final VPI-5 phase. Moreover, the conclusion concerning the role of water during the crosslinking of the aluminophosphate layers is supported by the ^{31}P cross-polarization MAS n.m.r. spectrum of an intermediate reaction product (Figure 6). Figure 6 clearly demonstrates that the cross-polarization experiment enhances the intensity only of the peaks at -18.7 and -29.0 ppm, which are representative for the layered phase. An interaction between water and the P atoms of VPI-5 cannot be derived.

The real crosslinking species are composed of both Al and P atoms because, in the case of a complete transformation of the layered phase into VPI-5, residual pseudo-boehmite is also totally converted (see Figure 5e and g).

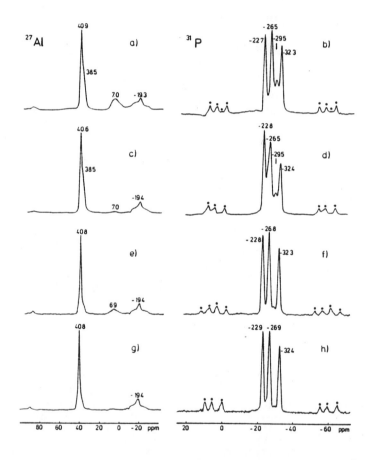

Figure 5. ^{27}Al and ^{31}P MAS n.m.r. spectra of samples of runs 5, 6, and 7: a) and b) run 5, 24- and 48-hour sample; c) and d) run 6; e) and f) run 7, 24-hour sample; g) and h) run 7, 48-hour sample.

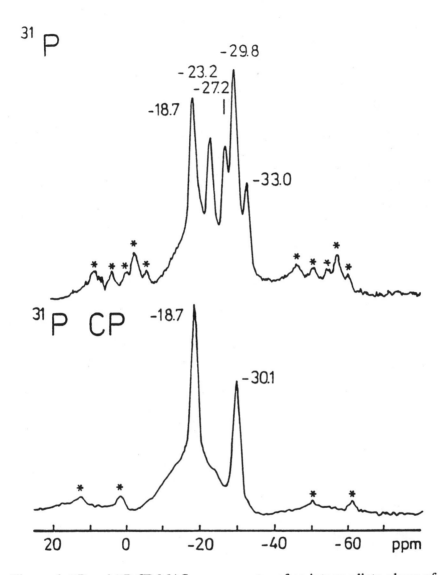

Figure 6. ³¹P and ³¹P CP MAS n.m.r. spectra of an intermediate phase of run 7. Asterisks indicate spinning side bands.

Since the crosslinking of the layers leads to the creation of a highly ordered system, it seems likely that the unique structure of the intermediate phase has favored places for the deposition of the pillars. The close dependence of the formation of VPI-5 on the development of the layered phase is evident in the following experiment. By the use of very diluted aqueous suspension of aluminum hydroxide and polyphosphoric acid, as Al and P starting materials, respectively, synthesis runs were performed without organic additives: In one case, VPI-5 seeds were used (Table 1, runs 8 and 9). Figures 7 and 8 show the X-ray diffraction patterns of the intermediate and final products of these two runs. Although in the course of both reactions layered intermediate phases are generated, only the layered phase that is characterized by a unique structure is transformed into VPI-5 (Figure 7b, c, and d). The latter layered phase, the X-ray reflections of which are denoted by triangles in Figure 7b, is obviously identical to the one in synthesis run 1.

CONCLUSIONS

The present investigation of the VPI-5 crystallization provides several indications that corroborate the proposed mechanism for this crystallization process. The supporting facts include the following:

• When pseudo - boehmite is used as the Al starting material, an essential feature of the VPI - 5 crystallization is the generation of an intermediate layered phase. This phase is composed of aluminophosphate sheets and polyphosphate chains containing species which occupy the interlayer space. The aluminophosphate sheets seem to be characterized by an unique structure.

• The intermediate layered phase is capable of swelling because of the hydrophilic properties of the polyphosphate chains.

• The transformation of the swelled layered phase to the three-dimensional network of VPI - 5 involves the crosslinking of the sheets by small aluminophosphate species acting like pillars. The latter are apparently created by the reaction of phosphate entities formed in a dissolution process of the polyphosphate chains and residual pseudo - boehmite. The unique structure of the aluminophosphate sheets promotes an organized deposition of the pillars.

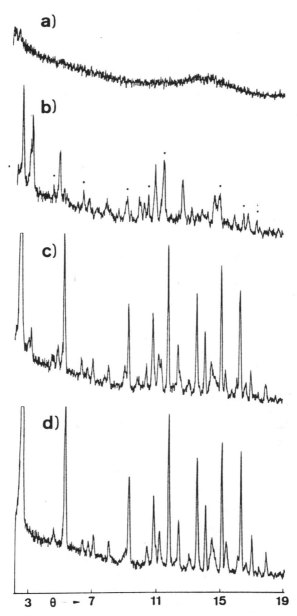

Figure 7. X-ray diffraction patterns of samples of run 8: a) after 6 hours; b) after 20 hours; c) after 30 hours; d) after 44 hours. Triangles in b indicate reflections due to the layered phase L.

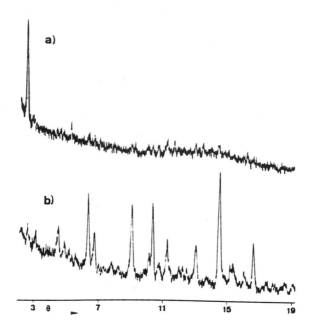

Figure 8. X-ray diffraction patterns of samples of run 9: a) after 12 hours; b) after 24 hours.

• Furthermore, the results of this study emphasize the role of the water as a structure - directing agent during the course of the crystallization of VPI - 5 (McCusker et al. 1991). The function of the organic additive seems to be restricted to its effect on the generation of the layered phase.

• These observations and their interpretation for the VPI - 5 crystallization process may allow further elucidation of the mode of action of both the pseudo - boehmite as a precursor for aluminophosphate layered phases and of polyphosphate compounds as pillars in other aluminophosphate syntheses.

REFERENCES

Davis, M. E. et al. 1989. Physico-chemical characterization of VPI-5. J. Am. Chem. Soc. 111: 3919-3924.

Grobet, P. J. et al. 1989. VPI-5: An aluminophosphate hydrate? Appl. Catal. 56: L 21.

Jahn, E. et al. 1989. On the synthesis of the aluminophosphate molecular sieve AlPO$_4$. Zeolites 9: 177-181.

McCusker, L. B. et al. 1991. The triple helix inside the large-pore aluminophosphate molecular sieve VPI-5. Zeolites 11: 308-313.

Müller, D. et al. 1983. Hochauflösende Festkörper-NMR-Spektroskopie an Aluminiumphosphaten Z. Anorg. Allg. Chem. 500: 80.

Tettenhorst, R. and Hofmann, D. A. 1980. X - ray diffraction investigation of boehmite. Clays and Clay Minerals 28: 373-380.

19

STRUCTURE AND CHEMICAL BEHAVIOR OF THE VPI-5 FAMILY OF COMPOUNDS

Abraham Clearfield and Jaime O. Perez
Texas A&M University

Introduction

The discovery of the wide pore 18-ring aluminum phosphate VPI-5[1,2] represents a milestone in the synthetic chemistry of molecular sieves. This compound was shown to have the framework structure illustrated in Figure 19-1.[2,3] In the original synthesis it was reported that VPI-5 did not contain the template amines used in the synthesis. Rather the channel was found to be filled with water and the unit cell contents were reported as $(AlPO_4)_{18} \cdot 36H_2O$.[3] The channel opening is ~12 Å, the largest recorded for a molecular sieve (except for cacoxenite which is a natural iron aluminum phosphate mineral). The illustrated structure shows strict Al, P alternation of AlO_4 and PO_4 tetrahedra in space group P6₃cm. A subsequent NMR study[4] however, revealed the presence of 6-coordinate aluminum and 3 equal ^{31}P NMR resonances. These data were interpreted to indicate that the aluminum atoms in every other 6-membered ring (6MR) are hydrated by two water molecules. Indeed the NMR data did show the presence of 6-coordinate Al in a 1:2 ratio to the 4-coordinate Al. There are two crystallographically distinct sites in space group P6₃cm, 12 atoms (P or Al) in the 6MR and 6 in the 4MR. Thus, the hypothesis of every other 6MR having 6-coordinate aluminum atoms appeared to adequately explain the NMR data as the phosphorus atoms would then have three distinct environments; 6P in the 4MR, 6P in the 6MR near 6-coordinate aluminum and 6P in the 6MR which were not near the 6-coordinate Al atoms.

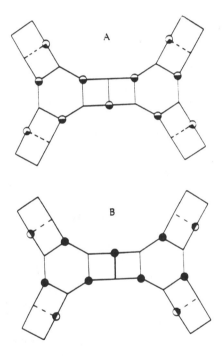

FIGURE 19-1. Schematic diagram representing the VPI-5 unit cell. Circles represent the phosphorus atoms which alternate with Al atoms located at the intersections of three lines. The fourth line completing the AlO₄ or PO₄ tetrahedra connect the A and B portions. The A section is at z=0 and the B section at z=1/2, the sequence being A B A. Shaded circles indicate the contribution of that atom to the total in the unit cell. Thus, O represents 1/2 a phosphorus atom, O an atom wholly in the cell and O, 1/4 atom. The dashed lines represent the unit cell boundaries enclosing 12 P atoms in the six membered rings and 6 P atoms wholly in the four membered rings.

Subsequent X-ray studies showed that at both low[5] and high[6] water contents, the 6-coordinate aluminium was in the 4MR not the 6MR. However, McCusker et al.[6] refined the VPI-5 structure in space group P6$_3$ which requires 3 distinct phosphorus sites again accounting for the three ^{31}P resonances observed. Yet it is not clear how the P atoms in the 6MR differ since they shifted only slightly from their original positions in space group P6$_3$cm.

Davis, et al.[7] have reported that the VPI-5 prepared with dipropylamine (DPA) is not stable in the mother liquor after approximately 24 h of heating, decomposing at longer heating times. In contrast they found the VPI-5 prepared in the presence of tributylammonium hydroxide (TBAH) to be stable at 150°C in the mother liquor. Very shortly after the announcement of the preparation of VPI-5 we undertook to synthesize it by variations in the synthesis conditions. At first we obtained mixtures of the type reported by d'Yvoire.[8] However, by controlling the gel viscosity and the aging process before and after adding amine, we were able to prepare pure wide pore phases. We will discuss three of them here. One phase gave an X-ray powder pattern which was identical to that reported for VPI-5. The second phase showed some minor differences in its X-ray powder patterns, but more importantly converted to AlPO$_4$-8 on heating at 300°C. As we shall see this compound is identical to the H1 synthesized earlier by d'Yvoire.[8] The third phase is AlPO$_4$-8 which has been reported to have a 14-ring uni-dimensional channel system.[9] We shall refer to the thermally unstable second phase as H1 to distinguish it from the more stable VPI-5.

Another point of difference between the H1 and VPI-5 preparations is their composition. The VPI-5 phase always had an Al:P ratio very close to 1 while the H1 solids contained phosphate deficiencies.[5] An average of 3 preparations gave Al:P = 1.08. However, in a subsequent study on H1[10] it was concluded that it was a form of VPI-5 that could be produced without an organic template agent and is not a separate phase. In the interim several studies have reported[11] that VPI-5, formed in the presence of dipropylamine, readily converts to AlPO$_4$-8. This paper presents new evidence bearing on these questions and details on the hydrothermal synthesis of AlPO$_4$-8.

Experimental

Sample Preparation. Three different amine template systems were used to prepare the wide pore aluminum phosphates including $AlPO_4$-8. In a previous publication[5] specific recipes for VPI-5, H1 and $AlPO_4$-8 were given. In this report we will provide some details on the important factors involved in the several synthetic procedures. In all cases the mole ratio of reactants was T:Al_2O_3:P_2O_5:40 H_2O, where T stands for the amine utilized in a particular preparation. Any deviations from this mole ratio will be stated explicitly.

The factors found to be important in the preparations are viscosity of the gel, aging time, temperature and choice of amine.[12] This is illustrated in the preparations described below.

$AlPO_4$-8 Synthesis. The viscosity of the initial gel could be altered from a very thin fluid one to a solid, rigid one by the way in which the phosphoric acid was added to the psuedoboehmite slurry. In general the procedure was as follows. 13.8 g of psuedoboehmite (Catapal B- alumina, Vista Chem. Co.) was slurried in 41.1 g of water with magnetic stirring for 10-15 min. A phosphoric acid solution consisting of 23.1 g of 85% phosphoric acid and 27.4 g H_2O was cooled to room temperature and added to the alumina slurry in different ways. These additions consisted of either (1) dropwise additions or (2) adding the acid in specific increments and homogenizing the slurry for a fixed length of time before the next addition. To obtain pure $AlPO_4$-8. the increments were 7 ml additions of phosphoric acid followed by 2 min of homogenization. The gel was then aged for times ranging up to 216 h. Gels aged for 10-15 h were found to yield the best results. To prepare $AlPO_4$-8. the amine used was dipropylamine (DPA). 10.1 g of DPA was added dropwise to the aged aluminum phosphate slurry with stirring, which was continued for an additional 20-40 min after the addition was complete. The gel was then heated at 125-130°C in a teflon-lined steel vessel for 18-22 h. The solid was then recovered by filtration, washed 6-10 times with 100 ml portions of distilled deionized water and dried in an air oven at 40-50 °C for 17 h.

Preparation of H1. The quantities of reagents were exactly the same as for $AlPO_4$-8. The major difference was that the phosphoric acid was added in 4 ml portions with a homogenization period of 2 min. This procedure produced a less viscous gel than that utilized in the synthesis of $AlPO_4$-8. The homogenized gel was left to age for 15-24 h at room temperature (23-26 °C) without stirring and then the DPA was added dropwise as before. This mixture was heated in a stainless steel vessel for 4-12 h at 120 °C. H1 could also be prepared by using a combination of tributylamine (80%) and dipentylamine (20%). A procedure for this method has been given previously.[5]

Preparation of VPI-5. For this synthesis the same quantities of reagents were used as for $AlPO_4$-8. However, the phosphoric acid was added dropwise[5] a procedure which produced a thin, fluid gel. After addition of the acid was complete, the mixture was transferred to a rotary evaporator and kept at 95 °C for 25 min while rotating the container, but with no vacuum applied. The gel was then aged at room temperature for 3 h and 15.9 g of dipentylamine added with stirring and the gel homogenized for 30-40 min after the addition was complete. The mixture was then heated at 150 °C and autogenous pressure for 20.5 h. The heating times given are at the required temperature and do not include the time required for the oven to achieve the operating temperature (usually 2 h).

X-ray Diffraction Analysis. X-ray powder patterns were recorded with either a Seifert-Scintag automated powder diffractometer (PAD II) or a Rigaku rotating anode RU200 unit using $CuK\alpha$ radiation and a variable scan speed. Routine X-ray patterns were obtained with a step size of 0.04° and count times of 2 seconds per step. For unit cell dimensions a step size of 0.01° at a count rate of 6 seconds per step was used and the 2θ range covered was 3-80°. Samples were ground together with 10% Si powder (NBS SRM #640) serving as calibrant. Dehydrated samples were run in an airtight sample holder of our own design. For the dehydration of H1, heating under vacuum at 50 °C was necessary to avoid formation of $AlPO_4$-8. For structure solution purposes the data was rerun without the addition of the silicon standard but with close attention to eliminate the zero point error. The data was processed by our GRAPH[13] series of programs. This set of programs carries out the necessary data processing and interfaces to prewritten programs such as TEXSAN, SHELXS, and GSAS.[14] For the VPI-5 data in the dehydrated state the non-centric space group P6₃cm was chosen based upon the systematic absences and the fact that this space group allows alternation of phosphorus and aluminum atoms. A direct method solution with 41 unambiguously indexed reflections in the TEXSAN program set yielded the positions of all the P and Al atoms and two oxygens. Normal Fourier methods were applied to locate the remaining oxygen atoms. This starting structure was then refined by the Reitveld method using GSAS.[14a] The refinement with isotropic temperature factors converged with $\omega Rp = 12.0$. A similar structure solution and refinement for the dehydrated form of H1 is almost complete. Full details of the X-ray study will be published separately but the results will be summarized here.

Solid State NMR. NMR spectra were obtained on a Bruker MSL-300 solid state spectrometer where proton, phosphorus and aluminum nuclei resonate at 300.1, 121.5 and 78.2 MHz, respectively. The ^{31}P chemical shifts were referenced to 85% H_3PO_4 aqueous solution. Samples at different states of hydration were sealed in a specially designed Kel-F capsule that snugly fit into the zirconia rotor.

Results

Effect of Synthesis Conditions. The total water content of the gel before heating has a marked influence on the products obtained. The major effect is on the viscosity of the gel, in general the higher the water content the less viscous the gel. However, this observation needs to be considered in relationship to the distribution of water between the alumina slurry and the phosphoric acid. The more concentrated the acid the more viscous the gel. However, if all the water is added to H_3PO_4 and then mixed with solid psuedoboehmite a very thick gel results. The choice of 40 moles of water per 0.1 mole of Al_2O_3 and the distribution of 40% to the acid and 60% to the alumina, after subtracting the water present in the 85% H_3PO_4, was arrived at after many trials.[12]

The most important influence on the viscosity of the gel is the way in which the phosphoric acid is added to the alumina slurry. Given the amount and distribution of the water between the alumina and the H_3PO_4 as described in the experimental section, the following qualitative observations hold. Addition of the acid dropwise yields a thin fluid gel while rapid addition of 3 ml portions of acid yields a more viscous gel. Increasing the acid portions give a progressively more viscous gel until addition of 20 or more ml yields a rigid non-flowing cake. Variation of the time between additions also may change the viscosity but we have fixed on 2 min intervals.

The way in which the H_3PO_4 is added also has an effect on the pH of the gel. For example, the pH of the gel was 1.19 when prepared by adding the acid in 3 ml portions for 10 additions each followed by a 2 min homogenization, followed by 4 additions of 5 ml each with 2 min homogenization periods. This pH rose slowly on aging at room temperature to 2.14 in 20 h and 2.26 in 47 h. Thus, changes in the aluminum phosphate gel in which protons are added to the solid or the OH⁻ groups of the psuedoboehmite slowly react with the acid continues at a slow pace. Even after 200 h the pH was still increasing. The pH, after amine addition and homogenization of the gel, is correspondingly higher the longer the gel has been aged. However, after the hydrothermal reaction the pH is approximately the same irrespective of the pH of the starting gel. For example, gels which varied in their initial pH from 3.1 to 5.4 had values of 6.4-6.6 after 20 h at 125 °C. Increasing the temperature of reaction generally resulted in an increase in the final pH and it should be noted that H1 could be prepared at 120 °C in the DPA system while VPI-5 required hydrothermal treatment at 150 °C (DPTA as template). Increased temperature was found to favor the formation of $AlPO_4$-11 in the DPA system. Therefore, we had to find a different amine which suppressed formation of this aluminum phosphate. Acid conditions favored the formation of H3 so that a delicate balance of all these factors was necessary to obtain the desired phase.

Composition and Identification of Phases. The phases were identified from their X-ray powder patterns.[5] The X-ray pattern for H1 is given in Figure 19-2 and that for VPI-5 is shown as Figure 19-3. These patterns show only minor differences which we will elaborate on in what follows. Elemental analysis of several of our preparations are given in Table 19-1. It is seen that there is a deficiency of phosphate in the H1 and $AlPO_4$-8 samples. TGA curves for some of these samples from which the amount of organic material in the pores may be deduced, are present in Figure 19-4. The amount of amine retained by the molecular sieve depended mainly on the amine used in the synthesis. When the template was DPA, very little or no amine was retained as shown in Figure 19-4, curve C. However, the larger amines, DPTA and TBA were retained in varying amounts. Reference to Figure 19-4 curve A shows that VPI-5 retained ~14.4% DPTA and about 4.5% H_2O, while H1 contained 14.3% of a TBA-DPTA mixture and about

8.6% water (curve B). The amines were removed from VPI-5 and $AlPO_4$-8 by calcination at 500 °C but even at 200 °C H1 transformed to $AlPO_4$-8. H1 does not form tridymite as reported by d'Yvoire.[8] Only when H1 contains H2 as a significant impurity does tridymite form. Other workers[10,11] also found that H1 thermally transforms to $AlPO_4$-8.

An interesting observation resulted from carrying out the hydrothermal preparation for different lengths of time. The amine chosen as template was DPTA and the temperature was 150 °C. The results are tabulated in Table 2. It is seen that 18-ring $AlPO_4$'s are formed over the entire time range of 6-26 h. However, on heating the washed product to 500 °C, decreasing amounts of $AlPO_4$-8 form. The sample prepared in 22.5 h (JP-66C) did not convert to $AlPO_4$-8 even at 650 °C. We interpret these results in the following way. After 6 h the hydrothermal reaction yields a mixture of VPI-5 and H1. On calcination only the H1 transforms to $AlPO_4$-8. More than half of the sample is H1. As the time of reaction

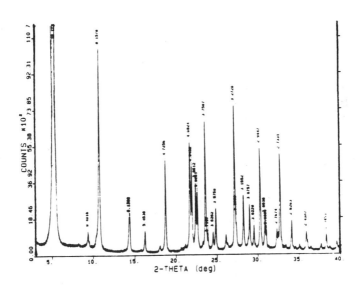

FIGURE 19-2. The X-ray powder pattern of H1(JP7-59A) from 3° to 40° 2θ. This sample was made with DPA and contains 24% H_2O. A trace of impurity, probably $AlPO_4$-11 is present.

increases so does the amount of VPI-5. In a narrow time range close to 22.5 h the entire product is VPI-5 but even at 24 h reaction time H1 forms once again (~50%) together with a small amount of H3 (<3%).

VPI-5 can be converted to AlPO4-8 via a hydrolytic process. Sample JP5-66C was slurried in water (25mg in 25ml H2O) at room temperature for 6 h after which time the aqueous phase contained 16.8µg/ml of P and only 1.49 µg/ml of Al. Continued stirring for 12 h more yielded a solid product with P/Al = 0.93 and consisted of 70% AlPO4-8 and 30% VPI-5.

X-ray Analysis. In Figure 19-5a we have reproduced portions of the X-ray powder patterns of the JP5-66 samples. They were prepared with DPTA as templating amine at different reaction times as listed in Table 19-2. Patterns A, B, and C were recorded for samples kept at 150 ºC for 6 h, 15 h and 22.5 h reaction time, respectively. The major differences seen in the X-ray patterns are in the region of 20-24º 2θ. The reflections in the neighborhood of 22º 2θ are doublets whereas in the pattern C, which represents pure

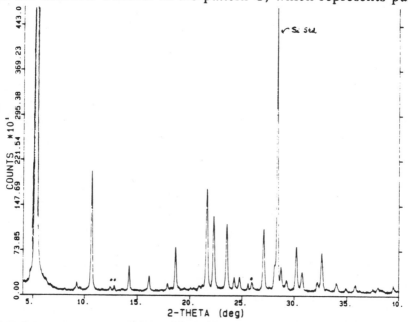

FIGURE 19-3. The X-ray powder pattern of VPI-5 (sample JP5-66C). This sample was prepared with DPTA at 150 °C. A small amount of an impurity is present as shown by the reflections at 7.14 and 6.92 Å marked with dots.

Table 19-1. Elemental Analysis of AlPO$_4$ Samples

Sample Identification	% P	% Al	P/Al	% Organic	Sample Description
JP4-91C	18.95	18.64	0.89	--	H1 (TBA)
JP5-21C	18.2	16.94	0.94	14.3	H1 (TBA-DPTA)
JP6-20B	17.66	17.85	0.86	--	H1 (DPA)
JP5-93B	18.51	17.37	0.93	--	H1 (DPA)
JP5-66C*	21.11	18.80	0.98	14.4	VPI-5 (DPTA), 22.5 h
JP5-69B	18.04	15.73	1.00	--	VPI-5 + ~5% H1(DPTA)
JP7-39A	20.14	20.0	0.88	--	AlPO$_4$-8 (DPA)
JP7-490	22.12	20.99	0.92	--	AlPO$_4$-8 (DPA)

*Prepared as described in ref. 5.

VPI-5, they are singlets. In the diffraction patterns in (b) these same samples were heated at 500 °C for 4 h. Curve A shows almost complete conversion to AlPO4-8 for the 6 h sample, less conversion to AlPO4-8 for the 15 h sample and none for the 22.5 h sample. This difference in stability is significant for potential use of these 18-ring molecular sieves as catalysts. One change that does take place in the VPI-5 X-ray pattern on heating is the splitting into doublets of the peaks in the region of 22° 2θ. This is shown in Figure 19-6A. Similarly some samples of H1 may yield X-ray patterns where these reflections look as though they are singlets (Figure 19-6C) but at higher resolution are seen to be doublets. A more typical example is shown in Figure 19-6D. This sample has retained the template (DPTA) but in Figure 19-6E we show a sample of H1, prepared in the presence of DPA, which retained no amine. Note the differences in relative intensities of the reflections in A and C and D and E.

In Table 19-3 are listed the unit cell dimensions for several samples of H1 and VPI-5 prepared by us, together with some d-

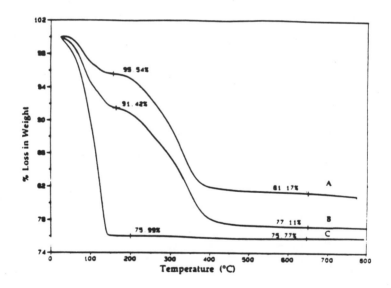

FIGURE 19-4. Thermogravimetric weight loss curves for as prepared samples. (A) VPI-5 containing 4.5% H_2O and 14.1% DPTA, (B) H1 (JB5-21C) containing 8.6% H_2O and 14.3% organic (TBA-DPTA), (C) H1 (JP5-93B) made from DPA and containing only water.

Table 19-3. Unit cell parameters for wide pore aluminum phosphate samples determined in this work with Miller indices and d-spacing for peaks in the region 21-24° 2θ

Sample	Unit cell parameters, hexagonal symmetry			Peaks (Miller indices/d-spacing (Å)[h]				
	a=b (Å)	c (Å)	Vol (Å^3)	(002)	(221)	(102)	(311)	(320)
H1[a]	18.9681(7)	8.1157(4)	2528.7		4103		3.978	3.770
H1[b]	18.927(2)	8.097(1)	2512.2	4.082	4.044	3.961	3.927	3.759
H1[c]	18.6576(7)	8.3284(5)	2494.2	4.152	4.057	4.023	3.936	3.695
H1[d]	18.9933(6)	8.1214(5)	2537.2	4.082	4.043	3.961	3.926	3.759
VPI-5[e]	19.036(5)	8.185(6)	2568.8	<u>4.093</u>	4.113	<u>3.972</u>	3.994	3.784
VPI-5[f]	19.036(6)	8.118(4)	2547.6	4.066	4.107	3.948	3.984	3.782
VPI-5[g]	18.6005(6)	8.3664(4)	2490.4	4.173	4.051	<u>4.040</u>	3.932	3.687
AlPO4-8	a=33.231(3)	b=14.749(1)	c=8.3796(3) Cmc2$_1$					

[a]JP5-21C, as made, contains 8.6% water and 14.3% organic (TBA-DPTA). Synchrotron data.
[b]JP5-44A, as made, contains 15.7% water and 9.1% organic (TBA-DPTA).
[c]JP5-44A, evacuated, contains ~0% water and 1.5% organic (TBA-DPTA).
[d]JP7-59A, as made (DPA system), contains 24% water and no organic.
[e]JP5-66C, as made, contains 4.5% water and 14.1% organic (DPTA).
[f]JP5-66C, calcined at 550°C for 5 h and 620°C for 2 h and rehydrated (23.2% water).
[g]JP5-66C, calcined at 550°C for 5 h and 620°C for 2 h and evacuated (~0% water and organic).
[h]Underlined numbers are those peaks not well resolved in the XRD pattern.

Table 19-4. Unit cell dimensions of VPI-5 and H1 reported in the literature

Sample	Space group	Unit cell dimensions			Ref.
		a (Å)	c (Å)	V (Å³)	
(1) Si-VPI-5, Si/(Si+Al+P)<0.1	P6₃cm	18.989	8.113		2
(2) Same sample No. 1	P6₃cm	18.9777(3)	8.1155(1)	2531.4 (1)	3
(3) Same sample No. 1	P6₃cm	18.989 (1)	8.113 (1)	2533	15
(4) DPA-VPI-5, Hydrated	P6₃cm	18.997 (1)	8.123 (1)		16
Dehydrated	P6₃cm	18.524 (1)	8.332 (1)		
(5) AlPO₄-54, as synth.	P6₃/mcm	19.009 (2)	8.122 (1)	2541.6	17
Dehydrated	P6₃/mcm	18.549 (1)	8.404 (1)	2504.1	
(6) AlPO₄-8	Cmc2₁	33.29 (2) b=14.76 (2)	8.257 (4)		9
(7) VPI-5	P6₃	18.9752(1)	8.1044(1)		6

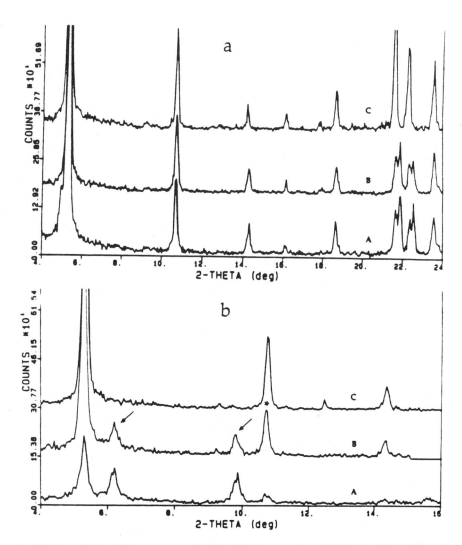

FIGURE 19-5. XRD patterns (4-24° 2θ) for samples JP5-66A, JP5-66B, and JP5-66C prepared in the DPTA system at 150° with different times of reaction. In (a), (A), (B), and (C) were reacted for 6, 15, and 22.5 h, respectively. These X-ray patterns represent mixtures of H1 and VPI-5 (A and B) and pure VPI-5 in pattern C. (b) Shows the X-ray patterns for the same series of samples calcined at 500°C for 4 h. Arrows indicate the appearance of the AlPO4-8 phase and the asterisk indicates a distinctive peak of VPI-5 used to estimate its percentage in the mixture.

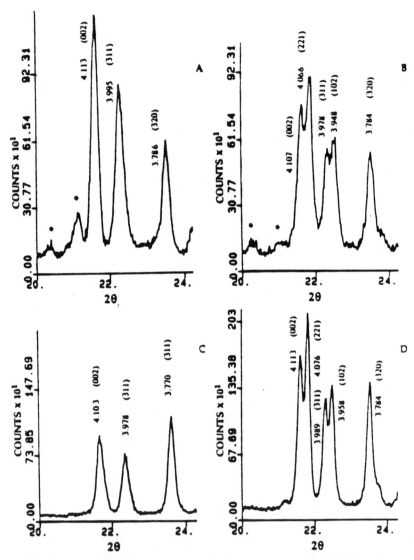

FIGURE 19-6.   A portion of the XRD patterns (at 20-24° 2θ) indicating changes in intensity and splitting of peaks. (A) VPI-5 as synthesized (JP5-21A, DPTA). Peaks marked with asterisks indicate AlPO₄-11. (B) Same VPI-5 sample (JP5-21A) calcined and rehydrated. (C) H1 as synthesized (JP5-21C, TBA-DPTA system). (D) H1 as synthesized (JP5-64A, DPTA system). (E) H1 organic free from synthesis (JP5-93B, DPA system). Asterisks indicate peaks of AlPO₄-11.

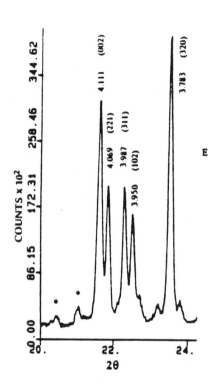

FIGURE 19-6 (continued).
(E) H1 organic free from synthesis
(JP5-93B, DPA system). Asterisks
indicate peaks of AlPO4-11.

spacings of prominent reflections in the 22-24° region of 2θ. We note that for both VPI-5 and H1, the a-axis decreases and the c-axis increases in length on dehydration. Occluded amines appear to make no difference in the cell parameters as observed by comparison of samples JP5-44A which contains 9.1% amine, and JP7-59A which contains only H2O. However, there does appear to be a significant difference in unit cell dimensions between H1 and VPI-5. The a- and c-axes for VPI-5 are slightly larger than those of H1 which may reflect the absence of some phosphate groups from the H1 lattice. This difference is not present in the dehydrated samples as the unit cell volumes become equal.

Listed in Table 19-4 are the unit cell dimensions for the wide-pore aluminumphosphates which have appeared in the literature. In general the samples referred to as VPI-5 have cell dimensions which lie between our values for H1 and VPI-5.

The X-ray powder pattern of a well crystallized sample of AlPO4-8, prepared as described in the experimental section, is given in Figure 19-7. It has been suggested that in this preparation H1 formed first and then this phase converted to AlPO4-8 on further contact with the hydrothermal solution. Strong evidence that this is not the case can be inferred from indirect observations. First, in no case where we prepared H1, in the DPA system up to 30 h of

Table 19-2. Effect of calcination of several samples with the H1/VPI-5 phase prepared in the DPTA system with different times of reaction at 150°C.

Sample	Reaction	Calcination		Thermal stability		
	time[a] (h)	Temp. (°C)	Time (h)	Initial phases	Final phases	Estimated %[b]
JP5-66A	6	500	2	H1/VPI-5	VPI-5	85
					$AlPO_4$-8	15
JP5-66B	15	500	4	H1/VPI-5	VPI-5	60
					$AlPO_4$-8	2540
JP5-69B	20	500	5	H1/VPI-5	VPI-5	90
					$AlPO_4$-8	10
JP5-66C	22.5	500	4	H1/VPI-5	VPI-5	100
		650	15		VPI-5	100
JP5-69C[c]	24	500	4	H1/VPI-5	VPI-5	-
				H3	$AlPO_4$-8	-
JP5-64A[c]	26	500	4	H1/VPI-5	VPI-5	50
				$AlPO_4$-11	$AlPO_4$-8	50
JP5-64B	26	500	5	H1/VPI-5	VPI-5	50
					$AlPO_4$-8	50

[a]Time of reaction is given as total time. For the type of reactors used (stainless steel 316) 2 h are required to reach the temperature in the range 150°C. Calcination times refer to steady state conditions.

[b]The estimated percentages of each phase were obtained from the areas under the peaks near 10° 2θ in figure 19-5 as $I_A = \dfrac{I_A}{I_A + I_B}$ where A and B refer to H1 and VPI, respectively.

[c]Samples JP5-69C and JP5-64A contain small amounts of H3 and $AlPO_4$-11, respectively.

reaction time, did we ever obtain evidence for the presence of $AlPO_4$-8. Second, $AlPO_4$-8 could be obtained by hydrothermal reaction in 6-12 h time, but with less crystallinity than shown in Figure 19-7. The real difference in which product is obtained by the hydrothermal synthesis has to do with the preparation of the gel. We found that a highly viscous gel, prepared by rapid addition of 7 or more ml portions of H_3PO_4 to the psuedoboehmite, followed by proper aging, yields $AlPO_4$-8 directly. This compound, unlike H1, is thermally stable.

Structure Analysis. We have initiated a structure analysis of dehydrated samples of VPI-5 and H1 from quantitative X-ray diffraction powder data. In the case of dehydrated VPI-5 the positions of all the Al and P atoms as well as those of two oxygen atoms were determined from an E-map prepared with 41 unambiguously indexed reflections. The remaining atoms were found in Fourier difference maps and the total pattern Rietveld refinement was carried out in space group $P6_3cm$. Attempts to further refine the structure in $P6_3$ produced poorer results. Full details will be published in a future paper. At this writing we present in a preliminary, way a comparison of results obtained for VPI-5 with those of a similar study on dehydrated H1. Dehydration of H1 was carried out at 150°C under vacuum to avoid formation of $AlPO_4$-8. The refinement has progressed sufficiently to reveal a significant difference in the two structures. The phosphate groups in the 6-membered rings in H1 were found to be disordered. The thermal parameters for these atoms were inordinately high and bond distances and angles were outside the range acceptable for tetrahedral coordination. By placing the phosphate group in the same position as found in dehydrated VPI-5 we were then able to find a second phosphate group, at half occupancy, in a new position in the difference map. This disorder may result from the fact that there is a deficiency of phosphate in the crystals. This is so because no extra framework aluminum was found and the Al-P ratio was greater than 1.

Solid State MAS NMR. In Figure 19-8 we show the NMR spectra of ^{31}P for VPI-5 at various stages of hydration. In the driest sample (less than 0.5% H_2O), six major resonances are present, as labeled by the letters a through f. As water is added the spectra change by disappearance of some peaks and the growth of others that were shoulders. The fully hydrated sample consists of three resonances which are now explained (McCusker et al. 1991) on the basis of the correct space group being P6₃. Two of these peaks appear as shoulders (a, d) to larger peaks in Figure 19-8 (at low water contents). We assume that in the fully dehydrated VPI-5 these shoulders would not be present, as they grow larger with increased water content. Their removal from the pattern leaves 3 major peaks remaining in the spectrum. However, as indicated in the previous section we were able to refine the dehydrated VPI-5 structure in space group P6₃cm, which requires only 2 unique ^{31}P peaks in a 2:1 ratio. Therefore, some drastic structural change is necessary to account for the three peaks. We have assumed that on complete dehydration some of the Al-O-P bonds break to form P=O or P-OH, Al-OH groups. The subsequent healing of these bonds to reestablish the Al-O-P linkage, probably through Al-OH and P-OH intermediates, in the presence of water accounts for the NMR spectral changes.[18]

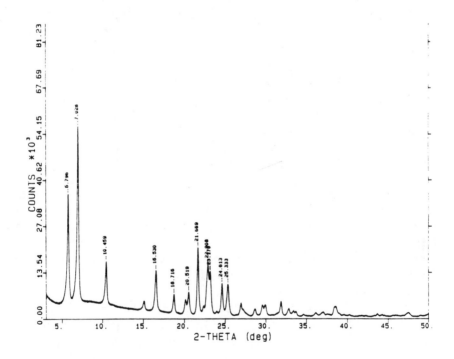

FIGURE 19-7. X-ray diffraction pattern of AlPO₄-8 synthesized from viscous gel containing DPA @ 125° for 22 h.

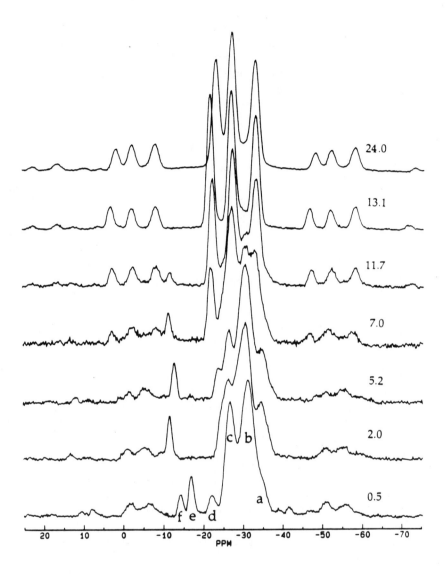

FIGURE 19-8.   The ^{31}P MAS NMR spectra of VPI-5 samples at different levels of water content. The percent water content is given at the right above the spectrum to which it refers.

The NMR spectra for H1 are similar to those for VPI-5 except that peak e is somewhat larger and peaks b and c are more equal in intensity. These differences may result from the phosphate deficiency in the H1 lattice and/or the disorder of the phosphate groups. Otherwise, the changes in the spectra with increasing water are similar.

Discussion

The number of variables to be controlled in order to obtain any one of the three aluminum phosphates discussed here is quite large. One may cite as the important variables to consider the gel viscosity, the choice of amine and the time and temperature of reaction. Some of the factors associated with gel viscosity are the concentration of H_3PO_4, the total water content, the way in which the alumina and acid are combined and aging of the gel. In our study we obtained excellent results with three amines; dipropylamine for synthesis of H1 or $AlPO_4$-8, a mixture of tributylamine (TBA) and dipentylamine for H1, and dipentylamine for VPI-5. Given the very large number of variables to control our approach was to choose a particular amine, fix the composition ratio and water distribution, and then vary gel viscosity, aging time, temperature and reaction time. H1 was prepared over a broad range of conditions and judging from the reported properties this phase appears to be the most common one prepared by other. For example, many workers have now reported the thermal transformation of what is referred to as VPI-5 to $AlPO_4$-8.[11] In contrast very few workers have obtained the thermally stable VPI-5. It is only formed at a more elevated temperature, ~150 °C, while H1 can be prepared in the temperature range of 120-150 °C. Other differences in the preparation methods are notable. For VPI-5 a thin, fluid gel, as prepared by dropwise addition of H_3PO_4 to psuedoboehmite, is best. This step is followed by a heat pretreatment before a short aging period and then amine addition, in our case DPTA.[5,12]

The X-ray study of dehydrated H1 and VPI-5 shows that both have the framework topology shown in Figure 19-1. Where they differ structurally is in the deficiency of phosphate groups in H1 and the disorder of the phosphate groups in the six membered rings. The thermal and hydrolytic instability of H1 apparently stems from this deficiency of phosphate. These differences in behavior between the two preparations are sufficiently important to require that a distinction be made in their naming. We have proposed[5] that the phosphate deficient, thermally unstable phase be referred to as H1 no matter what its origin, and the stoichiometric thermally stable phase be named VPI-5. This procedure will avoid confusion in the future literature.

In the dehydrated condition both H1 and VPI-5 X-ray patterns fully conform to the space group P6$_3$cm and not the lower symmetry group P6$_3$.[6] This observation is of consequence in connection with the interpretation of the ^{31}P NMR spectra of the dehydrated phases. There are at least 3 and possibly 4 resonance peaks in the spectrum of either phase. However, in space group P6$_3$cm only two separate environments are possible for phosphorus. Therefore, it is necessary to propose some structural change in the framework upon complete dehydration to account for the complexity of the NMR spectra.

It is altogether possible that H1 is a precursor of VPI-5. H1 can form under a variety of conditions in very short time of hydrothermal treatment. When DPTA was used at a temperature of 150 °C H1 was formed first and this phase was converted to VPI-5 over a 20 h period. More experiments along these lines need to be carried out.

AlPO$_4$-8 was prepared under conditions similar to those that would produce H1. The major difference is the greater viscosity of the gel. While H1 can be formed from gels with a broad range of viscosities, thicker, less fluid gels favor the formation of AlPO$_4$-8. The importance of the gel viscosity is such that uncovering the mechanism of the reaction between psuedoboehmite and phosphoric acid may be the key to controlling the synthetic procedures.

Acknowledgment. We wish to thank the Regents of Texas A&M University for support of this research under the Material Science and Engineering Program Commitment to Texas.

References

1.   Davis, M. E., Saldarriaga, C., Montes, C. , Garces, J. M. and Crowder, C. 1988. *Nature* , 331: 698 .
2.   Davis, M. E. , Saldarriaga, C. , Montes, C. , Garces, J. M. and Crowder, C. 1988. *Zeolites* , 8: 362 .
3.   Rudolf, P. R. and Crowder, C. 1990. *Zeolites* 10: 163.

4.   Grobert, P. J., Martens, J. A. Balakrishnan, I., Mertens, M. and Jacobs, P. A. 1989. *Appl. Catal.*, 56: L21 .
5.   Perez, J. O , McGuire, N. K. and Clearfield, A. 1991.*Catal. Lett.* 8: 145.
6.   McCusker, L. B., Baerlocker, Ch., John, E. and Bulow, M. 1991. *Zeolites* , 11: 308.
7.   Davis, M. E., Montes, C., Hathaway, P. E. and Garces, J. M. 1989. *Zeolites: Facts, Figures, Future*, Elsevier Sci. Publ. B.V., Amsterdam, pp 199-214.
8.   d'Yvoire, F., 1961.*Bull. Soc. Chim.*, Fr., 1762 .
9.   Dessau, R. M., Schlenker, J. L. and Higgins, J. B. 1990. *Zeolites*, 10: 522 .
10.   Duncan, B. Szostak, R., Sorby, K. and Ulan, J. G.1990. *Catal. Lett.* 7: 367.
11.   Stöcker, M., Akporiaye, D. and Lillerud, K.-P. 1991. *Appl. Catal.* 69: L7; Maistriau, L., Gabelica, Z., Derouane, E. G., Vogt, E.T.C. and van Oene, J. 1991. *Zeolites*, 11: 583; Szostak, R. Ulan, J. G. and Gronsky, R. 1990. *Catal. Lett..*, 6:209. Prasad, S., Gunjikar, V. G. and Balakrishnan, I. 1991. *Thermochim. Acta*, 191: 265; Liu, X., He, H. and Klinowski, J. 1991. *J. Phys. Chem.*, 95: 9924.
12.   Perez, J. O., Ph. D. Dissertation, Texas A&M University, August 1991.
13.   See Rudolf, P. R. and Clearfield, A. 1989. *Inorg. Chem.*, 28: 1706 for details.
14.   (a) Generalized Structure Analysis System (GSAS), A. Larson, R. B. Von Dreele, LANSCE, Los Alamos National Laboratory, copyright 1985-88 by the Regents of the University of California. (b) Sheldrick, SHELX76, University of Cambridge, Eng. 1976. (c) TEXSAN:TEXRAY Structure Analysis Package for Single Crystal Data; Molecular Structure Corp., The Woodlands, TX 1986; revised 1987.
15.   Crowder, C. E., Garces, J. M. and Davis, M. E. 1988. *Adv. X-ray Anal.*, 32: 507
16.   Annen, M. J., Young, D., Davis, M. E., Cavin, C. O. and Hubbard, C. R. 1991. *J. Phys. Chem.*, 95: 1380.
17.   Richardson, J. W., Jr., Smith, J. V. and Pluth, J. J. 1989. *J. Phys. Chem.*, 93: 8212.
18.   Perez, J. O., Chu, P.-J. and Clearfield, A. 1991. *J. Phys. Chem.*, 95: 9994.

20

In Situ Monitoring of the Degree of Transformation of VPI-5 to AlPO4-8 at Various Temperatures by Multinuclear Solid State NMR and DTA

Z. Gabelica, L. Maistriau and E.G. Derouane
Department of Chemistry, University of Namur, B-5000 Namur, Belgium

Combined TG-DTA-DTG with ^{31}P- and ^{27}Al-NMR provided choice techniques to follow the in situ dehydration and solid state transformation of VPI-5 to AlPO4-8, either under progressive heating or isothermally at different temperatures.

On a conventionally prepared (DPA)-VPI-5 (Type A synthesis) the transformation occurs readily and irreversibly as soon as about 95% of the total water structure is removed by heating, whatever the heating rate or the final temperature. The transformation can be visualized by a characteristic DTA exothermic effect. In the case of a more homogeneous sample prepared from the same ingredients mixed in a different way (Type B synthesis), neither DTA nor ^{31}P- and ^{27}Al-NMR detect any transformation. XRD confirms that the VFI topology remains unaltered even after complete dehydration and further heating at high temperatures. The stability of the initial structure was interpreted in terms of purity of the as-synthesized materials, probably closely related to their lattice defect density. These latter were supposed to induce the hydrolytic cleavage of the P-O-Al bonds when most of the water is removed.

Both materials also undergo a reversible destruction of the water triple-helix configuration around 60-70°C and independantly of the actual water content, as ascertained by NMR data but not seen by DTA.

INTRODUCTION

Since the first report of the synthesis procedure of the extra-large-pore molecular sieve VPI-5 by Davis et al. (1), a series of recent publications have dealt with detailed experimental conditions leading either to a highly crystalline VPI-5 (2-11) or to its silicon containing analogue (6, 11-16) or also to the organic-free VFI (9, 17, 18) presumed to be analogous to the aluminophosphate hydrate H1 (17), already prepared three decades ago by d'Yvoire (19).

The crystallization of VPI-5 is currently achieved in the presence of di-n-propylamine (DPA), as initially suggested by Davis et al. (2, 3), but a variety of other (poly)alkylamines (5, 7-10, 20) and quaternary alkylammonium ions, such as TBAOH (2, 3, 7), can also be efficiently used. Except in a recent work describing a dipentylamine-bearing VPI-5 phase (9), the organics are not found incorporated into the pore volume of the VFI structure, suggesting that they most probably act neither as templates nor as pore fillers in the synthesis. They may, for example, function as pH moderators (3, 21) or possibly have other functions (7, 10), such as to suppress completely co-crystallizing phases like H3 (17).

The structure of the as-synthesized (DPA)-VPI-5 actually confirms that we are dealing with an aluminophosphate hydrate (22). Seven water molecules are differently positioned in the pore volume of the unit cell: four of them form a H-bonded chain between the octahedrally coordinated framework Al atoms, so as to create a triple helix of H_2O molecules that follow the 6_3 screw axis in the 18 ring channel, and two others complete the octahedral coordination around the framework Al located between the fused 4-rings, while the last molecule, located near the center of the channel, is supposed to link the helices to one another.

Such a complex structuring of water is a priori related to the (hydro)thermal stability of VPI-5. In fact, the total amount of water (about 25 wt. %) can be totally removed as soon as the precursor is heated above 100°C (3, 7, 10, 16). However, the dehydrated material either still preserves its VFI framework topology that then remains remarkably stable at high temperatures (600-800°C) (1, 23), or undergoes a polymorphic transformation to $AlPO_4$-8, even under mild thermal treatments as low as 100°C (1, 6, 16, 17, 20, 24). $AlPO_4$-8, synthesized hydrothermally from gels having molar compositions similar to those used for the VPI-5 synthesis (25), is a microporous material involving pores circumscribed by 14 T atoms (26, 27) and structurally related to (but definitely different from) VPI-5 (15).

The ability of VPI-5 or H1 to convert into $AlPO_4$-8 by heating was shown to be a property unique to the VFI structure (18), so the straightforward hydrothermal preparation of $AlPO_4$-8, as reported by Wilson et al. (25) or by Perez et al. (9), might appear doubtful. Indeed, in both cases, $AlPO_4$-8, as identified by X-ray powder diffraction, was obtained after drying the corresponding "precursor", an operation now known to induce the phase transition (5, 16, 26).

Complete or partial conversion of VPI-5 into $AlPO_4$-8 achieved by heating the as-synthesized VPI-5 under various experimental conditions (16, 20, 26) seems also to depend markedly on the method by which the precursor is synthesized (7, 9, 17, 20) and, more particularly, on the nature of the organics used. For example, (DPA)-VPI-5 completly converts to $AlPO_4$-8 at 100°C (2, 16, 20), while the use of other amines such as triisopropylamine + TMAOH (15, 20) impedes in some extent the degree of conversion. The use of TBA^+ ions (2, 3, 7) or pentylamine (9) in the synthesis batch seems to completely prevent such a transformation, and only stable VPI-5 remains after heating.

Nevertheless, contradictory results were also recently mentioned; Vinje et al. (20) reported that (TBA)-VPI-5 can also undergo transformation under certain conditions, while, by contrast, (DPA)-VPI-5 may also show a great thermal stability, when prepared under carefully controlled conditions (5, 10, this work). Further experiments have shown that when extensively washed, VPI-5 does not undergo any transformation upon heating to at least 600°C (7, 10), thus suggesting that traces of residual organic material occluded within the VPI-5 crystals, and not detected by conventional analytical techniques such as thermal analysis or ^{13}C NMR, may play a role in the stability of the precursor. Furthermore it was confirmed that, for a given as-synthesized VPI-5 material, the extent of conversion very strongly depends on the various post-synthesis treatments and, more precisely, on the way under which water is removed (16, 20, 26). Preliminary experiments in our laboratory have shown that the VPI-5 structure can remain unaltered in aqueous medium even at temperatures close to 100°C, while rapid transformation is initiated when only traces of water remain at that temperature, confirming the observation that the presence of moisture is a factor favoring the transformation (20, 26). By contrast, when dehydration is achieved at low temperatures (20 to -40°C) either by evaporation or by washing with other solvents, the VFI topology stays unaltered, even in the presence of residual water (16).

The apparent contadictory observations that the degree of VPI-5 to AlPO$_4$-8 conversion depends either on the precursor synthesis conditions or on subsequent post-synthesis treatments can be rationalized if one admits that the prerequisite condition for the transformation to occur consists in the hydrolysis of the Al-O-P linkages followed by their further recondensation in forming new bridges from adjacent fused 4 rings, as proposed by McCusker et al. (22). The hydrolysis can be assisted either by the presence of impurities or traces of organic matter, or, more generally, can depend on the actual concentration of POH defects in the as-synthesized precursor. Obviously, the percolation of water through the VPI-5 framework at defined temperatures will induce the first hydrolytic cleavages.

The aim of this paper is to delineate conditions that favor or impede the VPI-5 to AlPO$_4$-8 transformation, by following in situ the behavior of water (by thermal analysis) in parallel with consequent short-range structural changes (by ^{31}P- and ^{27}Al-NMR), during a progressive dehydration under various conditions of two differently synthesized (DPA)-VPI-5 samples.

Upon finding out whether synthesis conditions or post-synthesis treatments are predominant factors that induce the polymorphic rearrangement, we wish, in a further step, to define conditions under which a thermally stable-extra-large pore VPI-5 can be prepared or stabilized.

EXPERIMENTAL

Synthesis

VPI-5-A was prepared by following a procedure derived from conditions described in.(1): 150 g of alumina (Catapal B) were slurried in 435 g of demineralized water, to which a solution of 237 g of H$_3$PO$_4$ (85 wt. %) in 218 g of demineralized water was added under stirring. The mixture was aged for 3 h before the addition of 7.7 g of SiO$_2$ (Degussa Aerosil 200) and 109 g of di-n-propylamine (DPA). The resulting gel was aged for another 2h and then left to crystallize at 130°C (temperature monitored inside the autoclave) for 18h. The washed (several times with portions of cold water) and dried (at ambient temperature) product contained 18.1 wt. % Al, 17.1 wt. % P and 0.5 wt. % Si (atomic absorption analysis of the hydrated precursor).

VPI-5-B was obtained from a gel having the same molar composition, except that no silicon was added to the mixture. The order of mixing of the reactants was also different. Alumina (Catapal-B) was added slowly under vigorous stirring to the phosphoric acid solution with the aim of preparing a more homogeneous $AlPO_4$ mixture. This latter was then aged for 3 h at 20°C before the addition of DPA. The final gel was further aged for 2 h. The crystallization time and temperature were respectively 16 h and 125°C. The final crystalline product was washed and dried as described above (VPI-5-A). It exhibited a pure 100% crystalline VPI-5 stucture (XRD).

Characterization

NMR spectra were obtained on a Bruker MSL400 spectrometer operating at a field of 9.39 T. The ^{31}P-NMR spectra were recorded at a frequency of 161.9 MHz with a magic angle spinning rate of 8.5 kHz. Chemical shifts were measured relative to $NH_4H_2PO_4$, used as an external reference. The shift correction relative to H_3PO_4 (85 wt. %) is -3.24 ppm. ^{31}P-NMR spectra were usually obtained by accumulating 50 free induction decays produced by 2.5 μs pulses, followed by a 10 s relaxation decay. All spectra were recorded with ^{1}H decoupling.

The ^{27}Al NMR spectra (104.25 MHz) were obtained with a magic angle spinning rate of 10 kHz. The chemical shifts were determined relative to aqueous $Al(NO_3)_3$, used as an external reference. In this case, the recycle delay and pulse length were 0.1s and 1μs, respectively. Semi-open spinners were used. For this purpose, the 4 mm rotor was closed with a cover in which a hole of 0.4 mm diameter was drilled. The rotor was placed in the preheated probe-head and maintained at the desired (variable) temperatures for various periods of time. NMR spectra were recorded at regular time intervals.

The nature of the products was determined by X-ray powder diffraction (XRD) using a Philips P.W. 1349/30 diffractometer (Cu Kα radiation).

Thermal analysis measurements were performed on a ST 780 Stanton-Redcroft thermobalance (combined TG-DTA-DTG system). The samples were heated in platinum open crucibles containing about 0.05 g of sample, in static air, at various heating rates. Variable heating programs, including isotherms, were chosen (see reults); α Al_2O_3 was used as the DTA reference.

RESULTS AND DISCUSSION

Sample VPI-5-A

^{31}P and ^{27}Al variable temperature NMR

^{31}P-NMR is presently considered as the most sensitive technique giving information on the local structure and structural modifications involving the T-framework elements (T = P or Al). In situ ^{31}P-NMR spectra of the sample VPI-5-A were recorded as a function of temperature and time (Fig. 20-1). From ambient temperature up to about 60°C, the spectra exhibit the three typical resonances located around -27, -31 and -37 ppm and showing an intensity ratio of 1:1:1 (see, e.g., spectrum A recorded at 40°C) as often reported in the literature (4, 8, 10, 14, 22, 24). These lines are currently assigned to the three crystallographically distinct P atoms in the structure that have been shown to possess a low (P6$_3$) symmetry (22) .

When the temperature reaches 60°C, slight modifications in the initial spectrum are noticeable. The three lines progressively move, the one at -27 ppm towards lower chemical shift values while the two other move downfields (spectrum B). The spectrum recorded at 80°C (in our conditions) shows only two resonances with an intensity ratio of 2:1. The resonance at -30.8 ppm is now probably the result of the coalescence of the two initial lines at -27 and -31.3 ppm, while the third line remains positioned at -36.9 ppm (spectrum C). We would agree with Cauffriez (10) in assuming that the triple-helix structure of water molecules is no longer rigidly held in the structure by the hydrogen bonds. Indeed, our thermal analysis data (see below) indicate that some of the total water is already released from the structure under these conditions. The remaining water, more randomly positioned in the channels because of the thermal agitation, probably induces an apparent symmetry increase, as far as the P atoms in the 6-MR are concerned, thus rendering them structurally "equivalent". One can also assume that under these "partly dehydrated" conditions, the framework finds a more stable conformation (29) so that NMR can only distinguish two types of sites (22). Note that the simple thermal agitation of all the water molecules (without any necessary water loss) can be a sufficient reason to cause this apparent increase in the framework symmetry (see below). Finally, when the sample preheated at 80°C within the semi-open rotor is cooled down to ambient temperature, the initial ^{31}P-NMR spectrum is completly restored, confirming the perfect reversibility of this dehydration/ rehydration effect.

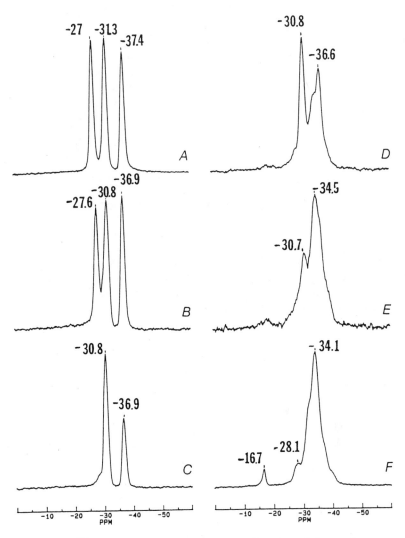

FIGURE 20-1 ^{31}P NMR spectra of VPI-5-A as a function of time and temperature.
A: 40°C, B: 60°C, C: 80°C, D: maintained at 110°C for 1 h, E: maintained at 110°C
for 2h, F: maintained at 110°C for 3 h.

Above 80°C, the ^{31}P-NMR spectrum remains unchanged until the temperature reaches 110°C (spectrum D). As the sample is maintained at that temperature for longer periods of time, the NMR pattern undergoes important modifications (spectra D, E, F). The two lines characterizing a VPI-5 structure progressively disappear while another strong line, identifying AlPO$_4$-8 (also detected by XRD), appears at about -34 ppm and increases in intensity with time. Finally, only one broad signal at -34.1 ppm with a shoulder at -28.1 ppm and a small additional resonance showing up at -16.7 ppm characterize the product heated at 110°C for 3 h (spectrum F), which thus corresponds to dehydrated AlPO$_4$-8.

On cooling down to ambient temperature, followed by a complete rehydration over a saturated NH$_4$Cl solution, the sample exhibits a ^{31}P-NMR spectrum showing three resonances with an intensity ratio 1:2:6 (Fig.20-2), which now fully characterize the structure of (hydrated) AlPO$_4$-8, as confirmed by XRD. This VPI-5 to AlPO$_4$-8 conversion, currently reported and described in the literature (see introduction) was already subject to a thorough ^{31}P-NMR investigation by our group (11). The assignment of all the resonances that progressively change along with the conversion was presented and a model for the transition, based on the evolution of the relative intensities of all the NMR lines as a function of the degree of conversion, was proposed (11).

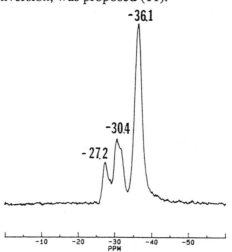

FIGURE 20-2 ^{31}P NMR spectrum of AlPO$_4$-8 rehydrated over a saturated NH$_4$Cl solution for 48 h, after a preliminary in situ NMR dehydration at 110°C for 3 h (experiment F, Fig.20-1).

The ^{27}Al-NMR spectra recorded in similar conditions confirm the ^{31}P-NMR results. They are shown in Fig.20-3. The as-synthesized sample (DPA)VPI-5-A is characterized by three lines, located at 40.4, 6.2 and -19.4 ppm (spectrum A). The line at 40.4 ppm is attributed to tetrahedrally coordinated aluminum atoms, while octahedral aluminum species are characterized by the broad line at -19.4 ppm. Some of the framework Al atoms have an octahedral configuration, involving four framework oxygens and two water molecules. Indeed, McCusker et al. (22) have recently shown that the water molecules from the triple helix interact with one third of the aluminum atoms of the unit cell, achieving thereby their octahedral coordination.

Recently, Wu et al. (29), using the new double-rotation ^{27}Al-NMR technique, confirmed the presence of the three resonances with equivalent intensities and belonging to two tetrahedrally and one octahedrally coordinated Al atoms in the as-synthesized (DPA)VPI-5-A structure.

When the sample is progressively dehydrated in the variable temperature ^{27}Al-NMR experiment, the line at -22.1 ppm disappears (Fig. 20-3, spectrum C, recorded at 80°C). A shoulder also appears at about 35 ppm and its intensity increases with time (spectra D and E). At the end of the VPI-5/AlPO$_4$-8 transformation (spectrum F), the +40 ppm resonance observed at ambient temperature and characterizing the as-synthesized VPI-5 completely dissapears. Only a broad signal is observed in that spectral region (at about 36.6 ppm) after 3 h of heating at 110°C (spectrum F). This line is assigned to Al(4P) configurations belonging to the dehydrated AlPO$_4$-8 structure.

Finally, the small resonance at 6 ppm always remains present and its intensity actually shows a slight increase during the heating experiment. Its assigment is not yet unambiguous. It possibly could correspond to Al atoms belonging to traces of unreacted pseudo-boehmite alumina, thus indicating that the crystallization of VPI-5-A was not 100% complete (thereby confirming the XRD data). Indeed, Catapal B alumina is characterized by a ^{27}Al-NMR spectrum consisting of one single resonance at 6.5 ppm. If our assignment is correct, the slight increase of this line during the variable temperature experiment could therefore indicate a progressive amorphization of the VPI-5/AlPO$_4$-8 material, with the consequent formation of some amorphous alumina.

Furthermore, the ^{27}Al-NMR data show that octahedral Al atoms are nearly totally converted to tetrahedral Al species, indicating that the AlPO$_4$-8 is totally dehydrated under these conditions.

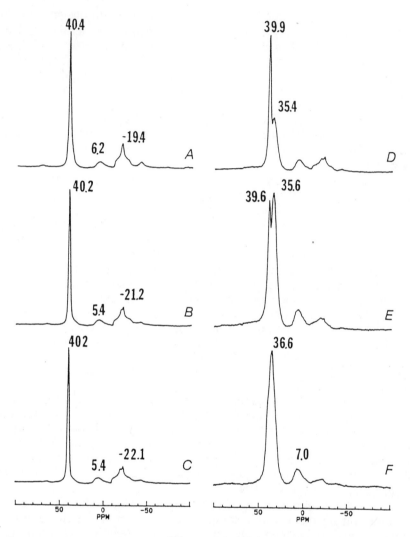

FIGURE 20-3   Evolution of the ^{27}Al NMR spectra of VPI-5-A as a function of temperature and time. A: 40°C, B: 60°C, C: 80°C, D: maintained at 110°C for 1 h, E: maintained at 110°C for 2 h, F: maintained at 110°C for 3 h.

In a completely closed rotor, the behavior of the sample is markedly different. The spectrum of the as-synthesized VPI-5 involving the usual three lines of equal intensity ratio (Fig. 20-4, spectrum A) is progressively replaced by the spectrum showing the two signals with a ratio 2:1, as soon as the temperature reaches 80°C (spectrum C). The spectrum remains unchanged in the range 80-120°C, even if the sample is maintained in these conditions for prolonged periods of time. When the sample is cooled down to ambient temperature, the intial spectrum A (spectrum F in the figure) is completely restored. This observation suggests that the apparent increase in the framework symmetry that renders the two phosphorus atoms in the 6 MR equivalent is directly or indirectly related to the increased motion of the water molecules at higher temperatures and is not necessarily due to a partial loss of water, as observed in the open system. The heating-cooling treatment in the closed system was repeated several times without showing any further changes. In particular, $AlPO_4$-8 was never detected, even in traces, at any stage of the heating-cooling cyclic process.

The comparison of both series of NMR data conducted in closed and semi-open containers suggests that the transformation of VPI-5 to $AlPO_4$-8 depends on the actual amount of water present in the system at a given temperature rather than on the temperature itself, as already suggested by our preliminary investigations (16).

A complementary ^{31}P-NMR investigation of the "stability" of VPI-5-A rapidly heated at 100°C in a completely open rotor, and maintained at that temperature for long periods of time, indicates that the conversion to $AlPO_4$-8 starts at least after one hour in these conditions, thus probably when enough water has been released from the structure, so as to induce the transformation.

The above conclusions needed experimental confirmation, namely a simultaneous monitoring of the amount of structural water as a function of heating temperature or time, in parallel with the corresponding structural changes. This was performed by combining the thermal analysis (TG-DTA) techniques with NMR.

Thermal analysis.

Sample VPI-5-A was heated in an open crucible under static air from ambient temperature (20°C) to 120°C at a heating rate of 1.5°C min[-1] and then maintained isothermally at 120°C. The corresponding TG (weight loss) and TDA (thermal effects) traces are shown in Fig. 20-5. On the same figure are reported the temperature profile used for the ^{31}P-NMR

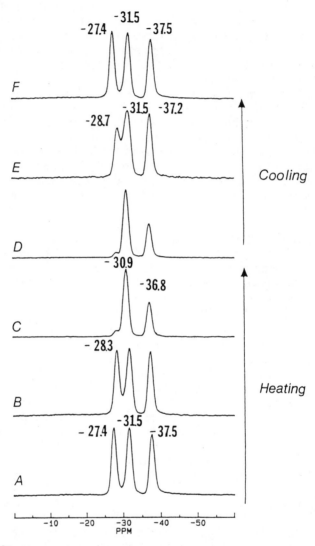

FIGURE 20-4 ^{31}P NMR spectra of VPI-5-A as a function of temperature (closed rotor). A: 20°C, B: 60°C, C: 80°C, D:120°C, E: 70°C, F: 20°C.

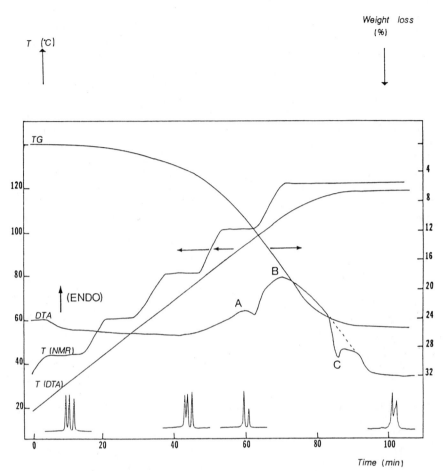

FIGURE 20-5   Comparison between TG and DTA characteristics of sample VPI-5-A heated in the thermobalance from 20 to 120°C (1.5°C min^{-1}) in static air and ^{31}P-NMR data recorded for the same sample heated in situ under similar conditions. Both temperature traces as indicated.

experiments conducted in a semi-open system as well as the corresponding spectral changes. Note that the temperature profile of the NMR experiment is slightly different from the continuous temperature increase of the thermobalance furnace. The apparent higher values recorded during the NMR experiments are actually those measured by a thermocouple in contact with the hot N_2 flow that is supposed to heat the sample inside the rotor. In fact the actual temperature experienced by the sample is probably a little lower, due to limited conductivity of the rotor walls, and may be close to the temperature measured by the thermobalance thermocouple in direct contact with the sample. Another difference is the stepwise profile of the NMR temperature illustrating the necessary upholding of the corresponding isotherm for about 10 min during the spectra recording. Except for these slight differences, we can suppose that the heating conditions in both systems are equivalent. The same should be postulated for the loss of water from holders (rotor and TG crucibles) that have different aperture diameters. In the worst case, the water release is somewhat retarded in the case of NMR experiments, with respect to the TG-DTA setup, which involves widely open crucibles.

The weight loss recorded is continuous and completely achieved after the furnace was heated at 120°C for about 30 min. Similar amounts of water lost by heating various as-synthesized VPI-5 were reported in the literature (2, 7, 23). We have also checked that no further detectable weight loss occurs when the sample is heated above 120°C, (e.g. up to 600°C), therefore confirming that little (if any) water or residual DPA remains in the framework after the 120°C treatment, in agreement with other data (24).

The DTA trace exhibits interesting features. Firstly, two endotherms (noted A and B in Figure 20-5) are observed between 90 and 100°C, i.e., when about 35-40 wt. % of the total water has been released. DTG data have confirmed that A and B are really endothermic effects and that the (reversed) pseudo-peak between them is not an exotherm. This behavior can be interpreted in various ways. Firstly, we can deal with water residing in several types of structural environments and therefore released in several (here two) steps, as already observed by Davis et al. using DSC measurements (24). Secondly, we can imagine that this behavior illustrates two different rates of the water release, namely before and after the triple-helix water structure destruction at around 60-80°C. This explanation was excluded for two reasons. Firstly, ^{31}P-NMR data show the presence of only two phosphorus lines (intensity ratio 2:1), before the endotherm A is recorded. On the other hand, a purer VPI-5 precursor (sample B, see later), which also undergoes the triple-helix destruction above 60°C (^{31}P-NMR data), never showed such a DTA pattern. Finally, we could imagine that traces of DPA are in some extent oxidized (thereby possibly accounting for the presence of the "exotherm" between peaks A and B) and/or released in this temperature range. This interpretation was not retained, as an analysis of the gaseous phase released during the whole heating procedure proved to be water; no traces of amine could be detected by the automatic titrimeter assembly, following the procedure of Kerr and Chester (31).

The position of these peaks in time and temperature was actually not reproducible for various VPI-5 type A samples. Moreover, for some of them, only one endotherm is observed. In fact, these two endotherms were found to be influenced by the amount of residual H3 phase, co-crystrallized with the as-synthesized VPI-5. It is possible that H3 undergoes dehydration at a different rate, thereby influencing the DTA endothermic effect due to the global water release. On heating pure phase H3 and various pure VPI-5 precursors of type B (see later), mechanically admixed with H3, we could confirm that H3 indeed dehydrates at a fairly slower rate and that the admixtures present a similar DTA pattern as the one shown in Fig. 20- 5.

A second "break" in the DTA curve was observed at a higher temperature, i.e., when about 95% of the total water content were lost. Peak C (Fig. 20-5) illustrates an exothermic effect. It is attributed to the solid state reorganization of the VPI-5 framework yielding ultimately AlPO$_4$-8. Indeed, the typical ^{31}P-NMR pattern characterizing AlPO$_4$-8 is recorded just after the completion of the DTA exotherm (Fig. 20-5). This effect is not reversed upon cooling the dehydrated AlPO$_4$-8, even in the presence of moisture (to allow a partial water readsorption), suggesting that the VPI-5 to AlPO$_4$-8 transformation is not reversible under our conditions. Actually VPI-5 could never be detected again in the AlPO$_4$-8 samples obtained after the various dehydration experiments and restabilized in ambient temperature and humidity conditions for a long time.

To check whether the transformation occurs at a constant temperature (here about 115°C) or is rather a function of the residual water content in the preheated VPI-5, we have submitted the same sample to various heating programs in similar conditions and recorded, in each case, the temperature and the water loss, in parallel with DTA peak C. The results show that while the temperature characterizing this exotherm is variable with the heating rate, the residual water content left in the solid at that moment was always about 5% of the total water amount present in the as-synthesized material.

To further confirm that the temperature is not important, we have preheated the thermobalance furnace at different constant temperatures chosen between 80 and 150°C and introduced the as-synthesized VPI-5-A into the crucible which was very rapidly brought to the desired temperature after the prompt closure of the furnace. A typical TG, DTA, DTG temperature and time trace recorded for the 140°C isotherm is shown in Fig. 20-6. The following observation should be noted. Firstly, no DTA effect (corresponding to peak C) is recorded when the temperature reaches 140°C, i.e., after about 7 min. By contrast, the exotherm appears when about 95% of the total water is released, i.e. after about 13 min of total heating. The effect is strong (due to increased sensitivity of the DTA combined with a rapid temperature increase) and sharp, another argument in favor of the assignment of peak C (Fig. 20-5) to a real exothermic effect. By contrast, endotherms A and B are now hardly detected, because of the high dehydration rate of the sample. Finally, DTG shows that the rate of the water release is markedly decreased when the transformation starts. The final release of the remaining 5% probably occurs as soon as the transformation is achieved, this being illustrated by the second (small) TDG bump.

Figure 20-7 shows schematically the rate of the water loss from the same VPI-5-A sample maintained isothermally at various temperatures under the above-described experimental conditions.

The double circle of each curve corresponds to the moment when the exothermic DTA effect C is recorded. Again, it is clear that the transformation depends neither on time nor on temperature, but on the actual water content of the sample. Moreover, when the sample is heated isothermally at 80°C for more than 2 h, the amount of water that is then released (about 50%) is not sufficient to allow the transition to start. XRD as well as ^{31}P-NMR investigations of the corresponding residues perfectly confirm that $AlPO_4$-8 is formed for samples heated between 100 and 150°C while the one heated at 80°C for 2.5 h essentially presents the characteristics of the VPI-5 structure with, nevertheless, traces of $AlPO_4$-8. This suggests that the transformation may have been partly induced at 80°C and possibly continues to proceed slowly with time (the XRD and ^{31}P-NMR were recorded several days after the sample was heated at 80°C).

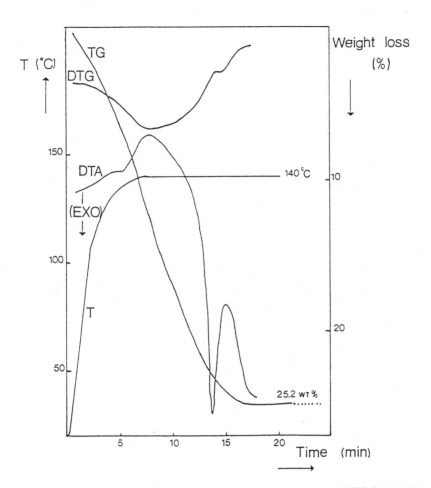

FIGURE 20-6   T (°C), TG, DTA and DTG traces recorded for sample VPI-5-A, held in the thermobalance furnace preheated at 140°C, for 25 min, in static air.

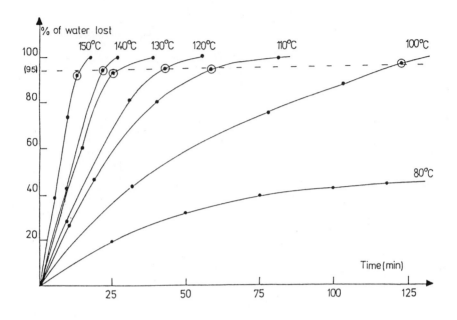

FIGURE 20-7   Rate of water loss for sample VPI-5-A maintained isothermally in the thermobalance preheated at various temperatures in static air. Double circles correspond to the moments at which the DTA exothermal effect due to the VPI-5 to AlPO$_4$-8 transition was recorded.

Sample VPI-5-B

^{31}P and ^{27}Al variable temperature NMR

VPI-5-B shows a very different behavior during the dehydration, as again investigated by in situ high-temperature NMR (semi-open rotor). Actually, the product was never found transformed into $AlPO_4$-8. The XRD pattern of the sample maintained at 110°C for 10 h in the open probe-head still corresponds to VPI-5.

Figure 20-8 shows the evolution, as a function of time, of the ^{31}P-NMR resonances of VPI-5-B maintained at 110°C. Initially, the ^{31}P-NMR spectrum recorded at ambient temperature (20°C) shows three lines with an intensity ratio of 1:1:1, similar to those observed in the case of VPI-5-A (spectrum not shown but similar to spectrum A, Fig. 1). At 110°C, the spectral changes are different than for VPI-5-A (Fig. 8, spectrum A). In addition to the two lines with a ratio close to 2:1 and located, respectively, at the chemical shifts of -31 and -36.3 ppm, an additional small peak, never observed so far, appears at -22.7 ppm. This peak progressively disappears with time while a very broad shoulder progressively arises in the same region of the spectrum (-25 ppm) (spectrum C). Another broad peak (-35 ppm) appears after 5 h of heating (spectra C and D) and progressively increases, as a function of dehydration (spectrum E). When the sample is completely dehydrated, only one broad asymmetrical line, peaking at about -34 ppm, is present (spectrum F). After this experiment, the sample was cooled down to ambient temperature (20°C). Part of the sample was then maintained at 20°C for a partial rehydration and the ^{31}P-NMR spectra were recorded after 86 h and 30 days, respectively. Another portion of the dehydrated sample was kept over a saturated NH_4Cl solution for three days, to achieve a complete rehydration. The ^{31}P spectra of the partially (86 h and 30 days) and totally (3 days over NH_4Cl) rehydrated sample are shown in Figure 20-9 (respectively spectra A, B and C). All spectra show three more or less separated lines, which are superimposed on a broad line. The chemical shifts of the three lines are characteristic of VPI-5 in agreement with XRD data, confirming that the VFI structure was always maintained, whatever the degree of rehydration. The X-ray patterns also indicate that the material is less crystalline.

Note that the initial 1:1:1 intensity ratio of the three ^{31}P-NMR lines is never completely restored, possibly because of a different organization of water in the structure, probably more random than in a triple helix configuration and thus affecting differently the three non-equivalent P atoms of the structure.

The actual amount of water reintroduced into the structure of VPI-5 after saturation over NH_4Cl for 3 days, as measured by TG, was as much as 28 wt.%, i.e., about 10% more than in the as-synthesized

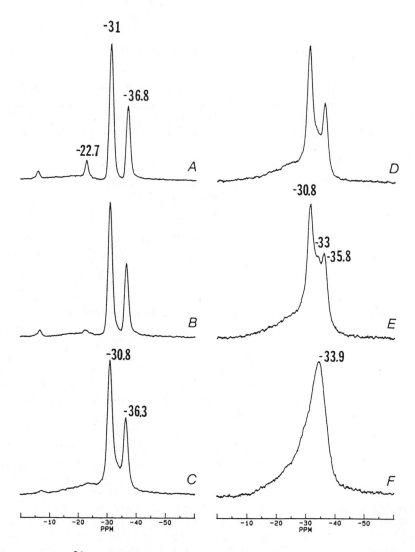

FIGURE 20-8 ^{31}P NMR spectra of VPI-5-B maintained at 110°C for various periods of time. A: 10 min, B: 1 h, C: 3 h, D: 5 h, E: 6 h, F:10 h.

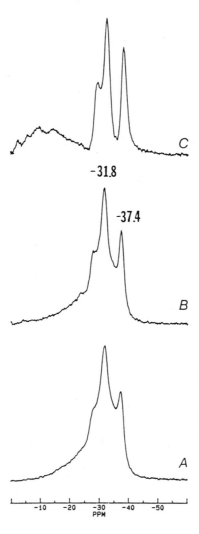

FIGURE 20-9 ^{31}P NMR spectra of sample VPI-5-B maintained in ambient air for 86 h (A) and 30 days (B), and of the same sample (B) completely rehydrated at ambient temperature over a saturated NH$_4$Cl solution for 3 days (C), after the variable temperature experiment described in Fig. 20-8.

material. Water is therefore probably more randomly (and more compactly) redistributed within the unit cell of VPI-5 than the water structurated in a triple helix configuration during the synthesis achieved at 125°C.

^{27}Al-NMR spectra were also recorded in situ after various periods of time, while the sample was maintained at 110°C (Fig. 20-10). Initially, the material is characterized by three lines at 39.2, -5.7 and -20.2 ppm. Upon dehydration, octahedral aluminum (-20.2 ppm) progressively disappears and the line at 39.2 ppm becomes asymmetrical (spectrum C) due to a shoulder appearing at higher field. For longer heating times, this line slightly shifts to higher fields and becomes symmetrical (spectrum F), but exhibits a larger width. At this stage, the octahedral aluminum species have almost completely disappeared. Tetrahedral Al species in the dehydrated material are therefore supposed to account for the broad signal located at 38.5 ppm. Similarly, the phosphorus species are also characterized by a broad ^{31}P-NMR signal centered near -31 ppm.

On the basis of these experiments, we cannot formulate a dehydration mechanism. We observe, however, that the removal of water leads to the disappearance of octahedrally coordinated aluminum atoms, while an additional broad line is observed in the ^{31}P-NMR spectrum (-33 ppm). This could be due to P species having a distorted environment but still belonging to a VPI-5 topology, as ascertained by XRD. This also demonstrates that NMR remains a powerful tool for detecting the slightest short-range structural changes.

Thermal analysis

Sample VPI-5-B was submitted to various thermal treatments. Figure 20-11 shows a typical TG-DTA trace as a function of temperature and time, for the sample heated from 20 to 120°C in static air (1.5°C. min^{-1}) and maintained at 120°C for about 1 h.

On comparing this pattern with the one recorded under similar conditions for sample VPI-5-A (Fig. 5), one can immediately notice that thermal effects related to the presence of H3 impurities in the sample (endotherms A and B) and to its transformation to AlPO$_4$-8 (exotherm C) are absent. Only a strong endothermic effect now characterizes the regular release of the 25 wt. % of water from the structure. NMR and XRD confirm the VPI-5 topology of the calcined residue.

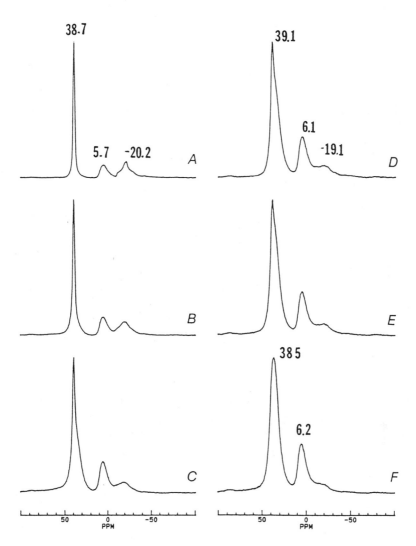

FIGURE 20-10 ^{27}Al NMR spectra of VPI-5-B maintained at 110°C for various periods of time. .A: 10 min, B: 1 h, C: 3 h, D: 5 h, E: 6 h, F:7 h.

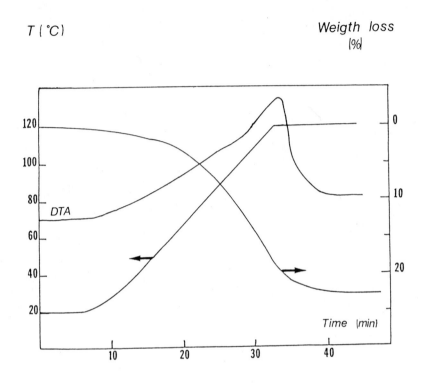

FIGURE 20-11   T (°C), TG and DTA traces recorded for sample VPI-5-B heated in the thermobalance from 20 to 120°C (1.5°C. min^{-1}) in static air and maintained isothermally at 120°C, as a function of time.

CONCLUSIONS

In this study, we have followed the in situ dehydration of two different VPI-5 precursors under similar conditions. Our data clearly show that, in this specific case, the variation of the heating conditions has little or no influence on the stability of the VFI structure. Conversely, the latter was shown to depend strongly on the synthesis procedure used to crystallize the precursors, directly related to their further purity and possibly to their lattice defect concentrations. This can be rationalized by considering that the polymorphic transformation of VPI-5 to $AlPO_4$-8 occurs by breaking of the Al-O-P linkages (11, 22, 26) via a hydrolytic process that is probably induced or enhanced by the presence of P-OH defects. These defects can be generated either during crystallization or during the loss of water during the first heating of the as-synthesized material. The presence of such defect groups was indeed detected by ^{31}P-NMR in both precursor or calcined (dehydrated) VPI-5 type A materials (11, 16). Obviously, a more homogeneous way of mixing the ingredients in procedure B would lead to less defected $AlPO_4$-type precursor species in the gel, during the aging stage at ambient temperature. Other synthesis variables, such as the nature of the amine, perhaps through its pH-moderating effect (7, 21) or by some not yet understood role, can also affect the final defect concentration. The apparent contradictory results or explanations reported in the recent literature about the influence of different synthesis variables on the purity and stability of the final VPI-5 materials (see introduction), can possibly be rationalized if one admits that different causes, namely the various synthesis parameters, can lead to the same result, namely the generation of appreciable amounts of defects that make the framework fragile and readily assist a hydrolytic process.

Our study also shows that in a defined (low) temperature range (80-150°C), the actual amount of water left in the structure has a predominant influence on the solid state structural transformation, rather than the temperature itself. Indeed, in such a temperature domain, one can imagine that the breaking of the P-O-Al structural bonds can start only if they were "fragilized" beforehand during the "history" of the framework building and also only when most of the loosely bound water molecules (about 95% in our case) have been eliminated. Hydrolysis is then probably only initiated by those water molecules directly anchored to the framework Al atoms. The transformation does not occur at 80°C, not because that temperature is too low for the hydrolytic reactions, but because too much water still remains in the framework and continues to stabilize the framework. Indeed, we have shown previously that when the

VPI-5-A precursor is maintained in contact with boiling water, the VFI framework remains unaltered (16). Similarly, when water is removed from the framework at low temperature, at which the hydrolysis rate is probably very low, or when the initial water is removed from the framework by washing with other solvents (10, 11), the original structure is also maintained stable.

The observation that stable VPI-5 frameworks can be prepared upon very rapid heating under various atmospheres or in vacuo (1, 7, 9, 10, 23) is also consistent with the above conclusions. In such cases, water is probably completely removed so rapidly that it has little ability to initiate the hydrolytic procedure. We have also tested the thermal stability of sample B at very high temperatures. Upon heating the precursor to 500°C at a rate of 5°C. min^{-1}, it still keeps its VFI framework topology, while sample A, treated in similar conditions, is completely transformed to AlPO$_4$-8 (XRD).

Finally, a partial or total loss of crystallinity due to a structural collapse has been observed in the case of various precalcined AlPO$_4$- and SAPO- molecular sieves, such as SAPO-11 (32) or SAPO-37 (33), when they are left in the presence of water or stored in a humid atmosphere for a prolonged time, even at ambient temperature. To that purpose, calcined sample B was stirred with water for 2 h at 20°C (0.1 g of solid in 15 ml of distilled water). After centrifugation, the solid was dried at 20°C for 12 h. The XRD spectrum indicates a partial decrease of the (still VPI-5) crystallinity. On the other hand, the as-synthesized VPI-5-B was also contacted with water (1 g solid/25 ml water) at 200°C in an autoclave, for 60 h. The VFI structure has now disappeared but the X-ray pattern of the residual product has not been identified so far. In this case, the combined action of water and temperature has completely destroyed the VPI-5 structure. These data confirm that a non-defected AlPO$_4$-type structure is also ineluctably a candidate for partial or total destruction by hydrolysis as soon as the stabilizing pore-filler molecules (here water) are disturbed or removed.

ACKNOWLEDGMENTS

L. Maistriau would like to acknowledge Akzo Chemicals B.V. (Amsterdam) for a doctoral fellowship. The authors are indebted for Dr. E. T. C. Vogt (Akzo Chemicals B.V.) for pertinent discussions and suggestions.

REFERENCES

1.   M.E. Davis, C. Saldarriaga, C. Montes, J.M. Garces and C. Crowder, Nature, 331, 698 (1988); Zeolites, 8, 362 (1988).
2.   M.E. Davis, C. Montes and J.M. Garces in "Zeolite Synthesis" edited by M.L. Occelli and H.E. Robson (ACS Symp. Series 398, American Chemical Society, Washington, D.C., 1989), pp. 291-304.
3.   M.E. Davis, C. Montes, P.E. Hathaway and J.M. Garces, Stud. Surf. Sci. Catal., 49A, 199 (1989).
4.   P. Grobet, J.A. Martens, I. Balakrishnan, M. Mertens and P.A. Jacobs, Appl. Catal., 56, L21 (1989).
5.   S. Prasad and I. Balakrishnan, Inorg. Chem., 29, 4830 (1990).
6.   E.G. Derouane, L. Maistriau, N. Dumont, J.B. Nagy and Z. Gabelica in "Synthesis and Properties of New Catalysts : Utilization of Novel Materials Components and Synthesis Techniques" edited by E.W. Corcoran and M.J. Ledoux (Mater. Res. Soc. Proc., Boston, MA, 1990), pp. 3-6.
7.   M.E. Davis and D.Young, Stud. Surf. Sci. Catal., 60, 53 (1991)
8.   X. Liu and J. Klinowski, ref. 6, pp. 33-36.
9.   J.O. Perez, N.K. Mc Guire and A. Clearfield, Catal. Lett, 8, 145 (1991).
10. H. Cauffriez, Ph.D. Thesis, Mulhouse University, France, May 1991.
11. L. Maistriau, Z. Gabelica, E.G. Derouane, E.T.C Vogt and J. van Oene, Zeolites, 11, 583 (1991).
12. E.G. Derouane and R. Von Ballmoos, Eur. Patent Appl. No. 146389 (1985).
13. E.G. Derouane, L. Maistriau, Z. Gabelica, A. Tuel, J. B.Nagy and R. Von Ballmoos, Appl. Catal., 51, L13 (1989).
14. P.J. Grobet, H. Geerts, J.A. Martens and P.A. Jacobs, Stud. Surf. Sci. Catal., 51, 193 (1989).
15. M.E. Davis, P.E. Hathaway and C. Montes, Zeolites, 9, 436 (1989).
16. L. Maistriau, Z. Gabelica and E.G. Derouane, Appl. Catal., 67, L11 (1991).
17. R. Szostak, B. Duncan, K. Sörby and J.G. Ulan, ref. 6., pp. 15-17.
18. K. Sörby, R. Szostak, J.G. Ulan and R. Gronsky, Catal. Lett., 6, 209 (1990).
19. F. d'Yvoire, Bull. Soc. Chim. France, 372, 1762 (1961).

20.   K. Vinje, J. Ulan, R. Szostak and R. Gronsky, Appl. Catal.,
        72, 361 (1991).
21.   X. Ren, S. Komarneni and D.M. Roy, Zeolites, 11, 142 (1991).
22.   L.B. McCusker, C. Baerlocher, E. Jahn and M. Bülow,
        Zeolites, 11, 308 (1991).
23.   M.E. Davis, C. Montes, P.E. Hathaway, J.P. Arhancet, D.L.
        Hasha and J.M. Garces, J. Am. Chem. Soc., 111, 3919 (1989).
24.   M. Stöcker, D. Akporiaye and K.P. Lillerud, Appl. Catal., 69,
        27 (1991).
25.   S.T. Wilson, B.M. Lok and E.M. Flaningen, U.S. Patent No. 4
        310 440 (1982).
26.   E.T.C. Vogt and J.W. Richardson, Jr., J. Solid State Chem.,
        87, 464 (1990).
27.   R.M. Dessau, J.L. Schlenker and J.B. Higgins, Zeolites, 10,
        522 (1990).
28.   P.R. Rudolf and C.E. Crowder, Zeolites, 10, 163 (1990).
29.   Y. Wu, B.F. Chmelka, A. Pines, M.E. Davis, P. Grobet and
        P.A. Jacobs, Nature, 346, 550 (1990).
30.   B.O. Brunner, Zeolites, 10, 612 (1990).
31.   G.T. Kerr and A.W. Chester, Thermochim. Acta, 3, 113
        (1971).
32.   N.J. Tapp, N.B. Milestone, M.E. Bowden and R.H. Meinhold,
        Zeolites, 10, 105 (1990).
33.   N. Dumont, T. Ito, J. B.Nagy, Z. Gabelica and E.G. Derouane,
        Stud. Surf. Sci. Catal., 65, 591 (1991).

21

Isomorphous Substitutions of Silicon And Cobalt in Molecular Sieve AlPO4–5

K. J. Chao, S. P. Sheu, S. H. Chen, J. C. Lin, and J. Lievens, *Tsinghua University*

The substitutions of Si(IV) for P(V) and Co(II) for Al(III) in AlPO$_4$ molecular sieves results in anionic tetrahedral frameworks coupled with exchangeable cations and Brønsted acid sites. The syntheses and characterizations of silicoaluminophosphate SAPO-5 and Co-aluminophosphate CoAPO-5 are reported. The cation sites were found to be close to the oxygen (O3) of the four-membered ring of the tetrahedral framework in the dehydrated Cs form of SAPO-5 by Rietveld analysis of powder XRD pattern. The oxidation-reduction chemistry of Co in CoAPO-5 was also investigated. The nature of 12 membered ring channels of SAPO-5 and CoAPO-5 were compared with that of AlPO$_4$-5 using ^{129}Xe NMR and xenon adsorption isotherm.

INTRODUCTION

The acidity and ion-exchange capacity of aluminophosphate molecular sieves (AlPO$_4$) can be improved substantially by the isomorphous substitution of tetravalent cations, e.g. Si^{4+} for P^{5+}, and divalent cations, e.g. Co^{2+} for Al^{3+}, in their tetrahedral framework (Lok et al. 1984; Tapp, Milestone, and Wright 1985; Flanigen et al. 1986). The incorporation of silicon and cobalt into the $AlPO_4$ framework leads to silico-aluminophosphate (SAPO) and cobalt-aluminophosphate (CoAPO) respectively. The framework structure of the number five of aluminophosphate molecular sieves, $AlPO_4$-5, has not been disturbed by partial silicon or cobalt substitution and consists of one dimensional channels bounded by twelve-membered rings with a pore opening of 0.8 nm (Bennett et al. 1983). In order to increase the Brønsted acidity and catalytic activity, the high-silicon containing SAPO-5 samples were prepared by increasing the concentration of the silicon reactant and varying the synthesis condition, the template agent, or the silicon source in the hydrothermal synthesis (Martens et al. 1988; Appleyard, Harris, and Fitch 1985; Wang et al. 1991). The chemical composition of SAPO samples indicates that two types of substitution mechanisms consist of a pair substitution of two Si for one P and one Al, together with one Si for one P (Lok et al. 1984; Appleyard, Harris, and Fitch 1985). The high-silicon SAPO-5 materials usually have low Brønsted acidity with a relatively high content of amorphous silica or silica patch. In our previous study (Wang et al. 1991), the relative amount of silicon substitution for phosphorus in the $AlPO_4$ -5 framework was found to be strongly dependent on the silicon source and the Brønsted acidity of SAPO-5 samples could be increased by using tetraethylorthosilicate $Si(OEt)_4$ instead of polymeric colloid silica, Ludox-AS, as the silicon source in preparation.

The net framework charge per silicon substitution for phosphorus or cobalt substitution for aluminum would be -1. Thus the SAPO and CoAPO materials have an anionic framework with a net negative charge coupled with exchangeable cations and Brønsted acid sites. Due to the similar x-ray scattering factors of Al, Si and P atoms and the similar bond distances of Si-O (0.162 nm) and P-O (0.153 nm), the location and occupancy of Si in the lattice cannot be easily extracted by x-ray diffraction analysis. However, the locations of exchangeable cations which play an important role in the catalytic and ion-exchange properties and also relate to the position of silicon substitution sites have not been characterized. In this paper, we determined the locations of cations in hydrated and dehydrated Cs exchanged SAPO-5 (CsSAPO-5) by using the x-ray powder diffraction method and Rietveld analysis. Furthermore, the ^{129}Xe chemical shifts and the adsorption isotherms of xenon on CsSAPO-5, HSAPO-5 and AlPO$_4$-5 were used to probe the cation effect and silicon incorporation as a function of the silicon source in synthesis.

Isomorphous subsitution of Co^{2+} for Al^{3+} can generate Brønsted acidity as Si^{4+} substitution for P^{5+} in AlPO$_4$-5. The incorporation of divalent cobalt ions into the tetrahedral sites of several aluminophosphate molecular sieves was identified by its d-d electron transition spectrum on UV-Vis spectroscopy (Schoonheydt et al. 1989; Montes et al. 1990; Goepper et al. 1989; Leu and Chao 1988). the presence of Brønsted acid sites was confirmed by temperature programmed desorption of ammonia on ammonium ion exchanged CoAPO-5 samples (Tapp et al. 1985; Shiralkar et al. 1989). In the case of cobalt, the concept of paired substitution was not accepted and Co^{2+} was assumed to substitute predominantly for Al^{3+} (Tapp, Milestone, and Bibby 1988). The framework Co^{2+} was found to be oxidized by heating under oxygen or air to Co^{3+} which was easily reduced to Co^{2+} localized in a distorted tetrahedral environment (Schoonheydt et al. 1989; Montes et al. 1990). Most of the synthesized CoAPO-5 was contaminated with an impurity phase. Where the cobalt ions are located in CoAPO-5 and how to synthesize CoAPO-5 of high purity and high cobalt content have not been fully determined.

The purpose of our work is to prepare high Co containing pure CoAPO-5 and to characterize quantitatively its physicochemical properties as well as its oxidation-reduction chemistry. In particular, the incorporation of cobalt ions into the AlPO$_4$-5 framework is confirmed by x-ray method, and the properties of its framework and intraframework space are studied by ^{129}Xe NMR and xenon adsorption isotherm. In addition; the thermal and acidic properties of CoAPO-5 before and after calcination are compared with those of SAPO-5. The conditions for synthesizing CoAPO-5 with triethylamine template are also discussed.

EXPERIMENTAL

Synthesis

The precursors of SAPO-5 were prepared by reacting orthophosphoric acid (Merck) pseudoboehmite (Versal 450, Kaiser), colloidal silica solution (Ludox As-40, Du Pont) or 40 wt% SiO$_2$ tetraethylorthosilicate Si(OEt)$_4$ (Merck, 98% pure) with tripropylamine, Pr$_3$N (Fluka), or triethylamine, Et$_3$N (Wako), with a molar composition of 0.85 Pr$_3$N or Et$_3$N: 1.0 P$_2$O$_5$: 1.0 Al$_2$O$_3$: 0.27 SiO$_2$: 30 H$_2$O in the hydrothermal condition outlined in Wang et al. (1991). After reaction, the reaction vessel was quenched immediately under tap water and the solid product was filtered, washed and dried at 80°C .

In the case of CoAPO-5, cobalt sulfate was added to an aqueous solution of orthophosphoric acid; then the alumina source (either Versal 450, surface area =300 m^{2}g^{-1}, Kaiser, or NG, surface area = 160 m^{2}g^{-1}, Condea) was added slowly with intense stirring. After the mixture had been thoroughly homogenized, the organic template Et$_3$N was added. The whole mixture was then stirred another 30 min., placed into a teflon-lined stainless steel autoclave and heated at 170°C for 24 hours.

After quenching to 25°C, the product was washed repeatedly with deionized water by stirring several minutes, letting solids precipitate, then discarding the supernatant liquid until the crystal product appeared clean under the microscope, and then dried at 70°C.

Ion-exchange

The synthetic SAPO-5 sample was placed into a U-type quartz cell to remove the organic template from the channel of the AlPO$_4$-5 based molecular sieves with flowing dry Ar at a flow rate of 60 mL min^{-1}. The temperature was increased at a heating rate of 1°C min^{-1} from room temperature to 150°C, and remained at this temperature for 4 hours. Same heating rate was used to increase the temperature from 150 to 350°C, remained at 350°C for 4 hours; and then increased the temperature from 350 to 600°C, maintaining the temperature at 600°C for 20 hours. The sample was calcined by flowing dry air at a flow rate of 60 mL min^{-1} for 20 hours. The template free white powder was obtained after cooling to 25°C.

The template of synthetic CoAPO-5 sample was removed by calcining in dry air flow to 550°C with a heating rate of 1°C min^{-1}, and held at 550°C for 20 hours. Then, the sample was treated with oxygen flow at 550°C for 8 hours, and the color of the CoAPO-5(III) sample was changed from blue to green. Evacuation was also used to remove the template from the channel of CoAPO-5. The synthetic CoAPO-5 sample was evacuated and heated with a heating rate of 0.2°C min^{-1} from room temperature to 400°C, and maintained at 400°C for about 30 hours. After evacuation and heat treatment, the color of CoAPO-5 was changed from blue to black. This black coked CoAPO-5 was used for xenon adsorption and acidity measurements.

$NH_4CoAPO-5$, $NH_4SAPO-5$, NaSAPO-5 and CsSAPO-5 molecular sieves were prepared by exchanging template-free CoAPO-5 or HSAPO-5 with 0.1N NH_4Cl, NaCl or CsCl solution at room temperature. In order to remove residual salt, the samples were washed with deionized water until free of Cl^- ions detected by $AgNO_3$ solution. All molecular sieve samples were then air-dried and stored over saturated NH_4Cl solution at room temperature.

Analysis

The phases of crystalline products in the synthesized samples were identified by powder XRD. The bulk Co, Al, Si, P, Na chemical compositions of SAPO-5 and CoAPO-5 samples which were dissolved in an acid mixture were determined by inductively coupled plasma-atomic emissioin spectroscopy (ICP-AES) on a GVM 1000. The cesium content of CsSAPO-5 was measured by neutron activation analysis. The variations of the Co content between CoAPO-5 and CoAPO-34 crystals coated with carbon were measured by electron probe microanalysis (EPMA) on a Jeol 840 A with EDX. Differential (DTA) and thermogravimetric (TG) thermal analyses were recorded by flowing argon, and a mixture of nitrogen with oxygen or air (30 mL^3min^{-1}) on a Seiko TG/DTA 300-10 apparatus.

Acidity Measurement

The acidity was determined by means of temperature programmed desorption of ammonia with on-line titration from ammonium form samples. The adsorbed water was removed by flowing dry argon (30 mL min^{-1}) at 150°C. Then, the temperature was increased at a rate of 5°C min^{-1} from 150 to 600°C. The desorbed ammonia was trapped in a solution of saturated boric acid and 1 M NH_4Cl, and titrated by sulphamic acid. The volume of titrant used per time interval vs. temperature was recorded and the concentration of Brønsted acid sites was estimated from the millimoles of ammonia evolved per gram sample.

The IR spectra of the hydroxyl group of CoAPO-5 were measured on a Bomem MB-100 FTIR. The self-supported water (20 mg) of Et_3N • CoAPO-5 or $NH_4CoAPO-5$ sample was fist evacuated at 150°C for 12 hours and then heated 10°C min^{-1} to 450°C, held at 450°C for 12 hours under vacuum and then cooled to ambient temperature for IR measurement. The evacuated wafer of Et_3N • CoAPO-5 was also oxidized under oxygen at 550°C.

XRD Measurement and Structure Refinement

A Seifert-Scintag PADII diffractometer using Ni filtered CuKα radiation at 50 kV, 25 mA was used. The powder diffraction intensities were collected in a step size of 0.01°C/10 sec from 2θ = 4-70°C at room temperature and 2θ = 5-80°C at 350°C for hydrated and dehydrated CsSAPO-5 respectively. The dehydrated sample pellet (15 x 12 x 1 mm) was evacuated and heated from room temperature to 350°C (1°C min⁻¹) and held at 350°C for 48 hours at 1 x 10⁻⁵ torr.

The coordinates and isotropic thermal parameters of the CsSAPO-5 framework atoms were fixed at the values of $AlPO_4$-5 (Bennett et al. 1983). The coordinates, isotropic thermal parameters and occupancy factors of Cs ions and water molecules in the intraframework void space were refined by DBW Rietveld program (Rietveld 1967) based on a pseudo-voigt profile function (Wiles and Young 1981). The space group of both hydrated and dehydrated CsSAPO-5 is P6CC, with the cell dimensions of hydrated CsSAPO-5 being a = 13.737(5) and c = 8.531(5) Å. The final agreement indices are R = 0.104 and Rw = 0.144 for 6500 steps. The cell dimensions of the dehydrated phase are a =13.752(2) and c = 8.411(2) Å and R = 0.109 and Rw = 0.151 for 7600 steps.

Xenon Adsorption and ^{129}Xe NMR Measurement

The sample (typically ca. 1.5 g) was loaded into a 10 mm NMR tube and evacuated to ca. 1x10⁻⁴ torr for 3 days at room temperature, then heated from room temperature to 150°C with a heating rate of 0.2°C min⁻¹, and maintained at 150°C for 8 hours, subsequently heated to 350°C for SAPO-5 and 400°C for CoAPO-5, and held at 350°C or 400°C for 30 hours (2 x 10⁻⁵ torr). All the xenon adsorption isotherms were measured by volumetric method at 22°C.

The ^{129}Xe NMR spectra of adsorbed xenon were obtained on a Bruker MSL-200 (Bruker MSL-300 for $AlPO_4$-5 only) spectrometer operating at 55.3 MHz and 295 K. Typically 2000-60000 signal acquisitions were accumulated for each spectrum with a recycle delay of 0.5 s between pulses. For $AlPO_4$-5 and SAPO-5 samples, a simple $\pi/2$ cyclic pulse technique was used while a $\pi/2$-τ-$\pi/2$ spin echo technique with a delay time τ=20 μs was employed in the CoAPO-5 sample. The chemical shifts were referenced to that of external xenon gas extrapolated to zero pressure using Jameson's equation (Jameson 1975) and taken to be positive in this report.

RESULTS AND DISCUSSION

Silicon Incorporation

The structures of hydrated and dehydrated CsSAPO-5 were determined by x-ray powder diffraction with the Rietveld analysis. The observed powder pattern are given in Fig. 21-1. The Cs^+ ion of dehydrated CsSAPO-5 is located close to the oxygen atom (O3) of the four-membered ring with a Cs-O distance of 0.327 nm and to the oxygen atom (O1) of the six-membered ring with a Cs-O distance of 0.333 nm. Both O1 and O3 oxygen atoms are in the coordination shell of the Cs^+ ion (Fig. 21-2). The silicon substituting phosphorus induces an anionic site, Si-O-Al. The cation or proton of dehydrated SAPO-5 is likely to be located nearby the anionic site in order to balance the charge. The present result indicates that the distribution of silicon is homogeneous in the tetrahedral framework. In the presence of water molecules, the Cs^+ ion was found to shift to the center of the main channel and coordinated to six water molecules with a Cs-O5 distance of 0.329 nm as shown in Fig. 21-2. The numbers of Cs and O5 per unit cell were refined to 1 and 6 respectively in hydrated CsSAPO-5. The Cs content is consistent with the value obtained by electron microprobe analysis.

Xenon atom, with a diameter of 0.44 nm, can only enter the twelve-membered ring channel of the $AlPO_4$-5 based molecular sieves and can be used as a sensitive probe for the geometric and electronic environments in the main channels. The equilibrium isotherms of xenon adsorption on $AlPO_4$-5, HSAPO-5, NaSAPO-5 and CsSAPO-5 are shown in Fig. 21-3. The isotherm is considered to follow Langmuir formulation, $\rho/\rho_\infty = KP/(1+KP)$, where ρ and ρ_∞ are xenon loadings at an equilibrium xenon pressure, P, and at a very high xenon pressure respectively. The values of Langmuir-form equilibrium constants, K, and ρ_∞ are listed in Table 21-1. Both the value of K constant and the amount of xenon adsorbed at the low xenon pressure depend strongly on the cation-xenon interaction and increase in the order of HSAPO-5 < NaSAPO-5 < CsSAPO-5. This is consistent with the results on Y zeolites (Chao, Chen and Liu 1991). Based on size of cation, the order of variation in the value of ρ_∞ is

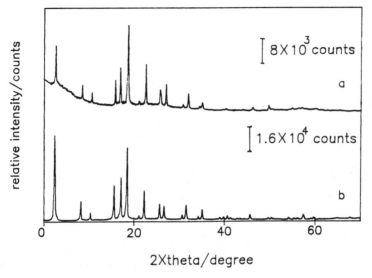

FIGURE 1. The observed powder XRD pattern of dehydrated (a) and hydrated (b) CsSAPO-5.

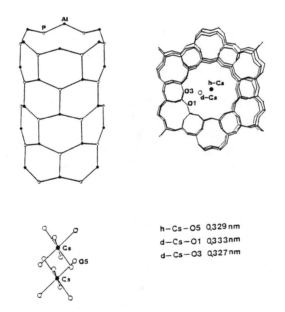

FIGURE 2. Stereoplot of channel and Cs ions inhydrated (h) and dehydrated (d) CsSAPO-5.

TABLE 21-1 The Simulated Langmuir Parameter (K and ρ_∞) for Xenon Adsorbed on AlPO$_4$-5 Based Molecular Sieves

Molecular Sieve	Anhydrous Unit Cell Composition	K $(10^{-4} \cdot torr^{-1})$	ρ_∞ (Xe atom/unit cell)
AlPO$_4$-5	$(PO_2)_{12}(AlPO_2)12.24$	7.4	3.0
HSAPO-5(L)	$H_{0.3}(SiO_2)1.96(PO_2)11.11(AlO_2)_{12}$	8.6	2.8
HSAPO-5	$H_{0.85}(SiO_2)2.06(PO_2)10.55(AlO_2)_{12}$	10.4	2.8
NaSAPO-5	$Na_{0.53}H_{0.32}(SiO_2)2.06(PO_2)10.55(AlO_2)_{12}$	8.1	2.4
CsSAPO-5	$Cs_{0.46}H_{0.39}(SiO_2)2.06(PO_2)10.55(AlO_2)_{12}$	22.9	2.0
CoAPO-5*	$(CoO_2)1.79(PO_2)_{12}(AlO_2)10.76$	8.7	2.5
CoAPO-5(III)	$(CoO_2)1.79(PO_2)_{12}(AlO_2)10.76$	9.0	2.8
CoAPO-5(II)	$(CoO_2)1.79(PO_2)_{12}(AlO_2)10.76$	9.0	2.8

*The coke content of 3.4 wt% was determined by TGA under O$_2$/N$_2$ flow.

expected to be HSAPO-5 > NaSAPO-5 > CsSAPO-5.

The silicon content of the SAPO-5 product which depends on the reactant composition and the relative amount of silicon substitution for phosphorus in the tetrahedral framework is controlled by silicon source (Wang et al. 1991). The silicon substitution in HSAPO-5 from $Si(OEt)_4$ preparation is higher than that in HSAPO-5(L) from Ludox-AS preparation. Therefore, the xenon loading at the same xenon pressure is in the order of $AlPO_4$-5<HSAPO-5(L)<HSAPO-5 as shown in Fig. 21-3. The saturated xenon loadings, ρ_∞, of both HSAPO-5(L) and HSAPO-5 samples (2.8 atoms (unit cell)$^{-1}$) are similar to that of $AlPO_4$-5 (3.0 atoms (unit cell)$^{-1}$). This indicates that the intraframework oxides should be very small in our SAPO-5 samples.

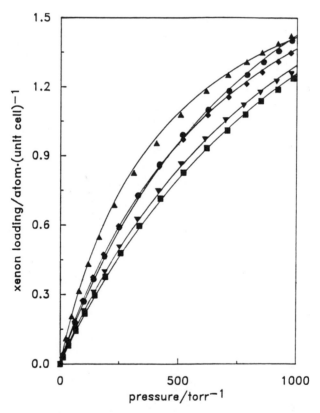

FIGURE 3. Comparison of the xenon adsproption isotherms of CsSAPO-5 (▲), NaSAPO-5 (♦), HSAPO-5 (●), HSAPO-5(L) (▼) and $AlPO_4$-5 (■) with the Langmuir simulation curves (–).

The influences of cation-exchange and silicon substitution on the framework were also detected by the chemical shift of xenon-129 NMR. Variations of the ^{129}Xe NMR chemical shift with the concentration of adsorbed xenon are shown in Fig. 21-4. The spectra of ^{129}Xe NMR display as a single signal, most symmetrical but somehow asymmetrical at low xenon concentration. According to the MAS-NMR results of Chen et al., (Chen, Springuel-Huet, and Fraissard 1989) this occasional asymmetry may be attributed to chemical shift anisotropy and not the superposition of two signals. The values of chemical shifts were assigned by considering the centre of gravity of the NMR signal as in the study on Nu-10 (Pellegrino et al. 1990). The previous study of Fraissard and Ito on zeolites (Fraissard et al. 1986) showed that the xenon-xenon interaction and the xenon-channel wall or xenon-cation interaction take account for the variation of the intercept and slope of chemical shift vs. xenon loading curves respectively. The intercept varies with amount of Si substitution and type of cation exchanged and are in order of AlPO$_4$-5 < HSAPO-5 (L) < HSAPO-5 $\cong$ NaSAPO-5 < CsSAPO-5. The reduction of intraframework void space by the occupation of Na$^+$ ions is also shown on the slope of NaSAPO-5, of which the value is larger than that of HSAPO-5. The slope of Cs form SASPO-5 is slightly smaller than those of H and Na forms due to the strong

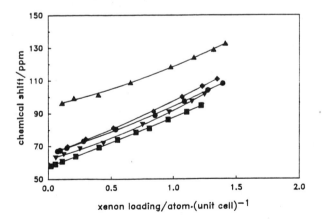

FIGURE 4, Dependence of the ^{129}Xe NMR chemical shift on the number of the xenon atoms adsorbed on CsSAPO-5(▲) , NaSAPO-5(♦), HSAPO-5(●), HSAPO-5(L)(▼) and AlPO$_4$-5 (■).

interaction between Cs^+ ion and xenon atom. The heteroatom substitution in the framework as well as the cation occupation in the intraframework void space of $AlPO_4$-based molecular sieves can also be detected by xenon adsorpiton isotherm and the chemical shift of ^{129}Xe NMR.

Cobalt Incorporation

The previous studies of Wilson et al. (Wilson and Flanigen 1989), Tapp et al. (Tapp, Milestone, and Bibby 1988) and Shiralkar et al. (Shiralkar et al. 1989) showed that triethylamine is the most suitable template for the production of cobalt substituted $AlPO_4$-5. Therefore, all the CoAPO-5 samples were prepared by using cobalt sulfate with triethylamine. High Co containing CoAPO-5 was formed together with the CoAPO-34 phase. Similar results were reported by Wilson et al. (1989) and Pyke et al. (1985). However, the CoAPO-5 and CoAPO-34 phases with high purity can be obtained by washing and separation, their electron microscopic pictures and x-ray powder patterns are shown in Figs. 21-5, 21-6. Thus, the relative amount of CoAPO-5/CoAPO-34 can be estimated by the intensities of the peaks at $2\theta=7.5°$ for CoAPO-5 and at $2\theta=9.5°$ for CoAPO-34. The amount of cobalt in CoAPO-5 crystals is less than that in CoAPO-34 crystals obtained from the same batch (Table 21-2).

The synthesis conditions for CoAPO-5 and the properties of crystalline and compounds are summarized in Table 21-3. The optimum pH value for synthesizing aluminophosphate based molecular sieves is 3-4. The pH value of the reactant mixture varies strongly with the amount of template (Et_3N) added. The molar ratio of 1.50 Et_3N: 0.2 CoO : 0.9 Al_2O_3 : 1.0 P_2O_5 : 50 H_2O was found to be the better condition for synthesizing pure CoAPO-5. The decrease of water content would lead to more CoAPO-34 phase. Under the microscope, the CoAPO-34 crystals appeared deep blue and in a cubic shape while the CoAPO-5 crystals were formed of lighter blue and hexagonal. The size of crystals was increased by increasing water content in the case of using boehmite as an aluminum source, which was also found in the synthesis of zeolite ZSM-5 by Muller et al. (1988). The dissolubility of alumina may be improved with increasing water content.

In the case of the reactant composition up to 0.4 CoO/P_2O_5, the product were contaminated by other crystalline phase. When the cobalt added was as low as 0.01 CoO/P_2O_5, some crystals of the CoAPO-5 showed

FIGURE 5. Scanning electron micrographs of CoAPO-5 (a,c) and CoAPO-34 (b,d) from Versal 450 (a,b) and NG (c,d) in 1.5 Et$_3$N:0.2 CoO:0.9 Al$_2$O$_3$:1.0 P$_2$O$_5$:50 H$_2$O.

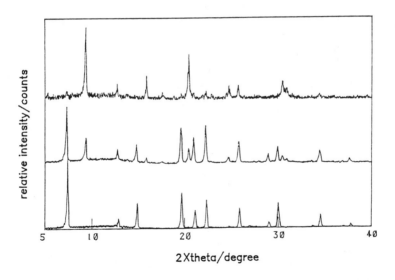

FIGURE 6. The observed powder XRD pattern of synthetic CoAPO-34(top), CoAPO-5 (bottom) and mixture of CoAPO-34 and CoAPO-5 (middle).

TABLE 21-2 Compositional Variations Across Different Size Crystals of Synthetic CoAPO-5 and CoAPO-34

	Crystal Size (μm x μm x μm)	Analysis Points	Average Co Elements %
Versal 450			
CoAPO-5	220x150x150	4	4.17(±0.5)
	180x80x80	3	5.44(±0.5)
	150x50x50	3	8.0(±1.0)
CoAPO-34	90x90x90	7	12.4(±0.8)
	50x50x50	4	11.9(±0.7)
NG			
CoAPO-5	450x100x100	4	7.7(±0.4)
	280x80x80	3	7.1(±0.2)
	80x50x50	3	4.0(±1.0)
CoAPO-34	200x200x200	3	9.8(±0.8)
	150x150x150	3	9.9(±1.2)
	90x90x90	2	10.4(±0.3)

degradation of color from the center to the end, a portion of transparent phase, presumely AlPO-5 was found to intergrow with CoAPO-5. This indicates that rapid uptake of cobalt occurs at the begining of synthesis and exhausts most of the cobalt concentration in the solution before all the alumina and phosphate participating in the type-5 structure formation. Besides, we found large crystals of 230 μm x 100μm x 100μm from the reactants of 1.50 Et_3N : 0.1 CoO : 0.95 Al_2O_3 : 1.0 P_2O_5 : 50 H_2O. The cobalt substitution for tetrahedral framework aluminum in CoAPO-5 had been confirmed by single crystal x-ray diffraction analysis and the structural data will be published elsewhere. The isomorphous substitution of cobalt for aluminum up to 13(4)% was observed by ICP-AES analysis and close to the result of x-ray method. The measured unit cell dimensions for CoAPO-5 are a=13.777(3)Å, c=8.429(7)Å, v=1386(12)Å^3; the larger cell dimensions in comparison with those of Montes et al. (1990); a=13.7Å, c=8.44Å, v=1372Å^3, indicate larger amounts of cobalt incorporated into our samples.

The cobalt substitution in the framework of $AlPO_4$-5 was suggested to occur exclusively on the aluminum tetrahedral sites. Therefore, the source of aluminum should play a certain role in Co incoporation. The morphology and the cobalt content of CoAPO-5 and CoAPO-34 crystals were found to vary with the reactivity of the source of

TABLE 21-3 Average Particle Size of Crystalline Products

| Reaction Composition (molar ratio) | | | | | | CoAPO-5[a] | | CoAPO-34 cubic |
Et3N	CoO	Al2O3	P2O5	H2O	pH	hexagonal[b] (µm)	spherical aggregates (µm)	(µm)
1.0	0.2	0.9	1.0	50	2.4	—	—	—
1.5					3.5	80 / 150 minor	10	
2.0	0.2	0.9	1.0		5.1	—	—	
1.5				30	3.7	25	<10	50
				48	3.5	40	40	30minor
				82	3.5	60	30	—
1.5	0.05	1.0	1.0	50	3.5	100 minor / 200 minor	40	—
	0.1	0.95			3.2	150 minor / 230 minor	30	20minor
	0.2	0.9			3.1	80 / 160 minor	30	—
	0.4	0.8			2.9	50	25	30minor
1.5	0.2	0.9	1.0	50[c]	3.3	80,150 / 400 minor	40	—

a. There are two distinct morphologies of CoAPO-5.
b. The tabulated values are the average crystal size along the (001) axis. Width = 1/2 - 1/3 length.
c. NG alumina was used instead of Versal 450.

aluminum; the product crystals from low surface area pseudoboehmite, NG, were usually contaminated by amorphous product, and their average size was larger than the CoAPO-5 and CoAPO-34 crystals from high surface area pseudoboehmite, Versal 450, in the same reactant composition (Fig. 21-5). The cobalt content varies between crystals and increasing with decreasing size on CoAPO-5 from Versal 450 and with increasing size on CoAPO-5 from NG, while the cobalt concentration seems to be independent on the size of CoAPO-34 crystals determined by EPMA as shown in Table 21-2.

The UV-Vis (Schoonheydt et al. 1989) and ESR (Montes et al. 1990) studies showed that the framework Co^{2+} (blue) can be oxidized to Co^{3+} (green) which can be subsequently reduced back to Co^{2+}. After calcination in air or O_2, the cobalt of CoAPO-5 remains in the tetrahedral framework or becomes octahedrally coordinated in the intraframework space that can be detected by xenon adsorption measurement. The xenon adsorption isotherms of both blue and green CoAPO-5 are compared with that of $AlPO_4$-5 in Fig. 21-7. In both CoAPO-5(III) and CoAPO-5(II), the simulated ρ_∞ value on the Langmuir model of 2.8 atom (unit cell)$^{-1}$ is slightly larger than the value for coked CoAPO-5 (Table 21-1). This shows that most of Co^{3+} obtained by framework Co^{2+} stay in the ramework. In addition, comparision of the chemical composition of coked NH_4CoAPO-5

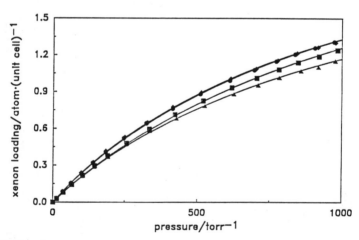

FIGURE 7. Comparison the xenon adsorption isotherms of coked CoAPO-5(▲), CoAPO-5 (III)(●) CoAPO-5 (II)(♦) and $AlPO_4$-5(■) with the Langmuir similation curves (–).

(Co/P=0.13) with oxidized $NH_4CoAPO-5$ (III) (Co/P=0.11) indicates most of Co in CoAPO-5 is incorporated into the framework and only a small portion of framework cobalt can be removed by oxidation and subsequent NH_4Cl exchange. After evacuation at 400°C for several days, the color of the oxidized CoAPO-5 sample was changed from green to blue. This indicates that green may be characteristic of the cobalt ion bonded to both four framework oxygen atoms and the extraframework oxygen ligands, and the extraframework oxygen species can be removed by prolonged evacuation at 400°C without open to air or moisture. The reduction of cobalt (III) may be carried out by H_2 (Schnoonheydt et al. 1989; Iton et al. 1989) with high temperature ($\geq$ 300°C), NO (Iton et al. 1989) and CH_3OH (Montes et al. 1990) at room temperature. In UV-Vis studies, the band at ca. 350 nm was assigned to be Co^{3+} of the green sample but did not tell whether cobalt in tetrahedral or octahedral coordination (Krourshaar-Czarnetzki 1991). In our case, the green calcined CoAPO-5 was reduced to blue color directly by evacuation at higher temperature. The association and dissociation of the extraframework oxygen species may play a role in the phenomena of color change. This indicates that the green sample might correspond to penta- or hexa- coordinated cobalt in the framework.

Variations of the ^{129}Xe NMR chemical shift and linewidth with the concentration of adsorbed xenon on the oxidized CoAPO-5 are shown in Fig 21-8. All the spectra of ^{129}Xe NMR display as a single, symmetrical but very broad signal. The variation in the range of low xenon loading, being characteristic of xenon-cage wall or xenon-cation interaction, depends only on the electronic and magnetic environment of the channel wall and intrachannel cations. The increased of chemical shift and linewidth with decreasing xenon coverage may be accounted for by the existence of cobalt on the framework or inside the channel cavities. The large values of chemical shift and linewidth may be interpreted as the magnetic field broadening due to the effect of unpaired electrons in the cobalt ion on CoAPO-5.

Acidity

The study of temperature programmed desorption of ammonia was used to provide the indirect evidence of silicon or metal substitution in the neutral aluminophosphate framework. With on-line titration, the number of Brønsted acid sites as well as their strength can be

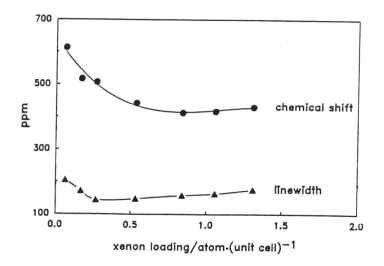

FIGURE 8. Dependence of the ^{129}Xe NMR chemical shift (●) and linewidth (▲) on the number of xenon atoms adsorbed on CoAPO-5 (III).

characterized on the NH_4^+ form sample. The TPD curves of the NH_4SAPO-5 and NH_4CoAPO-5 are given in Fig. 21-9. The maximum desorption temperature, (Td), varies proportionally to the strength of acid sites while the acid values are calculated from the volume of acid titrant used for neutralizing evolved ammonia.

The acid values of NH_4SAPO-5 (0.51 m mole g^{-1}) and NH_4CoAPO-5 (0.18 or 0.25 m mole g^{-1}) are about 1/2 of the heteroatom content in samples. The NH_4CoAPO-5 samples were prepared by NH_4^+ exchange with the synthetic sample preheated under vacuum to 400°C and with the synthetic sample calcined in dry air and oxygen flows to 550°C. After calcination, the color of CoAPO-5 turned from blue to green. Since the blue is typical of tetrahedral Co^{2+} and the green can be assigned to Co^{3+} ions, the acidity of CoAPO-5 may be different before and after calcination. Ammonia desorption peaks of NH_4CoAPO-5 from air treatment appear at 330 and 410°C instead of only one at 380°C for the vacuum treated sample. This implies that additional sites of medium acid strength have been generated with the strong acid sites on CoAPO-5 by calcination treatment. However, the generation of extraframework species may reduce the volume of intrachannel void space which can be detected by xenon adsorption

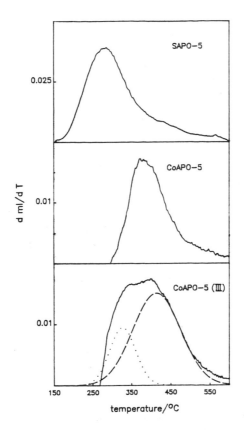

FIGURE 9. TPD of NH_3 from NH_4 form of SAPO-5, CoAPO-5 and CoAPO-5 (III) (the profiles were obtained by a least-squares fitting of titration data).

isotherm. The values of maximum xenon loading, ρ_∞, shown in Table 21-1 are the same for evacuated and calcined samples. This indicates that intraframework cobalt species are not formed upon calcination and Co^{3+} from structural Co^{2+} may remain in the lattice. It is not clear how the new acid sites are generated by the oxidation of framework Co^{2+}. The ammonium decomposition temperatures of both NH_4CoAPO-5 samples are higher than that of NH_4SAPO-5 ($Td=280°C$). This shows that the Brønsted acidity of CoAPO-5 is stronger than that of SAPO-5.

Evacuating to 400°C removes the NH_3 from both NH_4CoAPO-5 (II)and NH_4CoAPO-5 (III) samples and $Et_3N•CoAPO$-5 sample and induces a broad band of weak intensity at 3673 cm^{-1}. The similar band was observed by Tapp et al. (1985) and Shiralkar et al. (1989). This is the hydroxy stretch associated with framework Co^{2+} or Co^{3+} and higher than those of acidic OH group (3623 and 3520 cm^{-1}) brought by Si substitution for P on AlPO$_4$-5 (Wang et al. 1991).

For comparison, the temprature programmed decomposition of Et$_3$N•CoAPO-5 and Et$_3$N•SAPO-5 were studied under Ar flow or a gaseous mixture of N$_2$ with O$_2$ or air. The results of thermogravimetric (TG) and differential (DTA) thermal analyses are given in Fig. 21-10. The low-temperature (30-300°C) endotherms are characteristic of dehydration with weight loss of 4.69 wt% for SAPO-5, 6.30 wt% for CoAPO-5 and 8.78% for CoAPO-34. The weight loss of water indicates the order of hydrophilicity in CoAPO-34 > CoAPO-5 > SAPO-5. The water contents of our SAPO-5, CoAPO-5 and CoAPO-34 samples are larger than those measured by Goepper et al. (1989). It is interesting to note the Et$_3$N decomposition peak at 300-550°C. The exothermic peaks at 425-435°C on SAPO-5 and at 450-470°C

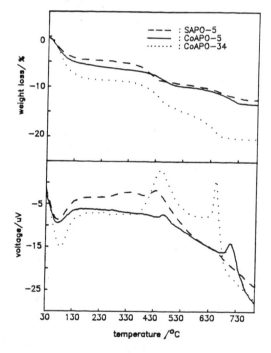

FIGURE 10. TG (top) and DTA (bottom) thermal analyses of SAPO-5, CoAPO-5 and CoAPO-34.

on CoAPO-5 under Ar flow may come from the Et_3NH^+ ions which are counterbalancing the anionic framework. The higher desorption temperature, the stronger acidity is consistent with the results of TPD of NH_3. There is only one DTA peak in the range of 300-450°C for CoAPO-34 which contains more cobalt in its framework than CoAPO-5 and all its organic template may be in a protonated form Et_3NH^+ and not in Et_3N. The third loss (500-800°C) is accompanied by the exotherm and relates to the organic species generated from Et_3N in the small intraframework void space. The DTA and TG curves under Ar flow can be employed to distinquish the product crystals in -5 or -34 structure. The temperature of the last stage of weight loss decreases under $O_2/N_2 = 0.13$ gas flow for both CoAPO-5 and SAPO-5.

References

Appleyard, I.P., Harris, R.K., and Fitch, F.R. 1985. Aluminum-27, phosphorus-31, and silicon-29 MAS NMR studies of the silicoaluminophosphate molecular sieve SAPO-5. Chem. Lett.: 1747-48.

Bennett, J.M. et al. 1983. Crystal structure of tetrapropyl-ammonium hydroxide-aluminum phosphate number 5. In Intrazeolite Chemistry, ed. G.D. Stucky and F.G. Dwyer, pp. 109-118. Washington, D.C.: American Chemical Society.

Chao, K.J., Chen, S.H., and Liu, S.B. 1991. NMR studies of cation location is zeolites. In Chemistry of Microporus Crystals, ed. Tomoyuki Inui, Seitaro Namba, and Takashi Tatsumi, pp. 123-132. Tokyo: Kodansha.

Chen, Q.J., Springuel-Huet, M.A., and Fraissard, J. 1989. ^{129}Xe NMR of xenon adsorption on the molecular sieves $AlPO_4$-5, SAPO-5, MAPO-5 and SAPO-37. Chem. Phys. Lett. 159 (1) : 117-121.

Flanigen, E.M. et al. 1986. Aluminophosphate molecular sieves and the periodic table. In New Developments in Zeolites Science and Technology, ed. Y. Murakami, A. Iijima, and J.W. Ward, pp. 103-113. Tokyo: Kodansha.

Fraissard, J. et al. 1986. Adsorption of xenon: a new method for studying zeolites. In New Developments in Zeolite Science and Technology, ed. Y. Murakami, A. Iijima, and J.W. Ward, pp. 393-400. Tokyo: Kodansha.

Goepper, M. et al. 1989. Effect of template removal and rehydration on the structure of $AlPO_4$ and $AlPO_4$ based microporous crystalline solids. Zeolites: Facts, Figures, Future, ed. P.A. Jacobs, and R.A.

van Santen, pp. 857-66. Amsterdam: Elsevier.

Iton, L.E. et al. 1989. Stabilization of Co(III) in aluminophosphate molecular sieve frameworks. Zeolites 9 : 535-38.

Jameson, C.T. 1975. An empirical chemical shielding function for interaction atoms from direct inversion of NMR data. J. Chem. Phys. 63 (12) : 5296-6301.

Karushaar-Czarnetzki, B. 1991. Characterization of Co^{II} and Co^{III} in CoAPO molecular sieves. J. Chem. Soc. Faraday Trans. 87(6): 891-95.

Leu, L.J., and Chao, K.J. 1988. Chemical modification of molecular sieve aluminophosphate number 5. Proc. Natl. Sci. Counc. ROC(A), 12(2): 91-96.

Lok, B.M. et al. 1984. Crystalline silicoaluminophosphates. U.S. Pat. 4 440 871.

Martens, J.A. et al. 1988. Synthesis and charactrization of silicon-rich SAPO-5. In Innovation in Zeolite Materials Science, ed. P.J. Grobet, W.J. Mortier, E.F. Vansant, and G. Schulz-Ekloff, pp. 97-106. Amsterdam: Elsevier.

Montes, C et al. 1990. Isolated redox centers within microporous environments 1. Cobalt-containing aluminophosphate sieve five. J. Phys. Chem. 94: 6425-30.

Muller, U., and Unger, K.K. 1988. Preliminary studies on the synthesis of alkaline-free large crystals of ZSM-5. Zeolites 8: 154-56.

Pellegrino, C et al. 1990. Characterization of zeolite NU-10 by ^{129}Xe NMR spectroscopy. Appl. Catal. 61: L1-L7.

Ryke, D.R., Whitney, P., and Houghton, H. 1985. Chemical Modification of crystalline microporous aluminium phosphates Appl. Catal. 18: 173-90.

Rietveld, H.M. 1967. Line profiels of neutron powder-diffraction peaks for structure refinement. Acta Cryst. 22 : 151-53.

Schnoonheydt, R. A. et al. 1989. Spectroscopy of cobalt in CoAPO-5. In Zeolites, Facts, Figures, Future, ed P.A. Jacobs, and R.A. van Santen, pp. 559-68. Amsterdam: Elsevier.

Shiralkar, V.P. et al. 1989. Synthesis and characterization of CoAPO-5, a cobalt-containing $AlPO_4$-5. Zeolites 9: 474-82.

Tapp, N.J., Milestone, N.B., and Bibby, D.M. 1988. Generation of acid sites in substituted aluminophosphate molecular sieves. In Innovation in Zeolite Materials Science, ed. P.J. Grobet, W.J. Mortier, E.F. Vansant, and G. Schulz-Ekloff, pp. 393-402. Amsterdam: Elsevier.

Tapp, N.J., Milestone, N.B., and Wright, L.J. 1985. Substitution of divalent cobalt into aluminophosphate molecular sieves. J. Chem. Soc. Chem. Commun. : 1801-2.

Wang, R. et al. 1991. Silicon species in a SAPO-5 molcular sieve. Appl. Catal. 72 : 39-49.

Wiles, D.B., and Young, R.A. 1981. A new computer program for Rietveld analysis of X-ray powder diffraction patterns. J. Appl. Crystal. 14: 149-151.

Wilson, S.T., and Flanigen, E.M. 1989. Synthesis and characterization of metal aluminophosphate molecular sieves. In Zeolite Synthesis, ed. Mario L. Occelli, and Harry E. Robson, pp. 329-345. Washington, D.C.: American Chemical Society.

22

Solid-State NMR Studies of a High-Temperature Structural Transformation of Hydrated Porous Aluminophosphate VPI-5

Heyong He, Waclaw Kolodziejski, João Rocha, and Jacek Klinowski

Department of Chemistry, University of Cambridge, Lensfield Road, Cambridge CB2 1EW, U.K.

A reversible structural transformation in the aluminophosphate molecular sieve VPI-5 has been observed in the temperature range 340-350 K by variable temperature ^{31}P, ^{27}Al and ^{1}H magic-angle-spinning (MAS) solid-state NMR, and ^{27}Al quadrupole nutation NMR with MAS. For the first time, using ^{2}H NMR of static samples, a direct evidence has been given for the presence of two kinds of water molecules in hydrated VPI-5.

INTRODUCTION

The aluminophosphate molecular sieve VPI-5 contains eighteen-membered rings of Al and P atoms linked via oxygens (Fig. 22-1).[1,2] The

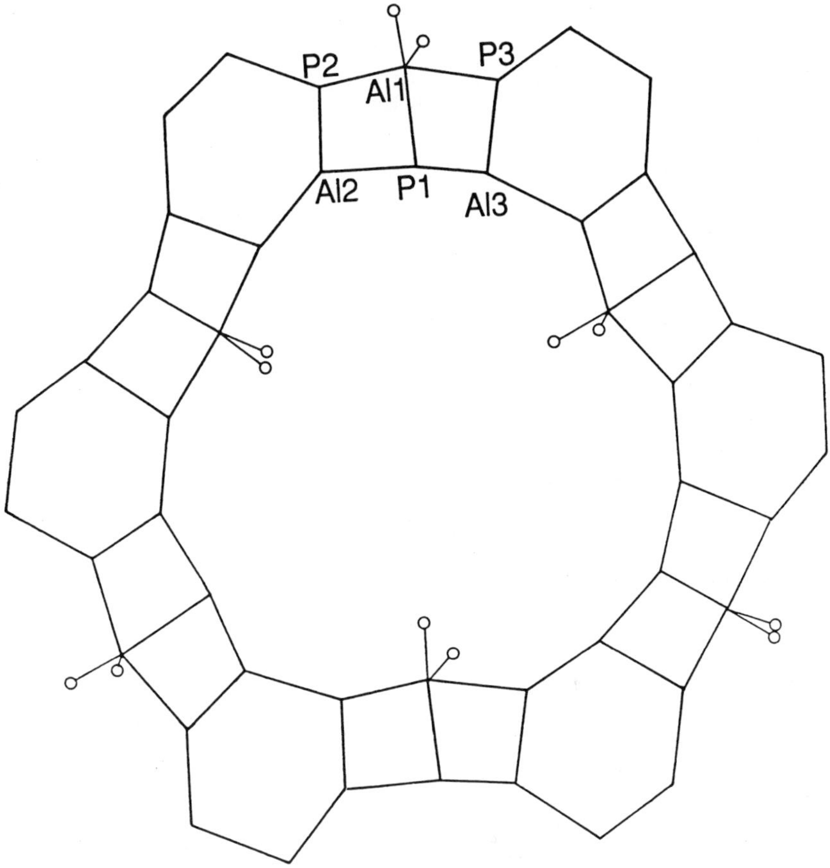

Fig. 22-1. One layer of the framework structure of hydrated VPI-5 taken from the stereoscopic view along the [001] direction according to McCusker et al.[6] showing the deviation from $P6_3cm$ symmetry. Aluminum and phosphorous atoms, linked via oxygen atoms (not shown for clarity) are located at the apices of the polygons. Sites located between two fused 4-membered rings are referred to in the text as 4-4 sites; those located between 6-membered and 4-membered rings are referred to as 6-4 sites. P2 and P3, and Al2 and Al3 sites are inequivalent as a result of the distortion. The Al1 site is 6-coordinated as a result of bonding to four bridging oxygens and two "framework" water molecules. Other intracrystalline water is not shown.

large channel diameter of ca. 12 Å gives the material potential for the separation of large molecules and for catalytic cracking of heavy fractions of petroleum. Several attempts have been made to determine the structure of VPI-5[3-5] and the latest X-ray refinement of the fully hydrated sample in the space group $P6_3$ by McCusker et al.[6] is finally consistent with the room temperature ^{27}Al and ^{31}P NMR spectra. Thus ^{31}P MAS NMR resolves three peaks in a 1 : 1 : 1 intensity ratio[7-9] and three [27]Al signals have been observed by double-rotation NMR.[10] Further, variable-temperature ^{31}P MAS NMR[9] indicates the presence of a high-temperature structural transformation in hydrated VPI-5. We have carried out a combined variable-temperature ^{1}H and ^{31}P MAS, ^{27}Al quadrupole nutation MAS NMR,[11] as well as ^{2}H NMR of static samples, in the range 294-370 K using fast spinning rates. These are the first variable-temperature quadrupole nutation spectra using MAS.

EXPERIMENTAL

VPI-5 was prepared according to Davis et al.[12] using di-n-propylamine (DPA) as a template with hydration for 12 h (25 wt % water) over a saturated solution of NH_4Cl. Elemental analysis indicates that the DPA content in the samples corresponds to about one molecule per three unit cells. For ^{2}H NMR a sample saturated with D_2O was prepared.

Variable-temperature ^{31}P, ^{27}Al, ^{1}H and ^{2}H NMR spectra were recorded on a Bruker MSL-400 spectrometer at 162.0, 104.2, 400.13 and 61.4 MHz, respectively, allowing 15 min for temperature equilibration. A MAS double-bearing probe was used and tightly closed 4 mm zirconia rotors were spun in nitrogen gas. [31]P spectra were measured with MAS at 8-12 kHz using 30° pulses and 30 s recycle delays. Single-pulse [27]Al spectra were measured with MAS at 12 kHz using very short, 0.6 μs (less than 10°), radiofrequency pulses and 0.4-0.5 s recycle delays. Quadrupole nutation spectra were recorded with a radiofrequency field $\omega_{rf}/2\pi$ of 115±5 kHz and with MAS at 8-12 kHz. Forty to 64 data points were collected in the t_1 dimension in increments of 0.5 or 1 μs, the damping of the t_1 FID being ca. 30 μs. A sine-bell digitizer filter and zero filling were used in the t_1 dimension and the FIDs were doubly Fourier transformed in the magnitude mode. [1]H spectra were measured with MAS at 7 kHz using 45° pulses and 2-4 s recycle delays. Static ^{2}H spectra were recorded with the two-pulse quadrupole echo pulse sequence using 20 μs echo delay and 2 s recycle delay.

RESULTS AND DISCUSSION

Variable-temperature ^{31}P MAS spectra of hydrated VPI-5 are shown in Fig. 22-2. The room temperature (294 K) spectrum contains three resonances in a 1 : 1 : 1 intensity (peak area) ratio with spin-lattice relaxation times of 26±2 s (MAS at 7.2 kHz). The ^{31}P resonances at -23 and -27 ppm come from P atoms in the 6-4 sites and the signal at -33 ppm to P atoms in 4-4 sites.[9] Above room temperature the resonance at -27 ppm grows at the expense of that at -23 ppm. The growing peak shifts only slightly from -27.7 ppm at 294 K to -26.6 ppm at 348 K, while the resonance decreasing in intensity shifts gradually from -23.6 to -25.0 ppm. The signal at -23 ppm disappears in the 340-350 K range, and the spectrum at 353 K contains only two signals with the intensity ratio of 2 : 1.

The ^{27}Al MAS NMR spectra of VPI-5 at 294-370 K (not shown) contain two signals at ca. 40 and -18 ppm in a 2 : 1 intensity ratio. The high-frequency signal is composed of two overlapping resonances from 4-coordinated Al in a 1 : 1 intensity ratio (6-4 sites), while the low-frequency resonance is due to 6-coordinated Al (4-4 site). A room temperature quadrupole nutation spectrum in the region corresponding to 4-coordinated Al (Fig. 22-3) has one signal at $\omega_{rf}/2\pi$ and two signals between $\omega_{rf}/2\pi$ and ca. $3\omega_{rf}/2\pi$. This indicates that the two signals from 4-coordinated Al, which overlap in the ordinary MAS NMR spectrum, have slightly different quadrupole interaction parameters. When the temperature is increased from 294 to 343 K, the resolution of the two 4-coordinated Al resonances decreases considerably and above the latter temperature only one signal is detected (Fig. 22-4).

At room temperature VPI-5 gives a broad proton signal from water (at 5 ppm from TMS), which narrows considerably (from 1220 to 240 Hz) in the 294 - 370 K range. This temperature effect is well described by a single-parameter exponential function, and an Arrhenius plot (not shown) yields an activation energy of -21.2 kJ mol^{-1}. We stress that this value should be regarded merely as an estimate of the magnitude of the interactions involved. Preliminary results show that $T_{1\rho}$ of the water signal increases substantially with temperature. The ^{2}H NMR spectrum of water contains two overlapping components, the shape and possibly relative intensities of which substantially change with temperature (Fig. 22-5).

CONCLUSIONS

Our NMR studies for *both* ^{27}Al and ^{31}P in the VPI-5 framework show that above ca. 343 K the two 6-4 tetrahedral sites collapse into one type of 6-4 site, while proton NMR simultaneously monitors a dramatic increase of water

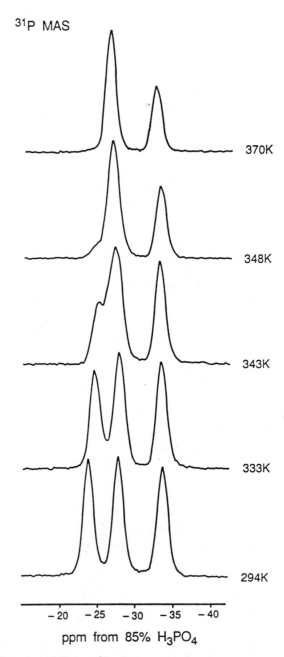

Fig. 22-2. Variable-temperature ^{31}P MAS NMR spectra of hydrated VPI-5.

^{27}Al quadrupole nutation

294K

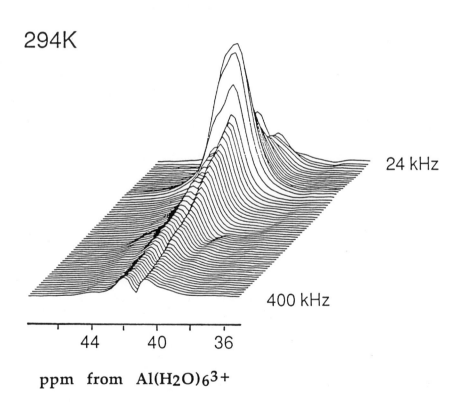

24 kHz

400 kHz

44   40   36

ppm from Al(H$_2$O)$_6^{3+}$

Fig. 22-3. Room temperature ^{27}Al quadrupole nutation spectrum in the region of 4-coordinated Al of hydrated VPI-5 measured with $\omega_{rf}/2\pi = 115\pm5$ kHz radiofrequency field.

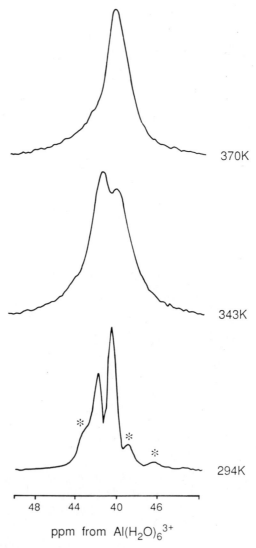

^{27}Al quadrupole nutation NMR
cross-sections at $2.5\omega_{rf}/2\pi$

370K

343K

294K

48   44   40   46

ppm from $Al(H_2O)_6^{3+}$

Fig. 22-4. Cross-sections parallel to the F_2 axis (taken at $2.5\omega_{rf}/2\pi$) of the ^{27}Al quadrupole nutation spectra of hydrated VPI-5 at temperatures given. Asterisks mark impurity signals.

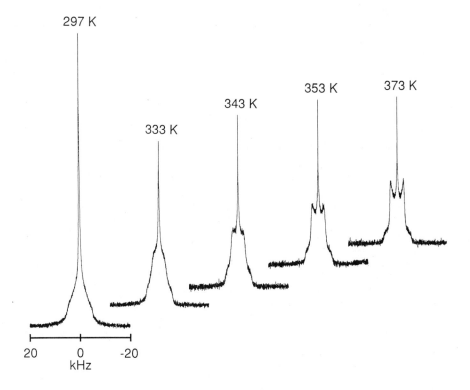

Fig. 22-5. Variable-temperature static ^{2}H NMR spectra of hydrated VPI-5. The spectra are plotted on the absolute intensity scale.

molecular motion. The results thus confirm the presence of a high-temperature structural transformation to a higher framework symmetry, possibly to a $P6_3cm$ space group, and support the suggestion[9] that it is caused by the breakdown of the hydrogen-bonded structure of water inside the pores. All the techniques used show that the phase transformation is fully reversible. Further, ^{2}H NMR detects two kinds of water molecules with different mobility, in accordance with the X-ray refinement.[6] Lineshape simulations, intended to extract details of this motion, are in progress.

Acknowledgments. We are grateful to Unilever Research, Port Sunlight, for support.

References

1 .   Davis, M.E.; Saldarriaga, C.; Montes, C.; Garces, J.; Crowder, C. *Nature* **1988**, *331*, 698.

2 .   Davis, M.E.; Saldarriaga, C.; Montes, C.; Garces, J.; Crowder, C. *Zeolites* **1988**, *8*, 362.

3 .   Crowder, C.E.; Garces, J.; Davis, M.E. *M. E. Adv. X-ray Anal.* **1988**, *32*, 507.

4 .   Richardson, J.W., Jr.; Smith J.V.; Pluth, J.J. *J. Phys. Chem.* **1982**, *93*, 8212.

5 .   Rudolf, R.R.; Crowder, C.E. *Zeolites* **1990**, *10*, 163.

6 .   McCusker, L.B.; Baerlocher, Ch.; Jahn, E.; Bülow, M. *Zeolites* **1991**, *11*, 308.

7 .   Davis, M.E.; Montes, C.; Hathaway, P.E.; Arhancet, J.P.; Hasha, D.L.; Garces, J.M. *J. Am. Chem. Soc.* **1989**, *111*, 3919.

8 .   Grobet, P.E.; Martens, J.A.; Balakrishnan, I.; Mertens, M.; Jacobs, P.A. *Appl. Catal.* **1989**, *56*, L21.

9 .   van Braam Houckgeest, J.P.; Kraushaar-Czarnetzki, B.; Dogterom, R.J.; de Groot, A. *J. Chem. Soc., Chem. Comm.* **1991**, 666.

1 0 . Wu, Y.; Chmelka, B.F.; Pines, A., Davis, M.E.; Grobet, P.J.; Jacobs, P.A. *Nature* **1990**, *346*, 550.

1 1 . Samoson, A.; Lippmaa, E. *J. Magn. Reson.* **1988**, *79*, 255.

1 2 . Davis, M. E.; Montes, C.; Hathaway, P. E.; Garces, J. M. *Zeolites: Facts, Figures, Future,* Jacobs, P. A.; van Santen, R. A. (eds.), Elsevier, Amsterdam, 1989, p. 199.

23

Synthesis of VPI-7: A Novel Molecular Sieve Containing Three-Membered Rings

Michael J. Annen and Mark E. Davis
Department of Chemical Engineering, Virginia Polytechnic Institute and State University, Blacksburg, VA 24061
Current address: Department of Chemical Engineering, California Institute of Technology, Pasadena, CA 91125

We report here various synthetic procedures used to crystallize VPI-7. The framework of VPI-7 contains 3, 4 and 5 T-atom rings which form unidimensional 8- and intersecting 9MR channels and is the first microporous zincosilicate to contain 3-membered rings (3MR). Unlike the other 3MR-containing molecular sieves, lovdarite and ZSM-18, VPI-7 can be synthesized in the absence of potentially toxic materials or highly specific organic molecules, respectively. Four different synthesis procedures are provided.

Zeolites and zeolite-like materials represent a large class of microporous solids. Novel molecular sieve structures, particularly those with void volumes lower than currently attainable, are of great interest for their potential use as catalysts and adsorbents. The void volume is inversely related to framework density (FD, the number of T-atoms per cubic nanometer) and recently the existence of a relationship between FD and ring size has been shown (1,2). This correlation indicates that materials which contain 3-membered rings (3MR) could have FDs lower than materials composed of 4 T-atom rings or larger, and thus yield novel, high void-volume molecular sieves.

We have initiated a project aimed specifically at the synthesis of novel molecular sieves which contain 3MR. Beryllosilicate and zincosilicate minerals with 3MR do exist (3). However, due to the toxic nature of certain beryllium compounds we chose zincosilicate gel chemistry as the starting point. VPI-7, the first microporous zincosilicate to contain 3MR, is the first molecular sieve to be synthesized in this project. VPI-7 (portion of the structure shown in Figure 23-1) is structurally related to the LOV topology (3) and contains 3, 4 and 5 T-atom rings which form unidimensional 8- and intersecting 9MR channels. VPI-7 possesses tetragonal I4m2 symmetry. As made, VPI-7 has unit cell dimensions of $\underline{a}$=7.265(1)Å and $\underline{c}$=40.33(1)Å and a FD=16.9.

We report here four different synthesis procedures used to crystallize VPI-7 and some physicochemical characterization data from the materials. The syntheses were chosen to give an indication of the broad range of reaction conditions and gel compositions over which VPI-7 can be synthesized. While these synthesis procedures are not necessarily optimized, they do show the utility of zincosilicate gel chemistry for the synthesis of 3MR-containing materials and the ease with which VPI-7 can be synthesized. The synthesis of VPI-7 does not require the use of potentially toxic materials or highly specific organic molecules, unlike the other 3MR-containing molecular sieves, lovdarite (4) and ZSM-18 (5), respectively.

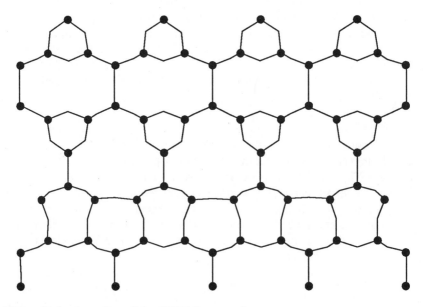

Figure 23-1. A portion of the VPI-7 framework.

Experimental Section

Zinc oxide (Aldrich), zinc acetate (Baker) and zinc sulfate (Alfa) were the zinc sources used. Ludox HS-40 (DuPont), amorphous silica (Syloid 74, Davidson) and tetraethylorthosilicate (TEOS) (Aldrich) were used as the silica sources. NaOH, NaCl and KOH, purchased from Aldrich, were used as the sources of sodium and potassium. Aqueous (40 wt%) tetraethylammonium hydroxide (TEAOH) was purchased from Alfa; ethanol and nitric acid were purchased from Fisher.

X-ray diffraction (XRD) powder patterns were collected on a Siemens I2 diffractometer using Cu Kα radiation. A step of 0.05 2-Θ and count of 2 seconds/step were used to collect the powder patterns. The ^{29}Si NMR spectra in Figure 23-3 were recorded on a Bruker MSL 300 spectrometer at 59.6 MHz and a spinning rate of 3 KHz.

Figure 23-2 shows unmodified experimental XRD powder patterns for VPI-7 synthesized by two different methods (vide infra). The differences in relative intensity between the two patterns are most likely due to preferred orientation effects which are present in pattern A. VPI-7 commonly crystallizes as thin rectangular plates which often agglomerate into fans. Such a crystal morphology frequently gives rise to preferred-orientation effects in the powder pattern.

Synthesis I

A gel of composition:

0.44 Na$_2$O :0.15 K$_2$O :0.04 TEA$_2$O :0.039 ZnO :SiO$_2$:22 H$_2$O is prepared by dissolving 18.6 g NaOH, 9.0 g KOH and 1.7 g ZnO in 160 ml of deionized H$_2$O with stirring for 10 minutes. The resulting solution is clear. 80.0 g of Ludox HS-40 are added and the solution is stirred for 5 minutes. Finally, 61.2 g of TEAOH (40 wt%) are added and the resulting reaction mixture stirred for an additional hour. The gel is charged into 45-ml Teflon-lined autoclaves and heated at 200°C under autogenous pressure for 3 days. To recover the product VPI-7, the autoclaves are quenched in cold water and the contents slurried in de-ionized H$_2$O. The supernatant liquid is decanted and the solid is recovered by vacuum filtration. The crystals are dried in ambient air or in an oven at 100°C. Aside from the substitution of beryllium by zinc, this synthesis procedure is very similar to that reported by Ueda et al. for the synthesis of lovdarite (4).

Figure 23-2A shows a representative XRD powder pattern for the product collected. Table 23-1 contains a listing of d-spacings, relatives intensities and Miller indices for as-made VPI-7.   A representative ^{29}Si NMR spectrum for VPI-7 from this synthesis is shown in Figure 23-3A.

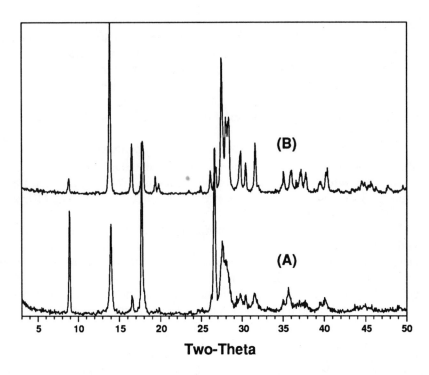

Figure 23-2. Representative X-ray powder diffraction patterns of VPI-7 synthesized by (A) method I and (B) method III.

For all VPI-7 syntheses in which TEAOH is added to the reaction mixture, it serves only serves to enhance the rate of crystallization and is not found within the micropores of VPI-7. Figure 23-4 shows a typical thermal trace for as-made VPI-7. All weight losses are due to the removal of water from the channels since differential scanning calorimetry indicates that all the loss events are endothermic. Elemental analysis also does not indicate the presence of TEAOH within as-made VPI-7.

Table 23-1. X-ray Powder Diffraction Data for VPI-7

2-Θ	d (Å)	I/Io	h k l
4.382	20.164	1	0 0 2
8.771	10.082	100	0 0 4
12.343	7.171	1	1 0 1
13.865	6.387	36	1 0 3
16.431	5.395	6	1 0 5
17.600	5.039	78	0 0 8
26.511	3.362	67	0 0 12
27.925	3.195	9	2 0 6
28.244	3.160	6	2 1 3
30.359	2.944	2	2 0 8
31.448	2.845	7	1 0 13
34.947	2.567	2	2 2 0
35.621	2.520	6	0 0 16
37.750	2.383	2	3 0 3
39.448	2.284	1	3 1 2
43.721	2.070	1	2 0 16

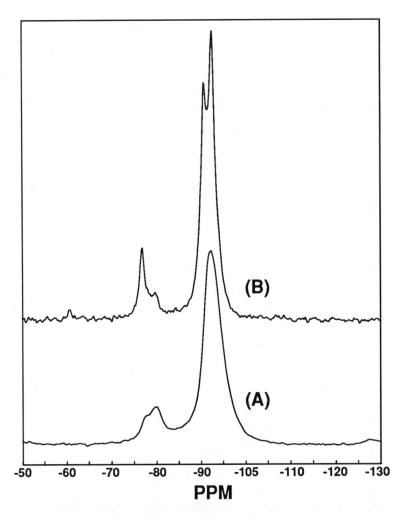

Figure 23-3. Representative ^{29}**Si NMR spectra for VPI-7 synthesized by (A) method I and (B) method III.**

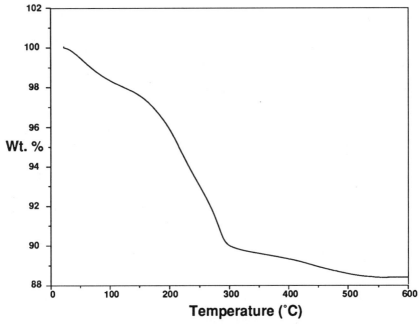

Figure 23-4. Thermogravimetric analysis for VPI-7.

<u>Synthesis II</u>
A gel of composition:
0.45 Na$_2$O :0.03 TEA$_2$O :0.03 ZnO :SiO$_2$:17 H$_2$O is prepared by dissolving 2.3 g of zinc acetate and 11.8 g of NaOH in 80 ml of deionized H$_2$O with stirring for 15 minutes. To the resulting clear solution, 31.7 g of TEAOH (40 wt%) are added, the solution is stirred for 2 minutes and 52.3 g of Ludox HS-40 are added. This reaction mixture is stirred for 30 minutes and then charged to a 100-ml and a 25-ml Teflon-lined autoclave. The autoclaves are heated statically at 200°C for 40 hours under autogenous pressure. The product VPI-7 is recovered by the process outlined in Synthesis I. The XRD powder pattern and ^{29}Si NMR spectrum of the product collected are essentially the same as those shown in Figures 23-2A and 23-3A, respectively.

The next two syntheses are included to emphasize the different gel-preparation methods and reaction conditions which can be used to crystallize VPI-7.

Synthesis III

VPI-7 can be crystallized from a reaction mixture with composition:
0.46 NaOH :0.15 NaCl :0.23 ZnO :SiO$_2$:71 H$_2$O. The gel preparation method and synthesis conditions are similar to those outlined by Ueda and Koizumi for the synthesis of beryllosilicate gismondine (6). Briefly the procedure is as follows: 0.6 g NaOH and 0.3 g NaCl are dissolved in 20 ml of deionized H$_2$O. 5.0 g of Ludox HS-40 and 0.6 g of ZnO are added and the solution is stirred for 10 minutes. The mixture is evaporated to dryness in an oven at 100°C for 30 hours. The resulting solid is ground into a fine powder and mixed with 42 ml of de-ionized H$_2$O and then stirred for 40 minutes. This reaction mixture is charged to a 45-ml Teflon-lined autoclave and heated at 150°C for 55 days under autogenous pressure. The product is recovered by the method outlined in Synthesis I.

The XRD powder pattern for VPI-7 crystallized by this procedure exhibits unusually distinct reflections in the region of 30°-50° 2-Θ, as shown in Figure 23-2B. The ^{29}Si NMR spectrum for the product of Synthesis III, shown in Figure 23-3B, exhibits a splitting of the broad resonance near -90 ppm which is not typically observed for VPI-7 synthesized by methods I or II.

Synthesis IV

A gel of composition: Na$_2$O :ZnO :3 SiO$_2$
is prepared by a method similar to that described by Hamilton and Henderson (7). A similar gel was used in the synthesis of disodium zincosilicate (8). The following is a short description of the procedure: 4.5 g zinc sulfate, 6.6 g NaOH and 150 ml of nitric acid (~70 wt%) are combined and stirred for 15 minutes. The solution is filtered and the remaining solid is dried in a vacuum oven at 50°C. The resulting nitrate salts are dissolved in 15 ml of de-ionized H$_2$O. To this salt solution 17.3 g of TEOS are added, after which sufficient ethanol is added to render the TEOS miscible. Ammonium hydroxide is the added dropwise to precipitate silica from the stirring solution. The mixture is allowed to age overnight and then the excess ethanol is evaporated slowly by heating the solution. Next, the mixture is dried under vacuum at 25°C then 110°C. The dry solids are fired to 800°C in 1 hour and held at this temperature for 10 minutes. The remaining dry solid or "gel" is ground into a fine powder.

The synthesis proceeds by adding 0.100 g of the solid described above, 0.026 g of amorphous silica and 0.021 g of de-ionized H$_2$O into a gold tube. The gold tube is sealed and heated for 7 days under a pressure of 1000 bar at 250°C. After reaction, the solid product is mixed with acetone and the slurry is ground into a powder. The dry product is obtained by evaporating the acetone in ambient air. The XRD pattern for the solid product is essentially the same as that in Figure 23-2A; however, in this case quartz is an impurity in the product VPI-7.

Table 23-2. Summary of Gel Compositions and Reaction Conditions.

Code	Gel Composition	Temp. (°C)	Time (hours)	Product
I	0.44 Na_2O: 0.15 K_2O: 0.04 TEA_2O: 0.04 ZnO: SiO_2: 21 H_2O	200	72	VPI-7
II	0.45 Na_2O: 0.03 TEA_2O: 0.03 ZnO: SiO_2: 21 H_2O	200	40	VPI-7
III	0.46 NaOH: 0.15 NaCl: 0.23 ZnO: SiO_2: 71 H_2O	150	1320	VPI-7
IV	Na_2O: ZnO: 5.6 SiO_2: 7.1 H2O	250	168	VPI-7 + quartz
V	0.44 Na_2O: 0.15 K_2O: 0.04 ZnO: SiO_2: 22 H_2O	200	56	VPI-7
VI	0.44 Na_2O: 0.15 K_2O: 0.04 ZnO: SiO_2: 22 H_2O	150	432	VPI-7
VII	0.44 Na_2O: 0.04 TEA_2O: 0.286 ZnO: SiO_2: 44 H_2O	200	284	VPI-7 + amorphous gel
VIII	0.44 Na_2O: 0.04 TEA_2O: 0.039 ZnO: SiO_2: 22 H_2O	190 stirred	120	VPI-7
IX	0.3 K_2O: 0.04 TEA_2O: 0.3 ZnO: SiO_2: 22 H_2O	200	624	Unknown A
X	0.3 Li_2O: 0.04 TEA_2O: 0.3 ZnO: SiO_2: 22 H_2O	200	48	Unknown B

Discussion

The four synthesis procedures described above indicate the broad range of reaction conditions and gel compositions over which VPI-7 can be synthesized and each is included to bring out particular points. Methods I and II are considered "classical" hydrothermal VPI-7 syntheses. Method I illustrates both the similarities between VPI-7 and lovdarite syntheses and that VPI-7 can be synthesized in the presence of mixed inorganic bases (1.e. NaOH and KOH).

Method II shows that VPI-7 can be synthesized in the absence of potassium. Synthesis III is an example of a VPI-7 synthesis where NaCl is used as a sodium source and neither TEAOH nor KOH are components in the reaction mixture. When compared with the first three methods, Synthesis IV exemplifies the broad range of temperature and pressure under which VPI-7 can be synthesized. Like method 3, it also illustrates the utility of different gel-preparation methods applicable to the synthesis of VPI-7. While four different synthesis procedures have been described in detail, VPI-7 has also been synthesized using many other gel compositions; a summary of VPI-7 syntheses are included in Table 23-2.

The four synthesis procedures described above are distinctly different, yet together indicate an important point: sodium, like zinc, silicon and water, is a necessary component for the synthesis of VPI-7. Synthesis runs in which sodium, potassium and TEAOH were absent resulted in the crystallization of willemite and/or hemimorphite, depending on the synthesis conditions. The replacement of sodium by other metal cations such as potassium or lithium has resulted in other new materials which are currently being studied. These materials are labeled A and B in Table 23-2.

Acknowledgments

MJA thanks Prof. David Hewitt for assistance with Synthesis IV of VPI-7 and the Mobil Research and Development Corporation for financial assistance.

References

1. Brunner, G.O., Meier, W.M. Nature 1989, 337, 146.
2. Stixrude, L., Bukowinski, M.S.T. Am. Mineral. 1990, 75, 1159.
3. Annen, M.J., Davis, M.E., Higgins, J.B., Schlenker, J.L., Synthesis/Characterization and Novel Applications of Molecular Sieve Materials, Mater. Res. Soc. Proc., Vol 233, p. 245.
4. Ueda, S., Koizumi, M., Baerlocher, Ch., McCusker, L.B., Meier, W.M. 7th IZC, Tokyo, 1986, Poster Paper 3C-3.
5. Lawton, S.L., Rohrbaugh, W.J. Science 1990, 247, 1319.
6. Ueda, S., Koizumi, M. Nature Phys. Sci. 1972, 238, 139.
7. Hamilton, D.L., Henderson, C.M.B. Miner. Mag. 1968, 36, 832.
8. Hesse, K.F., Liebau, F., Bohm, H., Ribbe, P.H., Phillips, M.W. Acta Cryst. 1977, B33, 1333.

24

Zeolite NU-87: Aspects of Its Synthesis, Characterization, Structure, and Properties

J. L. Casci, M.D. Shannon, P.A. Cox, and S.J. Andrews,

ICI Chemicals and Polymers Ltd., England

NU-87 is a novel high-silica zeolite with a 2-D pore system based on intersecting 10 and 12 T-atom channels. This paper describes its preparation by hydrothermal crystallisation and characterisation by powder x-ray diffraction, electron microscopy and chemical and thermal analysis. Information on the thermal stability and sorptive properties of NU-87 is presented and its unique pore system described by means of a "pore map".

INTRODUCTION

In the last decade there have been a number of significant advances in the science of zeolite molecular sieves. Arguably two of the most important of these have been the preparation of the family of aluminophosphate zeotypes (Wilson, Lok and Flanigen 1982; Wilson et al. 1982) and the work on structure solution, which has resulted in a dramatic increase in the number of known molecular sieve topologies (Meier and Olson 1987).

Of particular interest is that what were probably the first three high-silica zeolites to be prepared, zeolites Beta (Wadlinger, Kerr and Rosinski 1967; Treacy and Newsam 1988), ZSM-5 and ZSM-11 (Meier and Olson 1987), are the only materials of this type to have multi-dimensional channel systems based on 10, or 12, T-atom rings. Materials discovered subsequently, such as ZSM-12, ZSM-18 (Lawton and Rohrbaugh 1990), ZSM-23, ZSM-48 (Schlenker et al. 1985), EU-1 and Theta-1 all have uni-dimensional 10- or 12-ring channel systems. Other materials such as the recently discovered ZSM-57 (Valyocsik and Page 1986), while having a 2-D channel system, consist of intersecting 10- and 8-ring channels (Schlenker, Higgins and Valyocsik 1989).

For many catalytic, and sorptive, applications it is advantageous to have materials with multi-dimensional channel systems and consequently the discovery of such a material is, potentially, of great significance.

This paper will describe aspects of the preparation, characterization and properties of a new 2-D structure type: zeolite NU-87 (Casci and Stewart 1990).

EXPERIMENTAL

Syntheses were carried out in 1- and 2-liter stainless steel autoclaves with continuous agitation using pitched-paddle type impellers. Autoclave temperature was controlled to within 2°C. Samples could be withdrawn from the autoclaves to allow the progress of the crystallization to be monitored. Typical reagents used were colloidal silica (SYTON X30, Monsanto; 30% w/w SiO_2), SOAL (sodium aluminate solution, Kaiser; 22.0% w/w Al_2O_3, 19.8% w/w Na_2O, 58.2% w/w H_2O) and decamethonium bromide (Fluka).

Products were analysed by x-ray powder diffraction (Philips APD 1700), SEM (Hitachi S650), TEM (Philips EM 301) and chemical (AES+ICP) and thermal (Stanton Redcroft STA-780) analysis. TGA and DTA were carried out in flowing air (heating rate 10°C per minute) on a sample which had been equilibrated with water vapor. Sorption measurements were obtained using an automated microbalance (CI Robal).

RESULTS AND DISCUSSION

Synthesis

Zeolite NU-87 can be prepared from a range of reaction mixture compositions. It is necessary, however, to add a template, or structure-directing agent, to facilitate the crystallization. For NU-87 suitable templates are long-chain bis-quaternary ammonium compounds of general formula:

$$[(CH_3)_3N-(CH_2)_m-N(CH_3)_3]^{2+} \, 2 \, Br^-$$

with m preferably in the range 8 to 12, that is, octamethonium to dodecamethonium bromide. NU-87 is most readily prepared, in high purity, from reaction mixtures with SiO_2/Al_2O_3 ratios from about 25 to 60 and moderate base levels ($NaOH/SiO_2$ about 0.3).

In this paper discussion will concentrate on NU-87 materials prepared using decamethonium bromide as the template, that is, decane-1,10-bis (trimethylammonium bromide). Using this template a typical synthesis mixture will have a (molar) composition:

$$60\ SiO_2 - 1.5\ Al_2O_3 - 9\ Na_2O - 7.5\ DecBr_2 - 3000\ H_2O - 2\ NaBr$$

The reaction mixture was prepared as follows:

Solution 1: 5.42g NaOH and 2.06g NaBr together with 6.95g of SOAL solution (22.0% w/w Al_2O_3, 19.8% w/w Na_2O, 58.2% w/w H_2O) were dissolved/dispersed in 200g of water.

Solution 2: 31.4g decamethonium bromide were dissolved in 200g of water.

Solutions 1 and 2 were thoroughly mixed and the resulting solution added, with stirring, to 120.2g of colloidal silica (Syton X30, 30% w/w SiO_2). The remaining water (51g) was then added and stirring continued for 5 minutes to give a smooth/homogeneous gel which was transferred to the reactor. Time zero was taken from when the autoclave reached reaction temperature.

Figure 24.1 Progress of the crystallisation of NU-87 as followed by xrd and pH measurements.

Crystallisation, at 180°C, from a reaction mixture of (molar) composition:

$$60\ SiO_2 - 1.5\ Al_2O_3 - 9\ Na_2O - 7.5\ DecBr_2 - 3000\ H_2O - 2\ NaBr$$

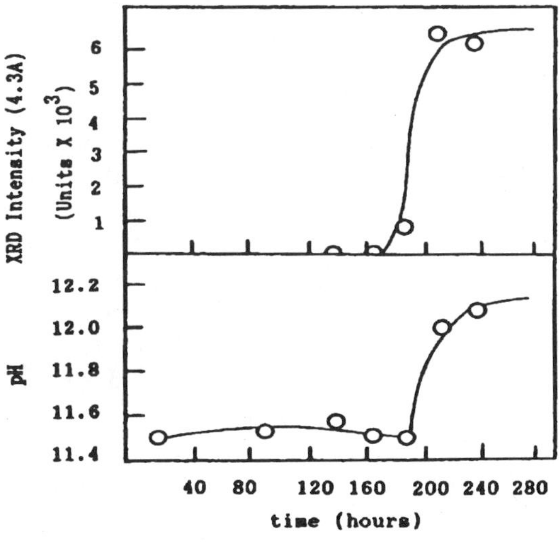

Figure 24.1 illustrates the crystallization of NU-87, at 180°C, from such a reaction mixture. The figure contains plots of both x-ray peak intensity and pH against time at reaction temperature. The interplanar spacing (d-spacing) at approximately 4.3 Angstroms in the NU-87 powder xrd pattern was used to follow crystallinity.

Examination of Figure 24.1 shows that, consistent with previous reports (Casci and Lowe 1983), measurement of pH gives a reasonable indication of the onset of crystallization, with the initial pH of about 11.5 rising to over 12 as crystallization proceeds. There is a suggestion that the pH lags behind x-ray crystallinity as previously predicted (Lowe 1983). The crystallization of NU-87 follows the standard format for zeolites with a relatively long induction period followed by a period of rapid growth during which the amorphous gel is converted to crystalline zeolite.

From Figure 24.1 it can be seen that maximum crystallinity is reached after about 9 days at reaction temperature. Compared to many other zeolites this is an unusually long crystallizsation time for such an elevated reaction temperature. However, attempts to shorten the synthesis time by significantly increasing reaction temperature tend to result in products contaminated with other phases, particularly mordenite.

Characterisation
XRD

Figure 24.2 contains the powder x-ray diffraction patterns for zeolite NU-87, both "as made" and after calcination and acid exchange, that is, H-NU-87.

The diffraction pattern for NU-87 acts as a fingerprint for this material, reflecting the fact that it has a new molecular sieve topology - see below. The pattern, however, does have some similarities with other zeolites, most notably EU-1 (Casci, Lowe and Whittam 1984). This, however, is a consequence of the structural relationship between these two materials (Shannon et al. 1991). NU-87 can be readily distinguished from EU-1 on the basis of the interplanar spacings at about 12.5, 8.3 and 4.2 Angstroms in addition to significant line shifts and changes in relative intensity.

Examination of Figure 24.2 shows that calcination, and acid exchange, of NU-87 results in some changes in the diffraction pattern with an increase in the intensities of low-angle lines relative to those in the mid-angle region. Similar changes occur in many high-silica zeolites following removal of the template, this effect being first reported for ZSM-5 (Wu et al. 1979).

Electron Microscopy
Examination of Decamethonium-NU-87 crystals by electron microscopy reveals an acicular morphology. Figure 24.3 contains both scanning (Figure

Figure 24.2   Powder x-ray diffraction patterns for "as made" and H-NU-87

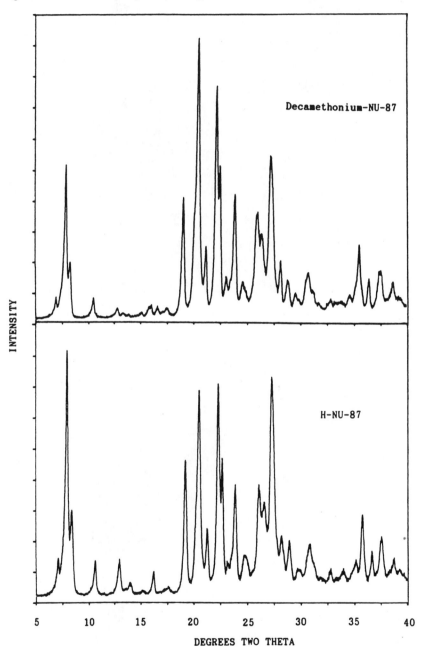

Figure 24.3   Electron photomicrographs of Decamethonium-NU-87

a (SEM)

1μ

b (TEM)

1μ

24.3a) and transmission (Figure 24.3b) electron photomicrographs from which a lath-like morphology can be clearly seen. Figure 24.3b, in particular, shows the crystals to have sharp, well-defined edges exhibiting the standard "cuboid-like" morphology for NU-87 samples made using the decamethonium template.

Typical crystals are approximately 0.5-1 micron in length and have an aspect ratio of about 5. There is considerable evidence for both multiple twinning and the formation of loosely bound aggregates of crystals.

Chemical and Thermal Analysis

Analysis of a sample of NU-87, prepared as described above, revealed the following (molar) composition:

$$27.9 \ SiO_2 - Al_2O_3 - 0.11 \ Na_2O$$

TGA and DTA results for the same material can be seen in Figure 24.4.

Figure 24.4   Thermal Analysis of Decamethonium-NU-87
(TGA/DTA; heating rate 10°C per min; air)

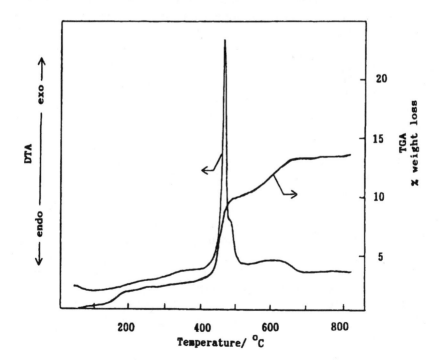

Examination shows a total weight loss of 13.5%. The initial weight loss of about 1.75%, completed by 250°C, is accompanied by a minor endotherm consistent with this being associated with water. The remaining weight loss can be attributed to occluded template (see below). From the TGA curve it can be seen that the template is lost in two distinct stages; while DTA indicates three separate exotherms. The first two exotherms occur at about 460 and 485°C (the higher temperature transition occurs as a shoulder on the 460°C peak) and are accompanied by rapid template loss. The final exotherm is very much broader and accompanies a much slower loss in weight. Of the total template lost, about one third occurs in this second stage.

Based on a unit cell containing 68 T-atoms (Shannon et al. 1991) this corresponds to a unit cell composition of (approximately):

$$63.5 \ SiO_2 - 4.5 \ AlO_2 - 0.5 \ Na - 2 \ Dec - 4 \ H_2O$$

Based on the assumption that all weight lost above 250°C was due to template, the exact number of moles of decamethonium calculated was 2.15. Such an assumption has dubious validity since it takes no account of water lost above 250°C, which may include water physically trapped by template or water of hydration associated with the occluded cations. This justifies the rounding down of the template content to 2 decamethonium molecules per unit cell. Template integrity was confirmed using ^{13}C cross-polarisation MASNMR spectroscopy, by comparing the spectra for a decamethonium-NU-87 with that for solid decamethonium bromide.

Interestingly, the above composition indicates complete charge balance. Analysis of several different samples of NU-87 prepared in the same way shows this to be a fairly consistent feature with samples in balance or showing a slight excess of cations (presumably associated with occluded hydroxide). In the samples examined, no evidence has been found for bromide.

Properties and Structure

Thermal Stability

The thermal stability of NU-87 was examined by taking a portion of NU-87, prepared as described above, and calcining to remove the template. (Calcination conditions were 450°C for 24 hours followed by 550°C for 16 hours, in static air.) Aliquots of the calcined material were then heated at progressively higher temperatures, in air, for 16 hours then examined by powder x-ray diffraction.

The absolute intensities of lines in the mid-angle region of the powder x-ray diffraction patterns were taken as a measure of crystallinity: the lines used corresponded to interplanar spacings of 4.6, 4.3 and 4.0 Angstroms. A plot of line intensities against calcination temperature is contained in Figure 24.5.

Figure 24.5   Thermal stability of NU-87.
                    Plot of XRD intensity against Calcination Temperature

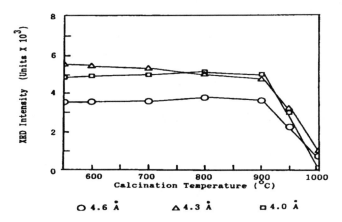

$\bigcirc$ 4.6 Å $\triangle$ 4.3 Å $\square$ 4.0 Å

From the figure it can be seen that the NU-87's x-ray pattern remains essentially intact until 900°C. At higher temperature there is a progressive loss in crystallinity until the sample is virtually amorphous to x-rays after 16 hours at 1000°C.

Not shown in Figure 24.5 is the effect of even higher temperatures. X-ray examination of a sample calcined at 1100°C showed alpha-quartz and alpha-cristobalite, indicating that at least some recrystallization had occurred.

Sorptive Properties

Sorption measurements were made on a sample of H-NU-87. The material was synthesized as described above and converted to the H-form by calcination (in air) to remove the template, then exchange with a 1M solution of hydrochloric acid.

Separate aliquots of H-NU-87 were used for each sorbate and samples were outgassed at 300°C before measurement. Although full isotherms, both adsorption and desorption, were obtained for all sorbates, only a summary of capacities will be presented here: results are shown in Table 24.1. Details of the complete isotherms will be presented elsewhere (Casci in preparation).

Table 24.1 contains information on the sorbates examined, the temperature at which the measurements were made and details of uptake. The sorbates are presented in ascending size, with methanol the smallest and neopentane the largest (Breck 1974). The figures for weight percent taken up by NU-87 were those recorded at p/p° of 0.4 and were used to calculate the values for "apparent voidage filled" assuming the sorbates have their normal (liquid) densities at adsorption temperature. With the exception of neopentane all of the isotherms were well behaved, "rectangular" in shape with little evidence for hysteresis (Gregg and Sing 1967).

Table 24.1. SORPTION DATA FOR H-NU-87

Sorbate	Adsorption Temperature/°C	Uptake %w/w at p/p° = 0.4	Apparent Voidage Filled/ cm³ g⁻¹
Methanol	21	11.75	0.15
n-Hexane	21	9.72	0.15
Toluene	27	12.18	0.14
Cyclohexane	27	11.73	0.15
Neopentane[1]	0	6.92[2]	0.11

1 2,2-Dimethylpropane.
2 Equilibration times very long; evidence for molecular sieving - see text.

Examination of the uptake figures in Table 24.1 shows that NU-87 has significant voidage available for methanol, n-hexane, toluene and cyclohexane. Indeed, the micro-pore volume figures calculated from the weight uptakes are remarkably constant.

It can be seen, however, that the uptake obtained for neopentane was significantly less than that measured for the other sorbates. In addition, time to reach equilibrium at each pressure was very much longer for neopentane than for the other hydrocarbons. This can be attributed to the size of neopentane, with a kinetic diameter of 6.2 Angstroms it is the largest of the sorbates (Breck 1974). This partial molecular sieving effect for neopentane indicates that sorption by NU-87 is controlled by windows close to 6.2 Angstroms.

Structure and Pore Map

The structure of zeolite NU-87 has been determined based on electron diffraction studies followed by high-resolution powder diffraction and lattice energy minimization (Shannon et al. 1991).

Based on a space group of P2₁/c NU-87 has unit cell dimensions:

a = 14.324 Angstroms
b = 22.376 Angstroms
c = 25.092 Angstroms
b = 151.515°.

There are 68 T-atoms in the unit cell
NU-87 has an interesting new framework topology based on a two-

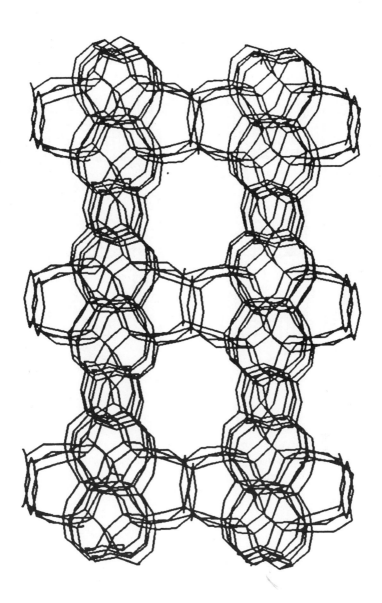

Figure 24.6   Framework structure of NU-87.   View is down the 10-ring channels

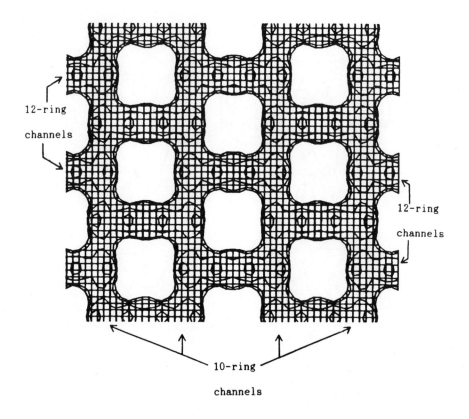

12-ring
channels

12-ring
channels

10-ring
channels

Figure 24.7　　　"Pore map" of NU-87 structure, illustrating the 10- and 12-ring channels

dimensional pore network; consisting of separate 10, and 12, T-atom channels which intersect at right angles.

Figure 24.6 contains a representation of the NU-87 structure. In the figure the T-atoms (Si or Al) are at the vertices. While the 10-ring channels can be clearly seen, the channels bounded by 12 T-atoms are more difficult to locate. Accordingly, Figure 24.7 contains what we would describe as a "pore map" of the NU-87 structure; in effect a representation of the Van der Waals surface of the channels without showing the framework atoms themselves. The figure shows the 10-ring channels running vertically with the intersecting 12-ring tunnels running in the horizontal direction.

From an examination of Figure 7 the offset "city block" type structure of NU-87 can be readily appreciated, with its intersecting 10- and 12-ring channels. What can also be seen is that, even though NU-87 contains 12-ring channels, access to, and movement through, the structure is effectively governed by 10-ring windows since molecules diffusing through the 12-ring channels have to pass through a 10-ring restriction before entering the next 12-ring section. This may explain the unusual sorption result obtained for neopentane.

CONCLUSIONS

NU-87 is a novel high-silica zeolite readily prepared from systems containing the template decamethonium (and its C-8 to C-12 analogues). Zeolite NU-87 has a new topology enclosing a 2-D pore system containing 10- and 12-ring channels. The material exhibits good thermal stability and has a high capacity for a variety of sorbates.

ACKNOWLEDGEMENTS

The authors gratefully acknowledge Ms. G. Von Bradsky (SEM/TEM) and Mr. S. Huntley, Mr. I. Robson and Mr. T. Muteen for technical assistance and ICI Chemicals and Polymers Ltd. for permission to publish.

REFERENCES

Breck, D. W. 1974. *Zeolite Molecular Sieves: Structure, Chemistry and Use.* New York: John Wiley and Sons, pp. 636-7.
Casci, J. L. and B. M. Lowe. 1983. *Zeolites* 3: 186.
Casci, J. L., B. M. Lowe and T. V. Whittam. 1984. *Proceedings of the Sixth International Zeolite Conference,* ed. D. Olson and A. Bisio,   pp. 894-94. London: Butterworths.
Casci, J. L. and A. Stewart. 1990. European Patent Application 377 291.
Gregg, S. J. and K. S. W. Sing. 1967. *Adsorption, Surface Area and Porosity.* London: Academic Press.
Lawton, S. L. and W. J. Rohrbaugh. 1990. *Science* 247: 1319.
Lowe, B. M. 1983. *Zeolites* 3: 300.

Meier, W. M. and D. H. Olson (Editors). 1987. *Atlas of Zeolite Structure Types (2nd Edition)*. London: Butterworths.

Schlenker, J. L., W. J. Rohrbaugh, P. Chu, E. W. Valyocsik and G. T. Kokotailo. 1985. *Zeolites* 5: 355.

Schlenker, J. L., J. B. Higgins and E. W. Valyocsik. 1989. *Zeolites for the Nineties:* Recent Research Reports, ed. J. C. Jansen, pp. 287-8.

Shannon, M. D., J. L. Casci, P. A. Cox and S. J. Andrews. 1991. *Nature*, 353: 417.

Treacy, M. M. J. and J. M. Newsam. 1988. *Nature* 332: 249.

Valyocsik, E. W. and N. M. Page. 1986. European Patent Application 174 121.

Wadlinger R. L., G. T. Kerr and E. J. Rosinski. 1967. U.S. Patent 3 308 069.

Wilson, S. T., B. M. Lok and E. M. Flanigen. 1982. U.S. Patent 4 310 440.

Wilson, S. T., B. M. Lok, C. A. Messina, T. R. Cannan and E. M. Flanigen. 1982. *J. Amer. Chem. Soc.* 104: 1146-7.

Wu, E. L. et al. 1979. *J. Phys. Chem.* 83: 2777.

25

Aluminum- and Boron-Containing SSZ-24: Inverse Shape Selectivity in the AFI Structure

Robert A. Van Nordstrand, Donald S. Santilli, and Stacey I. Zones

Chevron Research and Technology Company

The AFI structure ($AlPO_4$-5) is interesting in relationship to adsorption and catalysis. The structure comprises a one-dimensional 7.5 A diameter pore system constrained by 12-rings, with relatively smooth channels devoid of cavities. Some surprising adsorption preferences for branched chain hydrocarbons carry over into the catalytic properties of AFI materials such as MAPO-5 and SAPO-5.

SSZ-24 as an all-silica isostructural analog of $AlPO_4$-5 has been reported recently.[1] The catalytic utility of SSZ-24 has been limited by the extremely low aluminum content available from the original synthesis procedure. The synthesis procedure described here has produced SSZ-24 with significant boron and/or aluminum content, providing catalysts with some of the adsorption and catalytic features found with the $AlPO_4$-5 variants.

SYNTHESIS

The original synthesis procedure used Cabosil M-5 as the silica source (Cabot Corp., less than 2 ppm Al), and N,N,N-trimethyl-1-adamantammonium hydroxide (TMAA-1) as the template.[1,2] This is the only template found to produce SSZ-24 using this original procedure. As seen in Figure 1 the length of this template, omitting the hydroxide, is about 8.4 A, which corresponds to the c-dimension (the pore direction) of the SSZ-24 hexagonal unit cell. The diameter of the adamantyl group is 6.7 A, very close to the pore diameter of SSZ-24.[3] This provides a snug fit, satisfying the hydrophobic surface forces of both template and zeolite. The template is found in this organozeolite corresponding approximately to complete pore filling with one template per unit cell.

The general procedure for preparation of the boron-containing SSZ-24 is described below. A boron-beta zeolite provides both the boron and silicon. To prepare the aluminum-containing SSZ-24, the boron form (organic-free) is treated with an aluminum salt solution whereby part of the framework boron is replaced by aluminum.[4]

The boron-beta is prepared from a silica source (Cabosil M-5 or Ludox AS-30) plus sodium tetraborate with a template, either the diquaternary compound, bis-1,1'-diazabicyclo[2.2.2]octane 1,4-butane dihydroxide, or mixtures of this diquaternary salt with tetraethyl ammonium hydroxide.[5] The synthesis is carried out at 150°C, with or without stirring, usually for six days. The atomic or mole ratios used are: $B/Si = 0.03$ to 0.13; $RN/Si = 0.30$; $H_2O/Si = 12$ to 25; $Na/Si = 0.015$ to 0.065; $OH^-/Si = 0.30$. After synthesis, the organic template is removed from the boron-beta by calcining at 540°C in nitrogen with a small amount of air.

A detailed example of the overall synthesis of boron-SSZ-24 is provided here, starting with the synthesis of the boron-beta.[6] 10.85 g of a 0.90 M solution of the diquaternary template is diluted with 3.95 ml water. 0.23 g of $Na_2B_4O_7 \cdot 18H_2O$ is dissolved in this solution, followed by the addition of 1.97 g of Cabosil M5. This reaction mixture is heated in a Parr 4745 reactor at 150°C and rotated at 43 rpm on a spit in a Blue M oven for 9 days. The solid reaction product is filtered, washed repeatedly, and dried at 115°C. X-ray diffraction pattern of this product showed that it was boron-beta zeolite. This calcined product was then converted to the ammonium form by exchanging at 100°C in a solution of ammonium acetate.

The synthesis of the boron-SSZ-24 follows. 2.25 millimoles of the TMAA-1 and 0.1 g KOH [solid, 85%] in 12 ml H_2O are stirred until clear. Then 0.90 g Cabosil M-5 is stirred in and 0.60 g of NH_4-boron-beta (Al-free) is added. This mixture is heated at 150°C for 1 day without stirring. The product is boron-SSZ-24, free of beta. The atomic or mole ratios in this preparation are: template/$Si = 0.09$, $B/Si = 0.03$, $K/Si = 0.04$, $OH^-/Si = 0.13$, $H_2O/Si = 44$. The XRD pat-

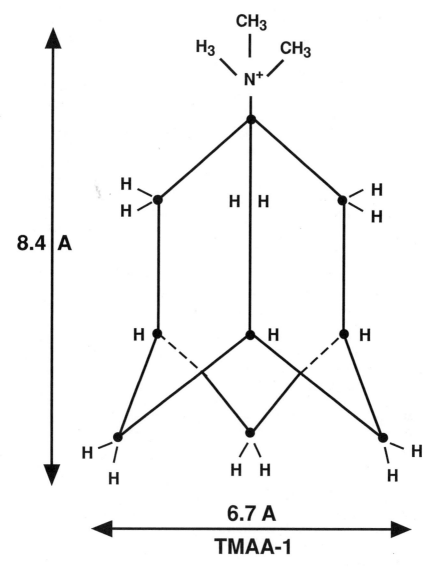

Figure 25-1. The original template used in synthesis of SSZ-24 and B-SSZ-24.

tern shows slightly reduced lattice constants as compared to the all-silica version, indicating the boron is incorporated, substituting for silica.

This synthesis may be carried out using boron-beta as the sole source of silicon, likewise a very rapid reaction. In this case about one-third of the boron present in the boron-beta is incorporated into the boron-SSZ-24. This synthesis procedure is an example of the hydrothermal conversion of a zeolite into an organozeolite as cited in previous work from this laboratory and from others.[7,8] As in our previous paper, the conversion of a zeolite (here boron-beta) to an organozeolite (boron-SSZ-24) is much more rapid than is the corresponding synthesis starting with the conventional oxide or gel sources (here Cabosil). The detailed example cited above, completed in 1 day, would require 7 to 10 days for the all-silica SSZ-24 case using Cabosil alone. And yet in the above synthesis more than half the silica is provided by the Cabosil.

Not only is this synthesis of the SSZ-24 much faster when starting with the boron-beta but it is much less demanding as regards the template. For example, in the original synthesis starting with Cabosil, only the 1-adamantammonium derivative (TMAA-1) was successful. In the synthesis starting from the boron-beta zeolite other templates have proven successful, including the 2-adamantammonium derivative (TMAA-2). In view of the "perfect fit" of the TMAA-1 in the SSZ-24 pore system, it is perhaps not surprising that the TMAA-2 also fits the pore system, as adamatane itself behaves as a freely rotating sphere in the solid state.[3]

Synthesis of boron-SSZ-24 from Cabosil plus conventional boron sources such as sodium tetraborate was not successful in making a pure SSZ-24. Instead the resultant product was usually a mixture SSZ-24, SSZ-23 and SSZ-13 (a high-silica chabazite).[9]

The Al-SSZ-24 is prepared from the calcined boron form following ammonium exchange. This NH_4-B-SSZ-24 is refluxed for 2 hours with a solution of aluminum nitrate. In this procedure a major portion of the boron is removed from the zeolite, being replaced by aluminum.[4] This has been shown by chemical analysis, XRD and NMR. Preparations using boron-beta as the sole source of Si and B have typical mole ratios as follows: B-beta, $SiO_2/B_2O_3 = 30$; B-SSZ-24, $SiO_2/B_2O_3 = 90$; Al-SSZ-24, $SiO_2/Al_2O_3 = 100$.

XRD RESULTS

XRD patterns of beta zeolites contain both sharp and broad peaks. In Table 1, three of the sharp XRD peaks are compared for the Al-beta and the B-beta in both the organozeolite and the calcined zeolite forms. These data reflect the difference between the Si-O and the B-O bond lengths.

X-ray diffraction patterns of the three forms of SSZ-24 are shown in Figure 2. They include an all-silica example plus a boron example and its aluminum derivative.

Hexagonal lattice constants for the various forms of SSZ-24 are given in Table 2. In general, in going from B to Si to Al there is the expected expansion of the lattice. Additional values cited for the all-silica SSZ-24 are from the recent XRD structure study of Bialek, Meier, Davis and Annen;[10] also from the neutron diffraction study of Richardson, Smith and Han.[11]

Table 25-1. Three XRD peaks in Al and B beta zeolites.
       (d/n spacings in A)

	As-prepared			Calcined		
Al-beta	3.97	3.30	3.03	3.97	3.30	3.03
B-beta B/Si = 0.10	3.94	3.24	2.99	3.90	3.26	2.98

Table 25-2. Lattice constants in A.

		a	c
SSZ-24 (all Si)	As-prepared Calcined	13.62 13.64	8.29 8.32
SSZ-24[10] (all Si)	As-prepared Calcined	13.60 13.67	8.28 8.33
SSZ-24[11] (all Si)	Calcined	13.64	8.31
B-SSZ-24	As-prepared Calcined	13.54 13.61	8.24 8.30
Al-SSZ-24	Calcined	13.68	8.34

An interesting feature seen in these XRD data is the expansion of the unit cell as the template is removed.

NMR RESULTS

Magic-angle spinning NMR study of the boron and aluminum versions of SSZ-24 show both these elements to be in tetrahedral sites.[12] (See Figure 3.) This applies to the boron-organo-zeolite, and to the latter after calcination and ammonium exchange. When the NH_4-B-SSZ-24 is refluxed with aluminum nitrate solution, the Al is inserted into tetrahedral sites, presumably those sites vacated by the boron. These measurements were made with a Bruker CXP 300 instrument.

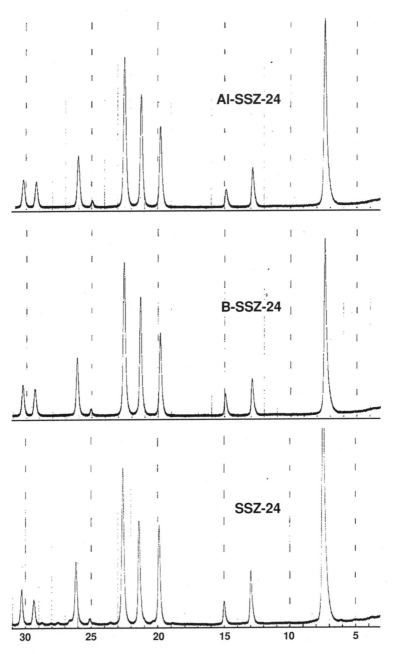

Figure 25-2. X-ray diffraction patterns of SSZ-24 in the all-silica form, the boron and the aluminum forms.

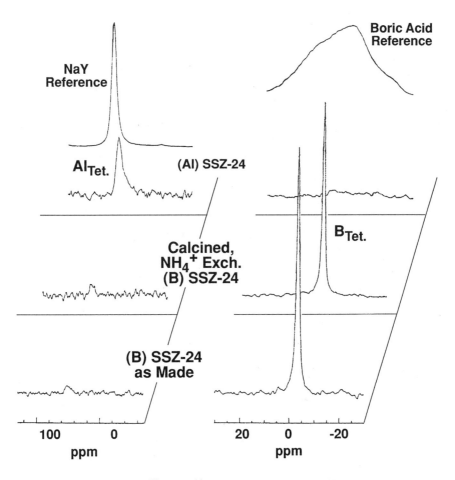

Figure 25-3. Quantitative ^{27}Al and ^{11}B MAS-NMR patterns, showing tetrahedral boron is replaced by tetrahedral aluminum in the AFI structure.

ADSORPTION RESULTS

A strong affinity for the adsorption of branched chain hexanes was pointed out previously for the three materials with the AFI structure (AlPO$_4$-5, SAPO-5 and the all-silica SSZ-24).[1] This same preference for branched chain over straight chain hexanes is also demonstrated in the new boron and aluminum versions of SSZ-24. This affinity for molecules such as 2,2-dimethylbutane is apparently characteristic of the AFI structure. The preference for branched chain hexanes has been found in competitive adsorption experiments both at room temperature and at 130°C.

Table 25-3. Pore probe data at 130°C.

MOL SIEVE	2,2-DMB	3-MP	n-C_6
Amorphous Si-Al	1.8	2.2	2.7
FAU (Y-82)	26	37	38
BETA	20	37	57
MFI	5	7	25
Al-SSZ-24	23	10	6
SAPO-5	27	13	6
ERIONITE	0	0	15

Data in Table 3 were obtained using the "pore probe" technique.[13] This technique includes flowing an equimolar mixture of n-hexane (n-C_6), 3-methylpentane (3-MP) and 2,2-dimethylbutane (2,2-DMB) over the molecular sieve at 130°C until a steady state adsorption is achieved, usually one hour. The loading is done at a flow rate of 1 ml of the hexane mix per hour with 20 ml per minute of helium. The flow is stopped and the sieve is stripped with a flow of helium and n-heptane. The amount of sieve used is about 0.5 g. Data in the table are in mg per g sieve.

The boiling points of these hexanes suggest that the branched chains have less mutual affinity in the liquid state than do the normals. The strong selectivity of the AFI structure for the branched hexanes apparently results from a more favorable fit of these branched hexanes in the AFI pores than of the normal hexanes in these pores.

CATALYSIS RESULTS

Data in Table 4 show the relative amounts of the various hexanes produced in the hydrocracking of n-hexadecane over several molecular sieves, each with 0.5% Pd.

The palladium is added to the sieve by aqueous ion exchange using $Pd(NH_3)_4 \cdot (NO_3)_2$ held at pH 9-10 with ammonium hydroxide. The Pd-sieve is dried, then calcined at 500°C.

The hydrocracking was carried out at 1200 psig over 0.5 g catalyst at 300°C, with flows of 160 ml/minute H_2 and 1 ml/hour n-hexadecane. Conversions were held in the range of 60 to 80% by tempering the more active catalysts with ammonia.

The same preference for adsorption of the branched hexanes in the AFI structures is exhibited in hydrocracking as a preference for production of these branched hexanes over the Pd/AFI catalysts. A recent patent by Ward points out

Table 25-4. Percents of the hexane isomers in products of hydrocracking of n-hexadecane.

MOL SIEVE	2,3-DMB	2- & 3-MP	n-C$_6$
FAU (Y-82)	3	72	25
MFI	0	33	67
Al-SSZ-24	14	69	17
SAPO-5	16	68	16

this feature of SAPO-5 as a cracking catalyst: increased branched chain products.[14]

This appears to be a novel type of catalytic shape selectivity. Previous types of shape selectivity, as reviewed by Csicsery, have been attributed to the limited pore dimensions of the molecular sieve;[15,16] each type favoring organic entities of smaller cross section. These previous types are attributed to exclusion of certain feed molecules, prevention of escape of certain product molecules, or prevention of formation of certain activated (transition) states.

The novel catalytic shape selectivity reported here, which is labeled "inverse shape selectivity", favors formation of molecules of larger cross section. Presumably this selectivity is caused by the same relative affinities shown in the adsorption experiments. However, in catalysis these relative affinities must pertain to affinities between the AFI pore system and the transition states or reaction intermediates. Thus the branched transition states or reaction intermediates are favored—resulting in branched chain products.

This inverse shape selectivity is not merely absence of shape selectivity. In the cases of larger pore zeolites such as faujasite and beta there is no enhancement of branched chain product formation compared, for example, to that with amorphous silica-alumina gel—no shape selectivity. Likewise there is not the preference for branched chains in adsorption experiments with these larger pore systems.

Somewhere in the catalytic transition from hexadecane to hexane products the preference for branched chains which is observed in adsorption experiments asserts itself in the cracking mechanism. In the adsorption experiments the hexanes themselves are involved. In the hydrocracking reaction, reactive intermediates or transition states are involved.

CONCLUSIONS

A synthesis procedure has been developed for making the aluminum- and the boron-containing SSZ-24. It is based upon the remarkably rapid conversion of the ammonium form of boron-beta zeolite into the boron-SSZ-24 organozeolite. The aluminum-SSZ-24 is formed by replacement of the boron by aluminum. Both the boron and the aluminum are shown to be in zeolite lattice sites.

These new forms of SSZ-24 also exhibit the surprising adsorption and catalytic behavior seen in previously known AFI materials. A marked affinity for branched hexanes in adsorption manifests itself in a significant preference for production of the branched chain hexanes. This relationship between adsorption and catalysis has suggested that in the AFI structures a new type of shape selectivity is operating, a type based not on repulsions between the organic entity and the zeolite pore wall, but rather, on attractive forces. This new catalytic shape selectivity is referred to here as "inverse shape selectivity".

ACKNOWLEDGEMENTS

Among the many at Chevron who assisted in this work we are especially indebted tp D. Wilson for the NMR, to L.-T. Yuen, D. A. Hickson and T. V. Harris for molecular sieve preparations, and to Y. Nakagawa, A. Bycraft and D. Dumelle for putting our ideas into respectable words and print. We appreciate the continuing support of Chevron for our zeolite studies.

REFERENCES

1. Van Nordstrand, R. A.; Santilli, D. S.; Zones, S. I. 1988. *Perspectives in Molecular Sieve Science*, ACS Symposium 368, ed. W. H. Flank and T. E. Whyte, Jr., pp 236-245. American Chemical Society.

2. Zones, S. I. 1987. U. S. Patent 4 665 110.

3. McCall, D. W.; Douglas, D. C. 1960. *J. Chem. Phys. 33*, 77.

4. Chu, P. 1990. U. S. Patent 4 912 073.

5. Zones, S. I.; Holtermann, D. L.; Jossens, L. W.; Santilli, D. S.; Rainis, A.; Ziemer, J. N. 1991. Patent Appl. WO 91/00777.

6. Zones, S. I. et al, 1991. Patent Appl. WO 91/00844.

7. Zones, S. I.; Van Nordstrand, R. A. 1990. *Novel Materials in Heterogeneous Catalysis*, ACS Symposium Series Series 437, ed. R. T. K. Baker and L. L. Murrell, pp. 14-24. American Chemical Society.

8. Dwyer, F. G.; Chu, P. 1979. *J. Catal. 59*, 263.

9. Zones, S. I.; Van Nordstrand, R. A.; Santilli, D. S.; Wilson, D. M.; Yuen, L.; Scampavia, L. D. 1989. In *Zeolites: Facts, Figures, Future,* ed. P. A. Jacobs and R. A. van Santen, pp 299-309. Elsevier Science Publishers.

10. Bialek, R.; Meier, W. M.; Davis, M.; Annen, M. J. 1991. *Zeolites,* 11, 438.

11. Richardson, J. W.; Smith, J. V.; Han, S. 1990. *J. Chem. Soc. Faraday Trans. 86,* 2341.

12. Turner, G. L.; Smith, K. A.; Kirkpatrick, R. J.; Oldfield, E. 1986. *J. Magn. Resonance 67,* 544.

13. Santilli, D. S. 1986. *J. Catal. 99,* 335-341.

14. Ward, J. W. 1987. U. S. Patent 4 695 368.

15. Csicsery, S. M. 1986. *Pure & Appl. Chem. 58,* 841-856.

16. Csicsery, S. M. 1976. *Zeolite Chemistry and Catalysis,* ACS Monograph 171, ed. J. A. Rabo, pp. 680-713. American Chemical Society.

26

Orienting Effect of Fluoride Anions in the Synthesis of a Novel Gallophosphate of the LTA Type

Abdallah Merrouche, Joël Patarin, Michel Soulard, Henri Kessler and Didier Anglerot[1] *Laboratoire de Matériaux Minéraux, URA CNRS 428, Ecole Nationale Supérieure de Chimie, 3 rue Alfred Werner, 68093 Mulhouse cedex, France, [1]Groupe ELF-Aquitaine, GRL, Lacq, BP 34, 64170 Artix, France*

By adding HF to the synthesis mixture of gallophosphates, it was possible to obtain two novel phases of the -CLO and LTA- structure types. The former which is the first 20-membered ring molecular sieve was obtained in the presence of quinuclidine as a template. The LTA-type crystallized from a mixture containing di-n-propylamine as a template and HF, whereas in the absence of fluoride, UCC $GaPO_4$-a was formed. The substitution of Ga and/or P in the former by Si or Al was also studied. The LTA-type gallophosphate is stable to at least 500°C in a dry atmosphere. The n-hexane adsorption capacity at $p/p_0 = 0.5$ is 14.8 wt%. The solid state MAS NMR results were compared with those obtained for the gallophosphate $GaPO_4$-a. In the LTA gallophosphate the ^{31}P MAS NMR spectrum shows two peaks at -5.3 and -11.7 ppm (peak area ratio 1:3), which might result from two chemically different tetrahedral environments for the P atoms owing to an interaction with the template. The ^{71}Ga MAS NMR spectrum of the new gallophosphate is poorly resolved. Only one broad peak at around -36.8 ppm was assigned; it might correspond to Ga atoms in a five-or six-fold coordination. All these results are in agreement with the structure refinement of the as-synthesized LTA $GaPO_4$ determined by the Rietveld procedure. The F- anions were found in the D4Rs of the LTA structure and presumably play a co-structuring role in addition to their mineralizing effect.

INTRODUCTION

In the eighties UCC developed a new generation of molecular sieves, $AlPO_4$-n, based on an aluminophosphate framework (Wilson, Lok, and Flanigen 1982). This initial discovery was followed by a number of reports describing partial isomorphous substitutions of aluminum and/or phosphorus by another element such as Si (Lok et al. 1984), Sn (Tapp and Cardile 1990), or Me (Me = Mg, Mn, Fe, Co, and Zn) (Wilson and Mercer 1988). These results showed that new open framework compositions of oxides outside of the then known aluminosilicate and silicate zeolites were possible. Many materials are counterparts of silica-based zeolites; however, for the LTA-type structure only the silicoaluminophosphate SAPO-42 (Lok et al. 1984) isostructural with ZK-21 and ZK-22 (Kühl and Schmitt 1990) was reported.

Recently a new family with a gallophosphate composition has been obtained (Parise 1985; Wilson et al. 1987; Feng and Xu 1987) and some of them show a novel structure (Yang, Feng, and Xu 1987; Wang et al. 1989). The use of a fluoride medium enabled us to prepare selectively two novel microporous gallophosphates of the -CLO (Joly et al. 1991) and the LTA-structure types (Merrouche et al. 1991). The former was obtained in the presence of quinuclidine as a template. The crystal structure (cubic, a = 51.712 Å, space group : Fm3c) comprises two non interconnected 3-dimensional channel systems with 20- and 8-membered ring openings (Estermann et al. 1991) The topological structure can be obtained by joining two types of double-four rings creating LTA α-cages and rpa cages. The main channels have a cloverleafed-shaped aperture described by 20 T atoms (Ga and P). All the double-four rings host a fluoride anion in their center. The Ga are pentacoordinated in a distorted trigonal bipyramidal environment and the P atoms remain tetracoordinated.

In this work, the preparation and the characterization by several techniques such as XRD, thermal analysis, adsorption measurements, and solid state NMR of the novel $GaPO_4$ with a LTA-type structure are described. The possibility of isomorphous substitution of Ga and/or P by Si and/or Al was also examined.

EXPERIMENTAL

Reactants

The reactants were di-n-propylamine (Fluka, purum 95%), 85% phosphoric acid (Prolabo, normapur), 40% hydrofluoric acid (Prolabo, normapur). The silicon and aluminum sources were, respectively, colloidal SiO_2 Ludox AS40 (40 wt% SiO_2, 60 wt% H_2O) and aluminum sulfate hydrate ($Al_2(SO_4)_3 \cdot 18H_2O$, Prolabo, rectapur 98%). Hydrated gallium sulfate was used as the gallium source. It was prepared by dissolution of gallium metal (Grade 1, Johnson Matthey) in an excess of hot concentrated sulfuric acid (Prolabo, rectapur). The mixture obtained was cooled to room temperature and diluted with 60% ethanol.

The hydrated gallium sulfate was then precipitated by adding ether ($(C_2H_5)_2O$). The solid was filtered, washed with ether, and slightly dried at room temperature (molar $H_2O/Ga_2(SO_4)_3$ in the final product ~25).

Synthesis

The LTA-type materials were obtained by hydrothermal synthesis at 80-200°C according to an extension (Guth 1989) of the procedure developed in our laboratory (Guth, Kessler, and Wey 1986) in which fluoride ions are used as mineralizing agents. The starting mixture was prepared as follows: the hydrated gallium sulfate was first dissolved in water and mixed with an aluminum or a silica source when a galloaluminophosphate (GAlPO) or a gallosilicophosphate (GaPSO) was prepared. H_3PO_4 was then added under stirring and finally di-n-propylamine (DPA) and hydrofluoric acid.

The obtained slurry was thoroughly mixed for 30 minutes and transferred into a PTFE-lined stainless steel autoclave. After heating, the solids obtained were filtered, washed with water until the pH of the filtrate was 5.5, and dried at 70°C.

Characterization

X-ray Powder Diffraction. The powder patterns were obtained with $CuK\alpha$ radiation on a Philips PW 1050 diffractometer. Fourteen unique peaks were selected for a least-squares refinement of the unit cell parameter. Lead nitrate was used as an internal standard. High temperature X-ray diffraction was performed on a Philips PW 1010 diffractometer ($CoK\alpha$ radiation) equipped with a variable temperature chamber in which the sample was kept in a flow of dry air.

Thermal Analysis. TG analysis was performed on a Mettler 1 thermoanalyzer by heating in air at a $6°C$ min^{-1} and DTA was carried out in air and in argon on a BDL-Setaram M2 apparatus with a heating rate of $6°C$ min^{-1}.

Adsorption Measurements (n-Hexane Adsorption). The adsorption measurements were performed by using a computerized thermogravimetric equipment TG92 from Setaram. About 100 mg of the as-synthesized solid were calcined by heating to 380°C at $5°C$ min^{-1}. After 1.5 h at 380°C, the solid was cooled to room temperature and subjected to a flow of a mixture of dry nitrogen and n-hexane. By mixing a second current of dry nitrogen it was possible to adjust p/p_0 between 0.1 and 0.9.

^{29}Si, ^{31}P, ^{71}Ga Solid State MAS NMR Spectroscopy. The spectra were recorded on a Bruker MSL 300 spectrometer for ^{29}Si and ^{31}P, and for ^{71}Ga a Bruker MSL 400 spectrometer was used. The recording conditions of the MAS spectra are given in Table 26-1. All samples were in their hydrated (ambient air) as-synthesized form.

Chemical and Electron Microprobe Analysis. Ga and P analysis was performed on the as-synthesized samples by inductively coupled plasma emission spectroscopy. The standards were high purity K_2HPO_4 and $Ga(NO_3)_3$ solutions

TABLE 26-1 Recording Conditions of the NMR MAS Spectra

	^{29}Si	^{31}P	^{71}Ga
Standard	TMS	85%H_3PO_4	$Ga(H_2O)_6^{3+}$
Frequency (MHz)	59.63	121.44	121.99
Recycle time (s)	10	5	0.2
Pulse width (10^{-6}s)	5.5	1.8	1.0
Spinning rate (Hz)	3400	8000	9000
Number of scans	1600	16	50 000

respectively. Scanning electron microprobe was also used for the analysis of Ga, P, and Si. The powder was first dispersed in an epoxy resin. After hardening, the surface of the pellet was polished mechanically using a diamond paste (D<3μm) washed with hexane and coated with a carbon film. The accelerating voltage was 15 kV; the smallest analyzed area was 1 μm^2. Ga, P, and Si analysis was performed on individual spots, the results were expressed as mol% Ga, P, and Si. Ga, P, and SiKα X-ray emission mapping of the crystals was also performed. F- was determined by using a fluoride ion selective electrode after mineralization (estimated accuracy = 2%). The amount of organic material and water was determined by thermogravimetry. Carbon and nitrogen were also analyzed by turning them into CO_2 and N_2 respectively, by combustion of 1-2 mg of sample at 1000°C in an excess of O_2 (C), and a He-3% O_2 mixture (N). The nitrogen oxides were reduced to N_2 by metallic copper at 500 °C. CO_2 and N_2 were titrated by a coulometric and a catharometric technique respectively.

RESULTS AND DISCUSSION

Synthesis, Crystal Morphology, and Chemical Composition

In Table 26-2 are given some typical synthesis examples having produced a LTA structure. The products obtained were checked by optical microscopy and by X-ray powder diffraction. As an example, the room temperature spectrum of sample B is given in Figure 26-1. The corresponding hkl indices and the unit cell parameter refined in the Fm3c space group (Meier and Olson 1987) are reported in Table 26-3.

The crystals are cubic (Figure 26-2) with a size ranging from 1 to 100 μm (see Table 26-2). The largest crystals were obtained with the lowest pH values. For instance, for a pH value of 3 (sample C), the crystal size is close to 80 μm. Such a result may be related to a slower nucleation rate.

With a similar composition, but without fluoride anions, only a phase type GaPO4-a (Wilson et al. 1987) was obtained (Table 26-2, example A). This sample comprises hexagonal prisms fairly regular in shape and approximately 120 μm in length (Figure 26-3). This result suggests that in addition to their mineralizing effect, fluoride anions play probably a co-structuring role beside the organic molecule.

For sample G obtained from an aluminum-containing mixture, beside the LTA- structure type, a novel GaPO4 or AlPO4 with an unknown structure was

TABLE 26-2 Typical Synthesis Examples

Sample	Reaction Mixture Composition	pH initial	pH final	T (°C)	Heating time (h)	Result	Crystal size (μm)
A	$1Ga_2O_3 : 1P_2O_5 : 70H_2O : 9$ DPA	5.5	5	140	24	$GaPO_4$-a[a]	120x120
B	$1Ga_2O_3 : 1P_2O_5 : 70H_2O : 9$DPA : 1HF	5	4.5	140	24	LTA	1-2
C	$1Ga_2O_3 : 1P_2O_5 : 65H_2O : 6.66$ DPA : 2.66HF	3	2.5	140	24	LTA	80
D	$1Ga_2O_3 : 1P_2O_5 : 65H_2O : 9$ DPA : 1HF	5	4.5	80	36	LTA	1-10
E	$1Ga_2O_3 : 1P_2O_5 : 65H_2O : 9$ DPA : 1HF	5	4.5	200	6	LTA	1
F	$1Ga_2O_3 : 1P_2O_5 : 1SiO_2 : 65H_2O : 9.8$ DPA : 2.66HF	4.5	4	140	24	LTA	70-100
G	$0.8Ga_2O_3 : 1P_2O_5 : 0.2Al_2O_3 : 65H_2O : 9$ DPA : 1HF	4.5	4.5	140	16	LTA ~80% +X[b]	30-40

a $GaPO_4$-a (Wilson et al. 1987)
b X = $GaPO_4$ or $AlPO_4$ with unknown structure (Guth 1989)

388

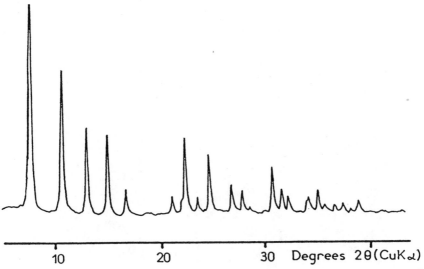

FIGURE 26-1 X-ray powder diffraction pattern of the LTA gallophosphate sample B.

TABLE 26-3 XRD Data of Sample B
Space group Fm3̄c   a = 24.040 (2) Å

hkl	d_{obs}	d_{calc}	I/I_o.
200	12.02	12.02	100
220	8.50	8.50	60
222	6.94	6.94	36
400	6.01	6.01	32
420	5.37	5.37	10
440	4.25	4.25	6
531	4.06	4.06	5
600/442	4.00	4.00	32
620	3.80	3.80	6
622	3.625	3.624	23
640	3.334	3.333	11
642	3.212	3.212	8
731	3.129	3.129	2
820/644	2.916	2.915	20
660/822	2.833	2.833	9
751	2.777	2.775	6
840	2.685	2.687	1
753	2.638	2.638	6
842	2.621	2.622	6
664	2.562	2.562	9
931	2.519	2.520	2
844	2.453	2.453	2
1000	2.403	2.404	3
1020	2.357	2.357	1
666/1022	2.312	2.313	4

389

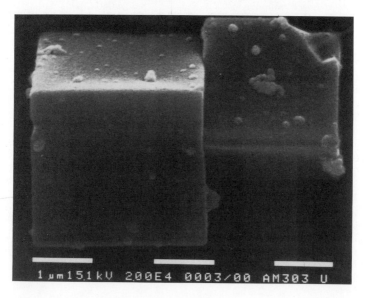

FIGURE 26-2 Scanning electron micrographs of the LTA gallophosphate sample B.

found; the same unknown type was reported previously for an $AlPO_4$ and SAPO phase prepared in fluoride medium (Guth 1989).

The chemical composition of the as-synthesized LTA samples is reported in Table 26-4. For the LTA gallophosphates (samples B, C, D, and E) the Ga/P molar ratio is very close to the expected value of 1. A small amount of Si was found in the GaPSO phase (sample F), it seems to substitute for P according to mechanism 1 proposed by Flanigen et al. (Flanigen, Patton, and Wilson 1988). SiKα X-ray mapping confirms that little Si was incorporated into the crystals, some silica impurities are also present on the outer surface of the crystals (Figure 26-4).

The molar C/N values found for the as-synthesized LTA-type samples B to E are in the range 5.9 - 6.3, in good agreement with C/N = 6 expected for dipropylamine. The amount of DPA present in the solids turned out to be variable. In samples C and F the number of moles of DPA per 24T (T = Ga, P, or Si) is close to 3, in agreement with the structure determination. The latter was performed on a GaPO$_4$ compound (Simmen et al. 1992). Three DPAF pairs per 24T were found and the fluoride anions were localized in the double-four-membered rings (D4Rs) of the structure as was already observed in octadecasil (Caullet et al. 1991). The DPA amount is higher for the three other samples, presumably due to the difficulty in washing out the intercrystalline excess DPA species because of their small crystal size, as opposed to samples C and F (large crystal size).

The amount of fluoride is variable too. The value found for sample B is close to 3, which was found by the structure determination. For the other samples, the excess fluoride which increases with decreasing pH of the starting mixture may be attributed to fluoride impurities according to the ^{19}F MAS NMR results (Simmen et al. 1992). In agreement with these results and the structure determination, the unit cell formula of the dehydrated LTA-type GaPO$_4$ sample is $[(C_3H_7)_2NH_2]_3\,Ga_{12}P_{12}O_{48}F_3$ for the pseudo-cell with a ~12 Å.

FIGURE 26-3 Scanning electron micrograph of the gallophosphate sample A (type GaPO$_4$-a).

TABLE 26-4 Chemical Composition of the LTA-Type Samples (As-synthesized)

Sample	wt%, Chemical Analysis					Moles per 24 T Atoms									
	Ga	P	C	N	F	Ga		P		DPA		F	Si	H$_2$O	
						a	b	a	b	a	c	a	b	c	
B	29.3	14.2	11.2	2.04	2.40	11.6	11.4	12.4	12.6	3.8	5.5	3.3	-	9	
C	29.9	13.1	7.8	1.53	3.03	12.1	11.7	11.9	12.3	3.1	3.2	4.2	-	17	
D	30.8	14.4	12.3	2.37	2.97	11.7	n.d.	12.3	n.d.	4.5	5.5	4.0	-	11.2	
E	29.5	13.7	11.2	2.20	2.65	11.7	n.d.	12.3	n.d.	5.0	5.4	3.9	-	18.7	
F	n.d.	n.d.	n.d.	n.d.	n.d.	n.d.	12.0	n.d.	11.4	n.d.	3.4	n.d.	0.6	19	

n.d. not determined
a chemical analysis
b electron microprobe analysis
c thermogravimetry

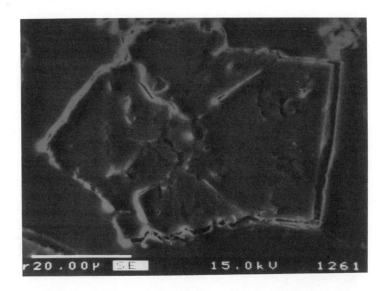

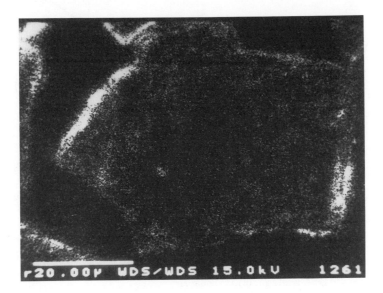

FIGURE 26-4 Scanning electron micrographs (top) and SiKα X-ray emission mapping (bottom) of the silicogallophosphate sample F.

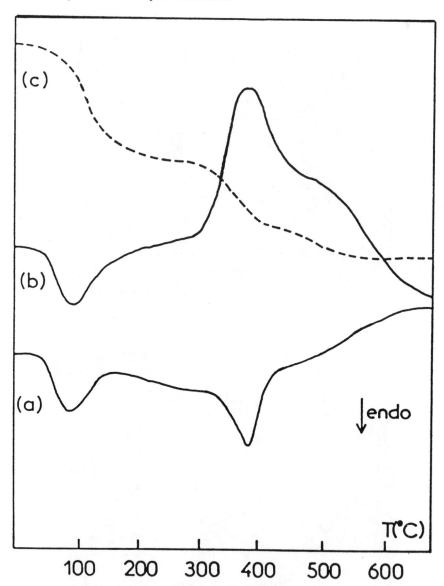

FIGURE 26-5   Thermal analysis of the LTA gallophosphate sample C. (a) DTA under argon, (b) DTA under air, (c) TG under air.

Thermal Analysis

The DTA and TG curves of the as-synthesized sample C are given as an example in Figure 26-5. Under inert atmosphere (Figure 26-5a) two broad DTA endotherms with a maximum, respectively, at 100°C and 390°C and a small endothermic drift near 500°C were observed. In air or in oxygen (Figure 26-5b), the DTA curves show at the same temperatures an endothermic (100°C) and two exothermic components (390°C and 500°C). The first peak (maximum at 100°C) corresponds in the TG curve (Figure 26-5c) to the loss of the hydration water (13 wt%), whereas the second at ~390°C and the third near 500°C can be attributed to the thermal decomposition of di-n-propylamine species (weight loss ~10 %). The thermal stability of the structure was studied by high-temperature X-ray powder diffraction. No significant change of the powder spectra was observed when the samples were heated in air to 500°C with a heating rate of 4°C min[-1] and cooled to room temperature in a flow of dry air. However, when the calcined samples were exposed to moist air at room temperature, as found for SAPO-37 (Briend et al. 1989), the crystallinity decreased significantly and after a few hours a complete amorphization was observed.

Adsorption Measurements (n-Hexane Adsorption)

The adsorption capacity found at 25°C for p/p_0 = 0.5 was 14.8 wt%. This corresponds to a n-hexane pore volume of 0.22 cm^3 g^{-1}, which is close to the value usually found for zeolite A, but significantly larger than that reported by Lok et al. (Lok et al. 1984) for SAPO-42 (0.11 cm^3 g^{-1} for p/p_0 = 0.4). The adsorption was fast, the half-reaction time was about 15 minutes, and the maximum of adsorption was reached after ~1.5 h. This demonstrates that there is no pore blockage and that the calcination did not deteriorate the structure.

Solid State NMR Results

29Si MAS NMR. The ^{29}Si MAS NMR spectrum (Figure 26-6) of the LTA-type silicogallophosphate consists of only one broad resonance at -112.2 ppm.

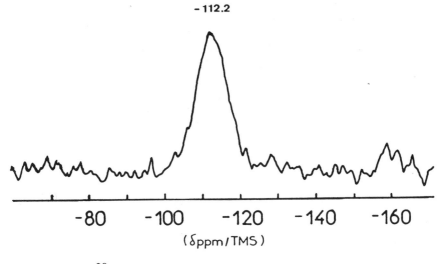

FIGURE 26-6 ^{29}Si MAS NMR spectrum of the silicogallophosphate sample F.

According to the SiKα X-ray mapping and to the chemical shift value (δ) found in SAPO's (Guth 1989), this broad peak maybe assigned to the SiO_2 impurity and, maybe, to some Si(4Si) islands resulting from a $2Si^{4+} = P^{5+} + Ga^{3+}$ substitution. Indeed, no peak that would correspond to Si(4Ga) resulting from mechanism no.1 suggested by Flanigen et al. (Flanigen, Patton, and Wilson 1988), and expected at δ ~ -70 ppm, is observed.

31P MAS NMR. The ^{31}P MAS NMR spectra of samples A ($GaPO_4$-a) and B ($GaPO_4$ with the LTA-type structure) are reported in Figures 26-7a and 26-7b. For both samples two lines are observed. The chemical shifts are given in Table 26-5 with those of different gallophosphate compounds where the phosphorus atoms are in a tetrahedral environment.

The chemical shift range (-6 to -12 ppm) is different form the one observed for ^{31}P in $AlPO_4$-type materials (δ range around -30 ppm) but is in good agreement with the value of -9.8 ppm which was found for $GaPO_4$ obtained by calcination of $GaPO_4 \cdot 2H_2O$ at 200°C (Turner et al. 1986). According to the XRD data, $GaPO_4$-a (Wilson et al. 1987) seems to be isostructural with the gallophosphate $Ga_9P_9O_{36}OH \cdot HNEt_3$ whose crystal structure was determined (Yang, Feng, and Xu 1987). Three types of crystallographic sites with an equal population were found for phosphorus. For sample A (i.e., $GaPO_4$-a) the observed peak area ratio is 1:2 (Figure 26-7a) and might correspond to the three crystallographic sites. According to the LTA structure, in which only one crystallographic site for P is present, only one NMR peak should be observed for phosphorus. The occurrence of two peaks (Figure 26-7b), with a peak area ratio 1:3, may be explained as follows. The structure refinement (Simmen et al. 1992) showed that the N-atom of DPA+ is shifted away from the center of the 8-membered ring by 0.6 Å. This results in closer contacts to 3 of the oxygen atoms forming the 8-membered ring. The distance of 2.7 Å to O(1) indicates a strong N—H--- O hydrogen bridge. Since there is only one DPA+ per 8-membered ring, only one hydrogen bridge can be present at any time. The H-bridge will pull some electron density away from the associated tetrahedra, thus

TABLE 26-5   ^{31}P MAS NMR Results

Sample	δ ppm	Reference
$GaPO_4$-a	-5.0 ; -11.0	This work
$GaPO_4$(LTA)	-5.3 ; -11.7	This work
$GaPO_4 \cdot 2H_2O$	-8.3 ; -12.9	This work
$GaPO_4$(quartz)	-10.7	This work
$GaPO_4$	-9.8	Turner et al. 1986

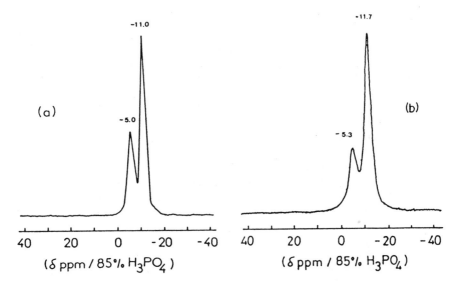

FIGURE 26-7 ^{31}P MAS NMR spectra of the gallophosphate sample A (type GaPO$_4$-a) (a), and of the LTA gallophosphate sample B (b).

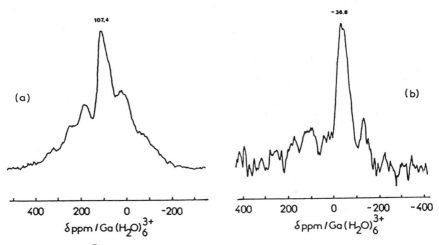

FIGURE 26-8 ^{71}Ga MAS NMR spectra of the gallophosphate sample A(type GaPO$_4$-a) (a), and of the LTA gallophosphate sample B (b).

deshielding the P atom. This has the effect of moving the chemical shift of the ^{31}P atom towards less negative values. Of the four PO$_4$ tetrahedra in the 8-membered ring, only one is affected in such a way, resulting in an intensity ratio of the ^{31}P signal of 1:3.

71Ga MAS NMR. To our knowledge, there is no published result on ^{71}Ga NMR in GAPO's. The ^{71}Ga MAS NMR spectra of samples A and B are given in Figure 26-8a and 26-8b. They are poorly resolved, presumably due to a quadrupolar effect. For $Ga_9P_9O_{36}OH \cdot HNEt_3$ (type $GaPO_4$-a), three types of gallium atoms have been found by the structure determination (Yang, Feng, and Xu 1987). Two are in a tetrahedral environment and the third is located in a distorted trigonal bipyramid. Hence, for sample A, the signal observed at 107.4 ppm can be assigned to tetracoordinated gallium. For sample B ($GaPO_4$ with LTA-type structure), the peak was found at -36.8 ppm (Figure 26-8b); thus, by comparison with the spectrum of sample A, the gallium atoms in the LTA gallophosphate are not tetracoordinated, but probably penta- or hexacoordinated. This result is in good agreement with the structure refinement (Simmen et al. 1992); indeed, the gallium atoms are pentacoordinated to four oxygen and one fluorine atoms.

CONCLUSION

The addition of F- to the synthesis medium of $GaPO_4$-type materials shows a specific effect due to the fact that F contributes to the stabilization of the structure by interacting with the framework. Thus, besides its mineralizing effect, the F- anion plays a co-structuring role and more particularly acts as a template for the D4R of the LTA-type structure. It can be assumed that for such a location the contribution to the stabilization of the structure is maximum.

ACKNOWLEDGMENTS

The authors would like to thank Ch. Baerlocher and J.L. Guth for fruitful discussions, L. Delmotte and Z. Gabelica for running the NMR spectra, and S. Einhorn, A. Eckhardt, and F. Muller for taking the photographs.

REFERENCES

Briend, M. et al. 1989. Thermal and Hydrothermal Stability of SAPO-5 and SAPO-37 Molecular Sieves. *Journal of the Chemical Society, Dalton Transactions:* 1361-62.

Caullet, P. et al. 1991. Synthesis, characterization and crystal structure of the new clathrasil phase octadecasil. *European Journal of Solid State and Inorganic Chemistry* 28 : 345-61.

Estermann, M. et al. 1991. A synthetic gallophosphate molecular sieve with a 20-tetraedral-atom pore opening. *Nature* 352 : 320-23.

Feng, S. and Xu, R. 1987. Studies on the syntheses and structures of a novel family of microporous gallophosphates. *Chemical Journal of Chinese Universities* 3 (10) : 867-68.

Flanigen, E.M., Patton, R. L., and Wilson, S. T. 1988. Structural, Synthetic and Physicochemical Concepts in Aluminophosphate-Based Molecular Sieves. *Innovation in Zeolite Materials Science*, ed. Grobet, P.J., Mortier,

W.J., Vansant, E.F. , and Schulz-Ekloff, G., pp.13-27. Amsterdam: Elsevier.

Guth, F. 1989. Synthèse et caractérisation de solides microporeux cristallisés contenant Al, P et Si. *PhD Thesis*, University of Mulhouse.

Guth, J.L., Kessler, H., and Wey, R. 1986. New Route to Pentasil-Type Zeolites Using a Non Alkaline Medium in the Presence of Fluoride Ions. *New Developments in Zeolite Science and Technology*, ed. Murakami, Y., Iijima, A., and Ward, J.W., pp. 121-28. Amsterdam: Elsevier.

Joly, J.F. et al. 1991. Nouveau phosphate de gallium microporeux cristallisés et dérivés substitués et leur procédé de préparation. *Fr. Patent*, registration number 91-03378.

Kühl, G.H. and Schmitt, K.D. 1990. A reexamination of phosphorus-containing zeolites ZK-21 and ZK-22 in light of SAPO-42. *Zeolites* 10(1): 2-7.

Lok, B.M. et al. 1984. Crystalline silicoaluminophosphates. *U.S. Patent* 4440871.

Meier, W. M., and Olson, D.H. 1987. *Atlas of Zeolite Structure Types*. London: Butterworths.

Merrouche, A. et al. 1991. Solides microporeux synthétiques renfermant du gallium et du phosphore, leur synthèse et leur utilisation comme catalyseurs et adsorbants. *Fr. Patent*, registration number 91-01106.

Parise, J.B. 1985. Some gallium phosphate frameworks related to the aluminium phosphate molecular sieves: X-ray structural characterization of {(i-PrNH$_3$)[(GaPO$_4$)$_4$·OH]}H$_2$O. *Journal of the Chemical Society, Chemical Communications*: 606-07.

Simmen, A., Patarin, J. and Baerlocher, Ch. 1992. The Rietveld refinement of a fluoride-containing gallophosphate with a LTA structure type. Proceedings of the Ninth International Conference, Montréal, 1992.

Tapp, N.J. and Cardile, C.M. 1990. Characterization of a tin-containing AlPO$_4$-5 molecular sieve (SnAlPO-5). *Zeolites* 10(7) : 680-84.

Turner, G.L. et al. 1986. Structure and Cation Effects on Phosphorus -31 NMR Chemical Shifts and Chemical- shift Anisotropies of Orthophosphates. *Journal of Magnetic Resonance* 70: 408-15.

Wang, T. et al. 1989. A novel mixed octahedral-tetrahedral framework: X-ray characterization of a microporous gallophosphate Ga$_2$P$_2$O$_8$ (OH) H$_2$O·NH$_4$·H$_2$O·0.16PrOH (GaPO$_4$-C7). *Journal of the Chemical Society, Chemical Communications*: 948-49.

Wilson, S.T., Lok, B.M, and Flanigen, E.M. 1982. Crystalline metallophosphate compositions. *U.S. Patent* 4310440.

Wilson, S.T., et al. 1987. Crystalline gallophosphate compositions. *Eur. Patent Application* 226219.

Wilson, S.T. and Mercer, W.C. 1988. Crystalline metal aluminophosphates. *Eur. Patent* 0293920.

Yang, G., Feng, S., and Xu, R. 1987. Crystal structure of the gallophosphate framework: X-ray characterization of Ga$_9$P$_9$O$_{36}$OH·HNEt$_3$. *Journal of the Chemical Society, Chemical Communications*: 1254-55.

27

Hydrothermal Synthesis and Structural Characterization of a Layered Zinc Phosphate, $RbHZn_2(PO_4)_2$

Edward W. Corcoran, Jr., Johanna B. Savader, and Meena Bhalla-Chawla
Exxon Research & Engineering Company, Annandale, NJ 08801

We have utilized hydrothermal synthesis techniques to produce a series of new zinc phosphate phases. These phases appear to be cation dependent in addition to being highly sensitive to both reaction temperature and pH. Most of the syntheses produce single-phase materials which form as crystals in a range of sizes (from ~μm to ~mm); crystal size can be controlled through slight variations of reaction pH. The synthesis and structural characterization of one of these phases, $RbHZn_2(PO_4)_2$, will be discussed.

INTRODUCTION

In an effort to identify new porous materials, we have utilized hydrothermal synthesis techniques to incorporate non-traditional framework components into oxide lattices. Our initial studies in this area resulted in a new group of materials (Corcoran et al. 1989a; Corcoran and Vaughan 1989b), the stannosilicates, which contained octahedral tin corner-shared through oxygen atoms to tetrahedral silicon.

We later extended this work (Corcoran 1990a; Corcoran et al. 1990b) to include another class of mixed framework oxides, synthesized at lower pH, consisting of octahedral molybdenum(V) and phosphate tetrahedra. The first member of this group, the layered compound $[N(C_3H_7)_4(NH_4)][(MoO)_4O_4(PO_4)_2]$ (Corcoran 1990a), is made up of extended sheets of Mo_4O_8 cubes linked through corner-sharing phosphate tetrahedra.

We have now continued our studies of phosphates to examine zinc-containing systems resulting in a series of new two- and three-dimensional lattices. We report herein the synthesis and single-crystal structural determination of one of these new zinc phosphates, $RbHZn_2(PO_4)_2$.

EXPERIMENTAL

The material was prepared as follows. A 19.23 g quantity of zinc dichloride (Aldrich) was dissolved in 10.0 g of distilled water; 16.27 g of 85% phosphoric acid was added and the mixture stirred. Next, 50.4 g of rubidium hydroxide (KBI, 50% aqueous solution) was slowly stirred into the solution until a homogeneous mixture resulted; distilled water was added to give a final weight of 100.0 g and the mixture blended.

Reaction of this gel in a polytetrafluoroethylene-lined autoclave (Parr) for 5 days at 200°C produced a white solid when the reactor was brought to room temperature. This solid was washed repeatedly with distilled water, filtered, and dried to yield a crystalline solid, $RbHZn_2(PO_4)_2$, with a characteristic powder X-ray diffraction pattern (Table 27-1). The pattern, which was indexed using a triclinic unit cell with the lattice constants $a = 5.23$ Å, $b = 8.89$ Å, $c = 9.70$ Å, $\alpha = 75.72°$, $\beta = 77.43°$, and $\gamma = 73.68°$, indicated that a single-phase sample was produced by the reaction.

Table 27-1   Powder X-Ray Diffraction Pattern for $RbHZn_2(PO_4)_2$

2θ	d	Relative I
9.53	9.37	100.0
15.60	5.68	57.3
19.12	4.64	69.0
1.97	4.44	10.7
21.27	4.17	57.6
22.75	3.91	6.5
23.34	3.81	10.4
23.69	3.75	38.5
24.05	3.70	6.0
25.04	3.55	5.6
28.29	3.15	20.6
28.84	3.09	54.4
29.51	3.02	5.1
29.84	2.99	7.6
31.65	2.82	76.4
32.40	2.76	12.6
32.80	2.73	6.4
34.12	2.63	11.7
35.32	2.54	6.8
37.52	2.40	5.4
38.79	2.32	30.7

Large, rectangular, colorless crystals were produced during the reaction having dimensions which were roughly 0.25 mm on each edge. A suitable crystal was selected for structural characterization by X-ray crystallography, which provided a structural solution (shown in Figure 27-1 and described below) consistent with chemical analysis (bulk elemental anal. calcd: Zn, 32.12; P, 15.21; H, 0.25; Rb, 20.99; found: Zn, 33.37; P, 14.99; H, <0.50; Rb, 21.51).

The structure of $RbHZn_2(PO_4)_2$ was solved and refined in the triclinic space group $P\bar{1}$ (No. 2) with $a = 5.248$ (1) Å, $b = 8.892$ (2) Å, $c = 9.702$ (2) Å, $\alpha = 75.72$ (2)°, $\beta = 77.43$ (2)°, and $\gamma = 73.68$ (2)°, $V = 415.7$ (2) Å^3, $Z = 2$, D(calcd) = 3.253 g/cm^3, λ (MoKα) = 0.71073 Å, $T = 20$°C. A Nicolet Autodiffractometer with a graphite monochrometer was used to collect 1916 (2θ > 25°) reflections from a colorless, rectangular parallelepiped crystal 0.2 X 0.25 X 0.3 mm. Of these, 1916 were unique and 1713 were observed (I > 3s(I)).

The data were corrected empirically for absorption effects using psi scans for 8 reflections having 2θ between 10.54° and 41.82°, which resulted in transmission factors between 0.527 and 1.000. The structure was solved by heavy-atom Patterson techniques and refined by full-matrix least-squares refinement. The hydrogen atom was located from a difference Fourier map and refined as an independent isotropic atom, $R = 0.027$, $R_w = 0.038$, GOF = 1.080. There were no peaks present above the background level (0.76 e$^-$/Å^3) in the final difference Fourier map.

RESULTS AND DISCUSSION

The structure of $RbHZn_2(PO_4)_2$ is a complex two-dimensional arrangement of corner-sharing zinc and phosphorus tetrahedra, utilizing both two- and three-coordinate oxygen (Figure 27-1). The anionic layers consist of a combination of three-, four-, and six-ring units and are separated by interlayer rubidium atoms. Each of the three-rings is made up of two zinc tetrahedra and a phosphate group with a three-coordinate oxygen atom. The three-ring phosphorus atoms, which comprise half of the phosphorus atoms within the structure, have hydroxyl groups which extend above and below each sheet, adjacent to the interlayer rubidium atoms. As expected, -P-O-P- linkages are excluded from the structure.

The $RbHZn_2(PO_4)_2$ layers (Figure 27-2) can be viewed as consisting of strands of connected double-six-ring cages with each strand flanked by a chain of edge-shared four-rings. The strands of cages alternate with the four-ring chains (with both running parallel to the a axis) to generate the ab plane layers, which are stacked down the c axis.

The four-ring chains contain only PO_4 phosphorus atoms. The O_3P-OH phosphorus atoms are all located within the double-six-ring cages with each cage contributing two hydroxyl groups, one below and one above, to the interlayer space.

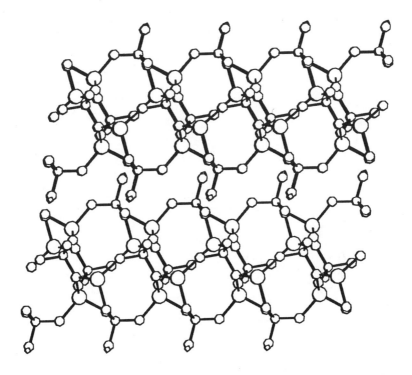

FIGURE 27-1. CHEMX (copyright Molecular Design, Ltd.) generated view of the the *bc* plane of RbHZn$_2$(PO$_4$)$_2$ looking at two layers of the structure. The P-OH hydroxyl groups extend into the interlayer space; rubidium atoms have been omitted for clarity. The larger circles represent zinc atoms.

Each four-ring in the chain is attached to the next through a shared -Zn-O-P- edge. The chains are "puckered," as if compressed along the *a* axis, and hinge on this edge to give the chain a staggered configuration. The chain connects to strands of double-six-ring cages on either side through a series of shared edges between the four-rings and the double-six-ring cages. Alternate four-rings in the chain contain two three-coordinate oxygen atoms, which are each shared between the ring and both a three- and four-ring of adjacent cages. The remaining four-rings in the chain (which have only two-coordinate oxygen atoms) each share an edge with a four-ring from a cage on one side and a cage six-ring on the other.

The strands (Figure 27-3) contain a series of double-six-ring cages, with each joined to the next through a shared four-ring window, forming a chain

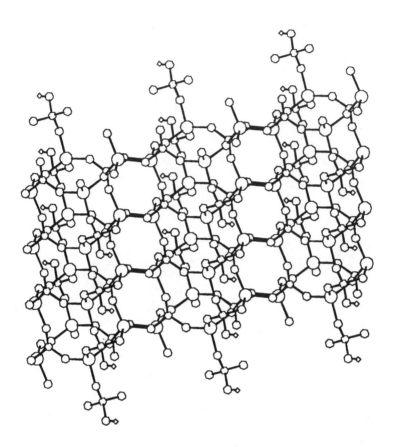

FIGURE 27-2. CHEMX (copyright Molecular Design, Ltd.) generated view of the layer plane (*ab* plane) of $RbHZn_2(PO_4)_2$. These layers are comprised of chains of four-rings which alternate with strands of double-six-ring cages (both of which propagate parallel to the *a* axis). The larger circles represent zinc atoms.

which runs along the *a* axis. The cages can each be viewed as consisting of a pair of parallel six-rings (consisting of six alternating zinc and phosphorus tetrahedra with one of the phosphorus atoms as O_3P-OH) connected to each other by four four-rings and two three-rings. The connecting four-rings are paired (through a common -Zn-O-P- edge) within each cage and coupled on either side by a three-ring unit.

The three-rings in the $RbHZn_2(PO_4)_2$ cages are unique in that they each contain a -Zn-O-Zn- linkage; the oxygen atom between the zinc atoms is

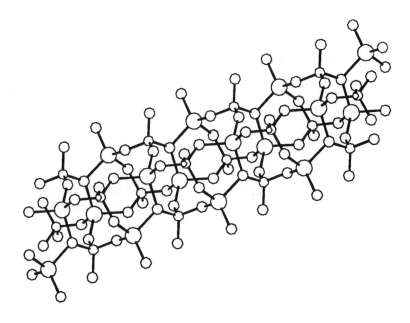

FIGURE 27-3. CHEMX (copyright Molecular Design, Ltd.) generated view of a strand (which runs parallel to the *a* axis) of double-six-ring cages in the layers of $RbHZn_2(PO_4)_2$. Each cage is comprised of two, parallel six-rings separated by both two pair of edge-sharing four-rings and two three-ring units. The larger circles represent zinc atoms.

three-coordinate and shared with alternate four-rings in adjacent four-ring chains. The remaining member of the three-ring is the O_3P-OH phosphorus which is also a member of the cage six-ring.

Though not unheard of in zeolite structures, three-ring units are relatively rare components in tetrahedral lattices. ZSM-18 was the first (Lawton and Rohrbaugh 1990), and remains the only example to date of an aluminosilicate zeolite containing three-rings within its framework structure. Three-ring units have also been observed in other silicates including the beryllosilicate lovdarite (Merlino 1984) and, more recently, the zincosilicate VPI-7 (Annen et al. 1991).

As expected, the T-O distances of the three-coordinate oxygen atom are slightly longer (Zn-O of 1.969 - 2.010 Å and P-O of 1.567 Å) than the average Zn-O and P-O distances in the framework (1.929 Å and 1.522 Å respectively). The P-O distance for the hydroxyl group is also slightly longer at 1.575 Å.

CONCLUSION

We have synthesized a new layered phosphate material, $RbHZn_2(PO_4)_2$, through the use of hydrothermal synthesis techniques, and conducted a structural analysis of the crystalline phase utilizing single crystal X-ray diffraction methods. The framework exhibits a unique three-ring sub-unit containing a -Zn-O-Zn- linkage.

Recent reports have also indicated that other structures are possible using hydrothermal synthesis techniques in the Zn/P(As)/O system illustrating the wide scope of available structural chemistry. In particular, zincophosphate and zincoarsenate analogs of the aluminosilicate zeolites RHO, Li-A(BW), SOD, and FAU are easily prepared (Harrison et al. 1991; Gier and Stucky 1991) under relatively mild conditions. Other new zincophosphates have also been synthesized (Chen et al. 1991). We have also produced additional new structures in this system which we will detail in a future publication (Corcoran et al. 1991).

REFERENCES

1. Annen, M. J.; Davis, M. E.; Higgins, J. B.; Schlenker, J. L. 1991. *J. Chem. Soc., Chem. Comm.* **17**: 1175.
2. Chen, J.; Thomas, J. M.; Jones, R. H.; Couves, J. W. 1991. To be submitted for publication.
3. Corcoran, E. W., Jr.; Newsam, J. M.; King, Jr., H. E.; Vaughan, D. E. W. 1989. From *Zeolite Synthesis*, ed. M. L. Occelli and H. E. Robson, chpt. 41. Washington, DC: ACS Symposium Series 398, American Chemical Society.
4. Corcoran, E. W., Jr.; Vaughan, D. E. W. 1989. *Solid State Ionics* **32/33**: 423.
5. Corcoran, E. W., Jr. 1990. *Inorg. Chem.* **29**: 157.
6. Corcoran, E. W., Jr.; Haushalter, R. C.; Lai, W.-Y. F. 1990. *U.S. Patent* 4,956,483.
7. Corcoran, E. W., Jr.; Savader, J. B.; Bhalla-Chawla, M. 1991. To be submitted for publication.
8. Gier, T. E.; Stucky, G. D. 1991. *Nature* **349**: 508.
9. Harrison, W. T. A.; Gier, T. E.; Moran, K. L.; Nicol, J. M.; Eckert, H.; Stucky, G. D. 1991. *Chem. Mater.* **3**: 27.
10. Lawton, S. L.; Rohrbaugh, W. J. 1990. *Science* **247**: 1319.
11. Merlino, S. 1984. From *Proceedings of the Sixth International Zeolite Conference*, ed. D. Olson and A. Bisio, pg. 747. England: Butterworths.

28

The Synthesis and Characterization of Some New Hydrated Sodium Zinc Phosphates

Thurman E. Gier, William T. A. Harrison, Tina M. Nenoff, and Galen D. Stucky   *Department of Chemistry, University of California, Santa Barbara, CA 93106-9510, USA*

The synthesis of several new hydrated sodium zinc phosphates, some of which are microporous, under mild conditions (ambient pressure; T $\leq$ 70°C) is described. The anionic frameworks of each are made up of ZnO_4 and PO_4 tetrahedra, with Na^+ ions in the pores for charge balance. The phase space is critically controlled by the pH of the medium as well as the concentrations of the ionic constituents of the final structures.

INTRODUCTION

Since the discovery of the class of microporous aluminosilicates known as zeolites (Cronstedt 1756; Breck 1984; Szostak 1989; Flanigen 1991), chemists have been intrigued by the possibility of introducing or completely substituting T-atoms (atoms tetrahedrally coordinated to oxygen) other than aluminum and silicon into open frameworks (Bennett and Marcus 1988). Several candidates have been proposed and many aluminophosphates (Bennett *et al.* 1986; Flanigen *et al.* 1988) and gallosilicates (Newsam and Vaughan 1986; Vaughan and Strohmaier 1990) are now well characterized, as well as materials which have essentially pure silica (*e.g.*, von Koningsveld 1990) and pure alumina (*e.g.*, Dempier and Bührer 1991) frameworks. Comprehensive recent reviews (Smith 1988 and references therein) and compilations (Meier and Olson 1987) of zeolite structure types are available.

For several years we have been interested in the possibility of synthesizing new "zeolites" containing neither aluminum nor silicon as the building-block framework tetrahedral atoms. The $AlSiO_4$ framework unit for many aluminosilicates (containing a 1:1 ratio of aluminum to silicon) has a formal net electronic charge of -1; the MXO_4-framework configuration beryllo(zinco)-phosphate(arsenate) or $(Be,Zn)(P,As)O_4$ unit has the same formal charge (but not the same formal *individual* atomic charges) and might be expected to form structures similar to the aluminosilicates, since these four cations commonly adopt tetrahedral coordination with oxygen.

Recently, Peacor and co-workers have reported the structures of several beryllophosphate minerals which are isostructural with aluminosilicates, including analogues of cancrinite (Peacor, Rouse and Ahn 1987) and zeolite-RHO (Rouse, Peacor and Merlino 1989). Harvey and Meier (1989) have synthesized five microporous zeolitic beryllophosphates, at least one of which has no known aluminosilicate zeolite analogue.

Synthesis of microporous beryllo(zinco)-phosphate(arsenate)s has proven to be a fertile field (Gier and Stucky 1991); here we report only on some new hydrated zinc phosphates and limit the counter-cation template necessary for charge balance of the framework to sodium. The use of other cations, including organic amino-derivatives and alkali metal mixtures, results in other structures which are now under investigation.

EXPERIMENTAL

All of the preparations described below were carried out in polypropylene or Teflon bottles at temperatures from -20 to $100\,^\circ C$. A typical synthesis consisted of rapidly mixing two homogeneous solutions of the reagents to form a gel/precipitate which was then aged at the desired temperature until crystallization was complete, generally a few hours to a few days. As with the aluminosilicates, different metastable structures may crystallize with time and following the course of the reaction by X-ray powder diffraction of sequential samples is a key to preparing pure, kinetically metastable materials. Since sodium phosphates are easily water soluble at all pHs, an excess of Na^+ and PO_4^{3-} over that required for the stoichiometry of the product was always employed.

Characterization techniques included X-ray powder diffraction (Scintag PAD-X automated diffractometer, θ-θ sample geometry, Cu Kα radiation, room temperature $(25\,(1)\,^\circ C)$), the results of which are reported below for each system. The instrumental Kα_1/Kα_2 profile was reduced to a single Cu Kα_1 peak position $(\lambda = 1.540568\,\text{Å})$ by a stripping routine, and d-spacings were established using silicon powder $(a = 5.43035\,\text{Å})$ as an internal standard, relative to this wavelength. Lattice parameters were optimized by least-squares refinements using Scintag software routines. For the low symmetry phases, strong peak assignments were confirmed by comparison with LAZY-PULVERIX (Yvon, Jeitschko and Parthe 1976) simulations. Thermogravimetric analysis (DuPont 9900 system), solid-state MAS NMR and, where possible, single-crystal X-ray methods (Huber automated diffractometer, $MoK\alpha$ radiation, $\lambda = 0.71073\,Å$) have also been used to characterize these phases. These results will be reported in detail separately.

$Zn_3(PO_4)_2 \cdot 4H_2O$: Solution-phase preparations from very acid conditions (pH < 1) yield the known phase hopeite, $Zn_3(PO_4)_2 \cdot 4H_2O$ (Whitaker 1975.) Thus, at very low pHs, the sodium cation is not incorporated in the product.

$NaH(ZnPO_4)_2$: A solution of 8.28 g of $NaH_2PO_4 \cdot H_2O$ (60 mmol), 3.43 g of 4 M NaOH (12 mmol) and 10 cc of water was prepared and 7.76 g of 2 M $Zn(NO_3)_2$ (12 mmol) was added rapidly. The initial white gel transformed to a "gummy ball" on shaking, but after overnight treatment at 70 °C, converted to a mass of small, transparent crystals. Reaction pH = 2–3.

$Na_6(ZnPO_4)_6 \cdot 8H_2O$: Gier and Stucky (1991) have reported the facile preparation of this aluminosilicate sodalite analogue as well as the arsenate congener. The sodium zincophosphate crystallizes rapidly at 70 °C at a pH of 6–8.

$NaZnPO_4 \cdot H_2O$: 10 cc of water was added to a clear solution of 0.49 g of ZnO (6 mmol) in 3 cc of 5 M H_3PO_4 (15 mmol), followed by 6.85 g of 4 M NaOH (24 mmol) (pH = 11.5) After 2 days at 70 °C about a gram of microcrystals was recovered. A few bi-capped rod prisms were selected out for single-crystal work.

$NaZnPO_4 \cdot nH_2O$: Increasing the Na/P ratio in the synthesis results in a higher final pH and the appearance of a new phase: 0.49 g of ZnO (6 mmol) was dissolved in 3 cc of 5 M H_3PO_4 (15 mmol) in a plastic bottle and diluted with 10 cc of water. The addition of 8.56 g of 4 M NaOH (30 mmol) yielded a gel which transformed to a sludge and then to a creamy milk on shaking (pH = 12.5–13). After two days at 70 °C the milk had crystallized to 0.2 mm cuboidal crystals.

$Na_6Zn_3(PO_4)_4 \cdot 3H_2O$: Increasing the pH still further and using relatively concentrated solutions gives rise to another new cubic phase. 7.76 g of 2 M $Zn(NO_3)_2$ (12 mmol) was added to 13.17 g of 4 M H_3PO_4 (44 mmol) in a plastic bottle, and 5.28 g of NaOH pellets (132 mmol) added slowly. The resulting hot, cloudy syrup was aged for 4 days at 50–70 °C (pH = 13) and the resulting 0.2 mm cubic crystals were recovered by filtration and washing. Thermal analysis showed 11% loss in 3 steps to 375 °C with loss of structure.

$Na_2ZnPO_4(OH) \cdot 7H_2O$: A clear, basic solution was prepared from 7.76 g of 2 M $Zn(NO_3)_2$ (12 mmol), 10.78 g of 4 M H_3PO_4 (36 mmol) and 37.65 g of

4 M NaOH (136 mmol). On standing at 70 °C, only ZnO precipitates, but if the pH is lowered slightly by the addition of 2.39 g of 4 M H_3PO_4 before holding at 70 °C (2 days), a suspension of needle-like crystals adequate for single-crystal characterization may be subsequently recovered by filtration.

ZnO: Very basic solutions of zinc, sodium and phosphate progenitors (pH > 13.5) only yield zinc oxide. Sodium cations and phosphate anions remain in solution.

RESULTS AND DISCUSSION

As noted above in the experimental section, combining homogeneous solutions containing Zn^{2+}, Na^+, and PO_4^{3-} ions results in different final structures depending principally on the pH but also on the concentrations of the reagents. Although the complete phase space (the $Na_2O-ZnO-P_2O_5-H_2O$ quaternary) has not been mapped, it is already clear that relatively minor synthesis changes are critical in determining the final products.

During the preparation of the initial gel suspensions, the viscosity and thixotropy of the mixtures undergo rapid changes, from gel to thick sludge to a creamy milk. Although amorphous at the beginning, this milk rapidly (compared to similar aluminosilicate preparations) crystallizes and crystalline or microcrystalline products settle out. We believe this is caused by the higher solubility of the reagents across the pH range employed.

We have also noticed that the nature of the anions (other than phosphate) present have an effect on the final products, implying that an anion structure-directing, or templating, effect exists; thus the results reported here are for nitrate (NO_3^-) and phosphate only.

NaH(ZnPO_4)_2: This material (NaZnP-Tri) is triclinic, space group $P\bar{1}$ (No. 2), with $a = 8.639(6)$, $b = 8.799(4)$, $c = 5.124(3)$ Å, $\alpha = 100.36(6)$, $\beta = 105.94(7)$ and $\gamma = 96.91(6)°$. Full details of the structure will be reported elsewhere (Nenoff et al. 1991a). It consists (Figure 28-1) of corrugated sheets of alternating, vertex-linked ZnO_4 and PO_4 tetrahedra: some of the sheet oxygens connect three tetrahedral nodes (2 Zn and 1 P), resulting in "bridged" tetrahedral 3-rings within the layers. Na^+ cations reside between the sheets, and NaZnP-Tri may be similar to the corresponding K^+ salt prepared, but not structurally characterized, by Frazier et al. (1966). NaZnP-Tri may also be related to the material with the same stoichiometry described by Kabalov et al. (1975) which was synthesized by high pressure hydrothermal methods. The powder pattern of NaZnP-Tri powder (Table 28-1) shows substantial preferred orientation in the [010] direction. Extensive grinding of NaZnP-Tri leads to an amorphous product.

Na_6(ZnPO_4)_6·8H_2O: This phase (NaZnP-SOD) is the exact analogue of the zeolitic aluminosilicate sodalite $Na_6(AlSiO_4)_6·8H_2O$, and, like other

TABLE 28-1: $NaH(ZnPO_4)_2$ Powder Data

Triclinic, space group $P\bar{1}$, $a = 8.639\,(6)$, $b = 8.799\,(4)$, $c = 5.124\,(3)$ Å, $\alpha = 100.36\,(6)$, $\beta = 105.94\,(7)$, $\gamma = 96.91\,(6)\,°$, $V = 362.5\,\text{Å}^3$.

h	k	l	$2\theta_{obs}$	$2\theta_{calc}$	$\Delta 2\theta$	d_{obs}	d_{calc}	Δd	I(rel)
0	1	0	10.345	10.380	-0.035	8.544	8.515	0.029	100
1	0	0	10.834	10.817	0.018	8.159	8.173	-0.013	2
1	1	0	16.335	16.311	0.025	5.422	5.430	-0.008	60
0	0	1	18.465	18.457	0.008	4.801	4.803	-0.002	< 1
1	1	-1	20.055	20.018	0.037	4.424	4.432	-0.008	7
0	2	0	20.835	20.847	-0.011	4.260	4.258	0.002	1
2	0	0	21.743	21.731	0.012	4.084	4.086	-0.002	28
2	0	-1	23.775	23.912	-0.137	3.739	3.718	0.021	2
1	0	1	24.015	24.153	-0.138	3.703	3.682	0.021	2
1	2	0	25.245	25.239	0.007	3.525	3.526	-0.001	6
1	2	-1	26.235	26.194	0.041	3.394	3.399	-0.005	2
2	-2	0	27.405	27.387	0.019	3.252	3.254	-0.002	13
1	1	1	28.725	28.713	0.013	3.105	3.107	-0.001	1
1	-2	-1	29.565	29.538	0.027	3.019	3.022	-0.003	9
0	3	0	31.425	31.493	-0.068	2.844	2.838	0.006	2
2	2	-1	31.827	31.776	0.051	2.809	2.814	-0.004	4
0	3	-1	32.782	32.853	-0.071	2.730	2.724	0.006	7
1	-3	1	34.683	34.684	-0.001	2.584	2.584	0.000	1
2	-3	0	35.096	35.083	0.013	2.555	2.556	-0.001	2
1	1	-2	35.295	35.260	0.035	2.541	2.543	-0.002	2
3	-2	0	35.775	35.776	0.000	2.508	2.508	0.000	1
1	2	1	36.045	36.012	0.033	2.490	2.492	-0.002	1
1	-3	-1	38.629	38.613	0.016	2.329	2.330	-0.001	1
2	-1	-2	39.994	39.958	0.037	2.252	2.254	-0.002	2
0	3	1	40.288	40.318	-0.030	2.237	2.235	0.002	2
3	-3	0	41.625	41.598	0.027	2.168	2.169	-0.001	< 1
1	-4	0	41.870	41.856	0.014	2.156	2.157	-0.001	2
0	4	0	42.435	42.427	0.008	2.128	2.129	0.000	4
1	4	-1	44.385	44.394	-0.009	2.039	2.039	0.000	3
1	4	0	45.915	45.900	0.015	1.975	1.975	-0.001	9
1	1	-3	53.793	53.800	-0.007	1.703	1.703	0.000	4
3	-4	-1	52.871	52.883	-0.012	1.730	1.730	0.000	3
4	-2	1	53.145	53.113	0.032	1.722	1.723	-0.001	2
5	-1	-1	53.295	53.321	-0.026	1.717	1.717	0.001	1

FIGURE 28-1: STRUPLO polyhedral representation (Fischer 1985) of the crystal structure of $NaH(ZnPO_4)_2$, showing the puckered tetrahedral-atom layers and inter-layer sodium cations (plain circles). Tetrahedral centers are labeled: the view is down the [001] direction.

sodalite-type materials with an ordered array of tetrahedral cations, it crystallizes in the primitive cubic crystal system (space group $P\bar{4}3n$ (No. 218) with $a = 8.8266(2)$ Å). NaZnP-SOD (Figure 28-2) contains the same ordered, $(Na_3(H_2O)_4)^{3+}$ "cubane"-like arrangement of three sodium cations and four water molecules within the cuboctahedral sodalite cage (Nenoff et $al.$ 1991b), like its aluminosilicate analogue (Felsche, Luger and Baerlocher 1986). Of interest here is the fact that crystallization in the presence of a large excess of sodium halide (chloride, bromide or iodide) does not generate the corresponding sodium-halo zincophosphate phases $Na_8X_2(ZnPO_4)_6$ (X = Cl, Br, I) as is the case with the well-known sodium-halo aluminosilicate sodalites, $Na_8X_2(AlSiO_4)_6$ (X = Cl, Br, I) (Henderson and Taylor 1978). A distance-least-squares (DLS) geometric simulation (Watkin, Carruthers and Betteridge 1990), subject to the crystallographic symmetry constraints of space group $P\bar{4}3n$ of the possible structure of $Na_8Cl_2(ZnPO_4)_6$, indicated that the framework would have to distort somewhat, but not unreasonably, to accommodate the chloride anion with reasonable Na–Cl contacts at the center of the sodalite cage. The powder pattern of NaZnP-SOD is reported in Table 28-2.

Unlike its aluminosilicate isostructure, which is stable to water loss, the sodalite-type framework of NaZnP-SOD converts smoothly to a new structure at 200°C upon dehydration, which has a powder pattern indexing well as hexagonal, $a = 8.82$, $c = 8.13$ Å. The systematic absences of this hexagonal pattern are consistent with space group $P6_3$. This new anhydrous NaZnPO$_4$ (NaZnP-H) phase slowly rehydrates to the sodalite form in boiling water and is the object of ongoing structure studies. NaZnP-H irreversibly converts to the stable stuffed-tridymite-type monoclinic form of NaZnPO$_4$ (Elammari et $al.$ 1987) at 600°C.

NaZnPO$_4$·H$_2$O: (NaZnP-Hex) Hexagonal, $P6_1$ (No. 169), $a = 10.4686(9)$, $c = 15.066(3)$ Å. A single-crystal structural study has confirmed the framework stoichiometry (Figure 28-3) and will be reported elsewhere (Harrison, Gier and Stucky 1991a). The composition seems to be essentially identical to that of the 15.2 Å cube reported below, but this synthesis requires a more concentrated solution of reagents. NaZnP-Hex consists of a new framework structure built up from 4- and 6-rings of Zn- and P-centered tetrahedral units. (See Table 28-3.)

NaZnPO$_4$·nH$_2$O, $n \approx 1.67$: Powder data for this phase can be indexed as primitive cubic (pseudo face-centered), with $a = 15.202(2)$ Å (Table 28-4.) The structure of this phase has not yet been determined. Water loss is complete at 300°C (observed weight loss 14%, calc. = 14% for $n = 1.67$) with concomitant transformation to a new structure, different from monoclinic NaZnPO$_4$. Analogous phases exist in the K/Zn/As/O and Rb/Zn/(P,As)/O systems, the latter apparently a tetragonal distortion of the parent cube.

TABLE 28-2: $Na_6(ZnPO_4)_6 \cdot 8H_2O$ Powder Data

Cubic, space group $P\bar{4}3n$, $a = 8.8266\,(2)$ Å, $V = 687.67\,(3)$ Å^3.

h	k	l	$2\theta_{obs}$	$2\theta_{calc}$	$\Delta 2\theta$	d_{obs}	d_{calc}	Δd	I(rel)
2	1	0	22.467	22.467	-0.001	3.947	3.947	0.000	11
2	1	1	24.646	24.648	-0.002	3.604	3.603	0.000	98
2	2	0	28.541	28.542	-0.001	3.121	3.121	0.000	27
3	1	0	32.001	32.001	-0.001	2.791	2.791	0.000	87
2	2	2	35.155	35.154	0.001	2.548	2.548	0.000	100
3	2	0	36.641	36.642	0.000	2.448	2.448	0.000	12
3	2	1	38.078	38.079	0.000	2.359	2.359	0.000	32
3	3	0	43.422	43.424	-0.002	2.081	2.080	0.000	16
4	2	0	45.903	45.906	-0.002	1.974	1.974	0.000	8
4	2	1	47.106	47.108	-0.002	1.926	1.926	0.000	14
3	3	2	48.286	48.287	-0.002	1.882	1.882	0.000	9
4	2	2	50.581	50.584	-0.002	1.802	1.802	0.000	6
4	3	1	52.806	52.807	-0.001	1.731	1.731	0.000	10
5	2	0	56.024	56.025	-0.001	1.639	1.639	0.000	11
5	2	1	57.069	57.071	-0.001	1.612	1.612	0.000	5
4	4	0	59.126	59.126	0.000	1.560	1.560	0.000	42
5	3	0	61.139	61.138	0.001	1.514	1.514	0.000	12
4	4	2	63.115	63.112	0.003	1.471	1.471	0.000	3
6	1	0	64.087	64.086	0.001	1.451	1.451	0.000	3
5	3	2	65.051	65.052	-0.001	1.432	1.432	0.000	6
6	2	0	66.964	66.962	0.002	1.396	1.396	0.000	1
5	4	1	68.846	68.846	0.000	1.362	1.362	0.000	4
6	2	2	70.726	70.705	0.021	1.330	1.331	0.000	21
5	4	2	71.630	71.627	0.003	1.316	1.316	0.000	< 1
6	3	1	72.545	72.544	0.001	1.301	1.301	0.000	3
4	4	4	74.369	74.365	0.004	1.274	1.274	0.000	< 1
7	1	0	76.169	76.170	-0.001	1.248	1.248	0.000	3
5	5	2	79.737	79.739	-0.002	1.201	1.201	0.000	3
6	4	2	81.505	81.508	-0.003	1.180	1.180	0.000	2
7	3	0	83.268	83.269	-0.001	1.159	1.159	0.000	3
7	3	2	86.775	86.773	0.002	1.121	1.121	0.000	3
8	0	0	88.519	88.520	-0.001	1.103	1.103	0.000	2
7	4	1	90.268	90.266	0.001	1.086	1.086	0.000	4
8	2	0	92.006	92.012	-0.006	1.070	1.070	0.000	1
7	4	2	92.894	92.885	0.008	1.063	1.063	0.000	< 1
6	5	3	93.760	93.760	0.000	1.055	1.055	0.000	< 1
8	2	2	95.506	95.511	-0.005	1.040	1.040	0.000	< 1
8	3	1	97.266	97.268	-0.002	1.026	1.026	0.000	1
6	6	2	99.029	99.031	-0.002	1.012	1.012	0.000	3
8	3	2	99.911	99.916	-0.005	1.006	1.006	0.000	< 1

TABLE 28-3: $NaZnPO_4 \cdot H_2O$ Powder Data

Hexagonal, space group $P6_1$, $a = 10.4686\,(9)\,\text{Å}$, $c = 15.066\,(3)\,\text{Å}$, $V = 1429.9\,(3)\,\text{Å}^3$.

h	k	l	$2\theta_{obs}$	$2\theta_{calc}$	$\Delta 2\theta$	d_{obs}	d_{calc}	Δd	I(rel)
1	0	0	9.736	9.748	-0.012	9.077	9.066	0.011	5
1	0	1	11.376	11.382	-0.006	7.772	7.768	0.004	100
1	0	2	15.278	15.280	-0.002	5.795	5.794	0.001	32
1	1	0	16.924	16.925	-0.001	5.235	5.234	0.000	19
1	1	2	20.647	20.647	0.001	4.298	4.298	0.000	26
2	0	2	22.878	22.878	0.001	3.884	3.884	0.000	20
2	1	0	25.975	25.982	-0.007	3.428	3.427	0.001	6
2	1	1	26.658	26.657	0.001	3.341	3.341	0.000	30
1	1	4	29.185	29.187	-0.002	3.057	3.057	0.000	48
3	0	0	29.543	29.534	0.009	3.021	3.022	-0.001	12
1	0	5	31.256	31.257	0.000	2.859	2.859	0.000	27
2	1	3	31.593	31.583	0.010	2.830	2.831	-0.001	21
3	0	2	31.885	31.881	0.004	2.804	2.805	0.000	36
2	2	0	34.241	34.234	0.007	2.617	2.617	0.000	21
3	0	3	34.615	34.613	0.002	2.589	2.589	0.000	3
2	1	4	35.408	35.385	0.023	2.533	2.535	-0.002	3
0	0	6	35.720	35.730	-0.009	2.512	2.511	0.001	6
3	1	1	36.213	36.188	0.025	2.479	2.480	-0.002	5
3	0	4	38.149	38.149	-0.001	2.357	2.357	0.000	8
2	1	5	39.804	39.805	-0.001	2.263	2.263	0.000	8
3	1	3	40.081	40.070	0.010	2.248	2.248	-0.001	3
2	2	4	41.995	42.005	-0.009	2.150	2.149	0.000	3
3	2	1	43.914	43.908	0.006	2.060	2.060	0.000	5
2	1	6	44.706	44.707	-0.001	2.025	2.025	0.000	3
3	1	5	47.038	47.031	0.006	1.930	1.931	0.000	9
2	2	6	50.314	50.318	-0.004	1.812	1.812	0.000	21
4	1	4	52.185	52.182	0.003	1.751	1.751	0.000	9
3	3	2	53.886	53.895	-0.009	1.700	1.700	0.000	5
5	1	1	56.794	56.825	-0.031	1.620	1.619	0.001	5

TABLE 28-4: $NaZnPO_4 \cdot nH_2O$ Powder Data

Cubic, $a = 15.202(2)\,\text{Å}$, $V = 3513.4(6)\,\text{Å}^3$.

h	k	l	$2\theta_{obs}$	$2\theta_{calc}$	$\Delta 2\theta$	d_{obs}	d_{calc}	Δd	I(rel)
2	0	0	11.655	11.633	0.023	7.586	7.601	-0.015	100
2	2	0	16.479	16.479	-0.001	5.375	5.375	0.000	1
2	2	2	20.211	20.218	-0.007	4.390	4.389	0.002	11
4	0	0	23.379	23.387	-0.008	3.802	3.801	0.001	5
4	2	2	28.749	28.746	0.003	3.103	3.103	0.000	36
5	1	1	30.549	30.530	0.019	2.924	2.926	-0.002	< 1
4	4	0	33.317	33.313	0.004	2.687	2.687	0.000	2
5	3	1	34.899	34.887	0.012	2.569	2.570	-0.001	6
6	0	0	35.394	35.398	-0.004	2.534	2.534	0.000	16
6	2	0	37.379	37.382	-0.003	2.404	2.404	0.000	5
6	2	2	39.265	39.279	-0.015	2.293	2.292	0.001	4
4	4	4	41.109	41.103	0.006	2.194	2.194	0.000	< 1
6	4	0	42.819	42.862	-0.043	2.110	2.108	0.002	< 1
6	4	2	44.555	44.565	-0.011	2.032	2.031	0.000	3
5	5	3	45.849	45.810	0.039	1.978	1.979	-0.002	1
8	0	0	47.859	47.827	0.032	1.899	1.900	-0.001	1
6	4	4	49.387	49.396	-0.008	1.844	1.844	0.000	1
8	2	2	50.924	50.928	-0.005	1.792	1.792	0.000	3
7	5	1	52.029	52.056	-0.027	1.756	1.755	0.001	1
6	6	2	52.449	52.428	0.021	1.743	1.744	-0.001	< 1
8	4	0	53.888	53.899	-0.011	1.700	1.700	0.000	< 1
8	4	2	55.337	55.342	-0.005	1.659	1.659	0.000	3
6	6	4	56.787	56.761	0.026	1.620	1.621	-0.001	< 1
9	3	1	57.789	57.810	-0.021	1.594	1.594	0.001	< 1

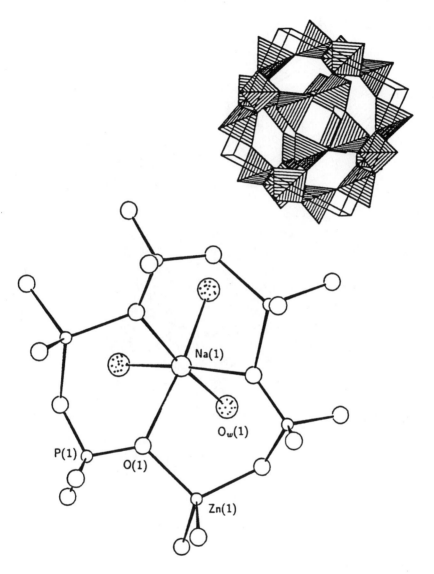

FIGURE 28-2: ORTEP view (Johnson 1976) of the sodalite-type phase $Na_6(ZnPO_4)_6 \cdot 8H_2O$, showing detail of the alternating ZnO_4/PO_4 tetrahedral 6-ring and the coordination sphere (3 framework O (plain circles), 3 water molecules (speckled circles)) of the sodium cation occupying the 6-ring window. The tetrahedral-atom topology is indicated in the small polyhedral plot.

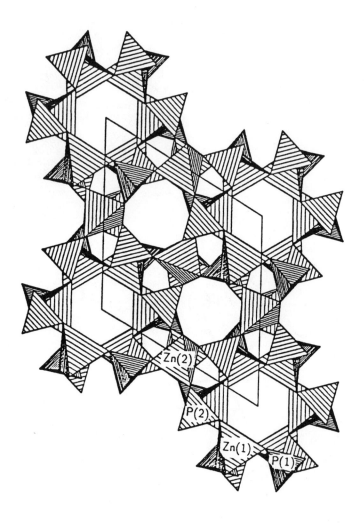

FIGURE 28-3: Polyhedral representation of the framework of hexagonal NaZnPO$_4$·H$_2$O, viewed down [001], showing the tetrahedral-atom labeling scheme.

$Na_6Zn_3(PO_4)_4 \cdot 3H_2O$: Body-centered cubic(?) or primitive cubic, $a = 11.938\,(2)$ Å, possible space groups $I\bar{4}3d$ (No. 220), $P4_132$ (No. 213) or $P2_13$ (No. 198): see below. This phase is analogous to the recently described $Na_6Zn_3(AsO_4)_4 \cdot 3H_2O$ (space group: $P2_13$) (Grey *et al.* 1989) and also has a seemingly unlikely topological analogue in "$11CaO \cdot 7Al_2O_3 \cdot CaF_2$" (or $2 \times Ca_6Al_3(AlO_4)_4 \cdot F$, body-centered cubic: space group $I\bar{4}3d$) (Williams 1973). Both these semi-condensed phases contain anionic 6-/8-ring tetra-hedral-atom networks (ZnO_4/AsO_4 in the former, just AlO_4 in the latter), again indicating the structural relationship between framework structures with different tetrahedral atoms. An 8-ring subunit surrounding a water molecule is illustrated in Figure 28-4.

Powder data (Table 28-5) were consistent with a primitive-cubic unit cell, while two different batches of single crystals appeared to be body-centered and primitive-cubic, respectively. This difference is as yet unex-plained; however, the body-centered model in $I\bar{4}3d$ was successfully refined to excellent agreement factors ($R \approx 2.5\%$), while the primitive-cubic data set (apparent space group: $P4_132$) has not yet been successfully modeled. Twinning, as was found for the arsenate phase (Grey *et al.* 1989) may be causing problems in the refinement. The structure is lost on heating to $375\,^{\circ}C$(11% weight loss in three distinct steps.) This corresponds to loss of the extra-framework water in three stages, even though there is only one crystallographically distinct water molecule in the as-synthesised material.

$Na_2ZnPO_4(OH) \cdot 7H_2O$: (NaZnP-Mon) is monoclinic, space group $P2_1/a$ (No. 14), with $a = 6.410\,(4)$, $b = 21.597\,(9)$, $c = 8.670\,(3)$ Å and $\beta = 109.91\,(3)\,^{\circ}$. NaZnP-Mon is not a framework structure, but contains novel one-dimensional tetrahedral stacks (ZnO_4 and PO_4 moieties) made up only of 3-rings (Figure 28-5) which include hydroxy-bridges (Zn–(OH)–Zn) be-tween adjacent zinc atoms in the chain. Adjacent tetrahedral stacks are interconnected *via* a complex arrangement of sodium cations and water molecules. For full details see Harrison, Gier and Stucky (1991b). A mono-clinic phase with this stoichiometry was reported by Kabalov *et al.* (1972) but seems to be different. Powder data are listed in Table 28-6.

CONCLUSIONS

Several new hydrated sodium zinc phosphates have been synthesized. Most of these materials are characterized by linked, alternating, zinc and phos-phorus tetrahedra, with sodium ions and water in the void spaces within the frameworks, akin to aluminosilicate zeolite. An important test of the "zeolite"-like nature of these zincophosphates will be the determination of their sorption properties, which are currently being investigated. Two observations concerning synthesis are of immediate interest. One is the wide variability in the zincophosphate frameworks formed, using only Na^+

TABLE 28-5: $Na_6Zn_3(PO_4)_4 \cdot 3H_2O$ Powder Data

Cubic, $a = 11.938(2)$ Å, $V = 1701.5(5)$ Å^3.

h	k	l	$2\theta_{obs}$	$2\theta_{calc}$	$\Delta 2\theta$	d_{obs}	d_{calc}	Δd	I(rel)
1	1	1	12.775	12.774	0.001	6.892	6.893	0.000	1
2	1	1	18.134	18.128	0.006	4.872	4.874	-0.002	100
2	2	0	20.955	20.972	-0.017	4.224	4.221	0.003	11
3	0	0	22.275	22.263	0.011	3.977	3.979	-0.002	1
3	1	0	23.505	23.488	0.017	3.772	3.775	-0.003	5
3	2	1	27.885	27.882	0.003	3.190	3.191	0.000	39
4	0	0	29.866	29.855	0.011	2.984	2.985	-0.001	38
3	3	1	32.596	32.611	-0.015	2.740	2.739	0.001	2
4	2	0	33.466	33.484	-0.018	2.671	2.669	0.001	90
3	3	2	35.162	35.174	-0.011	2.546	2.545	0.001	49
4	2	2	36.797	36.795	0.001	2.437	2.437	0.000	45
5	1	0	38.344	38.358	-0.014	2.342	2.341	0.001	3
3	3	3	39.107	39.119	-0.012	2.298	2.298	0.001	2
5	2	1	41.340	41.333	0.007	2.179	2.180	0.000	4
4	3	3	44.148	44.142	0.006	2.047	2.047	0.000	3
6	1	1	46.818	46.816	0.002	1.937	1.937	0.000	12
6	2	1	48.769	48.747	0.022	1.864	1.864	-0.001	1
5	4	1	49.383	49.378	0.005	1.842	1.842	0.000	6
6	3	1	51.846	51.845	0.000	1.760	1.760	0.000	5
4	4	4	53.060	53.048	0.012	1.723	1.723	0.000	6
7	1	0	54.230	54.232	-0.002	1.688	1.688	0.000	6
6	4	0	55.400	55.398	0.002	1.655	1.656	0.000	15
6	3	3	56.551	56.549	0.002	1.625	1.625	0.000	10
6	4	2	57.681	57.684	-0.004	1.595	1.595	0.000	12
7	3	0	58.791	58.806	-0.015	1.568	1.568	0.000	1

TABLE 28-6: $Na_2ZnPO_4(OH)\cdot 7H_2O$ Powder Data

Monoclinic, space group $P2_1/a$ (No. 14), $a = 6.410\,(4)$, $b = 21.597\,(9)$, $c = 8.670\,(3)\,Å$, $\beta = 109.91\,(3)°$, $V = 1128\,Å^3$.

h	k	l	$2\theta_{obs}$	$2\theta_{calc}$	$\Delta 2\theta$	d_{obs}	d_{calc}	Δd	I(rel)
0	2	0	8.144	8.156	-0.011	10.813	10.798	0.015	29
0	0	1	10.814	10.819	-0.005	8.155	8.152	0.003	3
0	1	1	11.564	11.568	-0.004	.7.629	7.626	0.003	100
0	2	1	13.604	13.574	0.031	6.491	6.506	-0.015	84
0	4	0	16.409	16.379	0.030	5.389	5.399	-0.010	6
1	3	-1	19.395	19.409	-0.015	4.567	4.564	0.003	2
0	4	1	19.695	19.681	0.014	4.498	4.501	-0.003	12
0	0	2	21.765	21.762	0.003	4.075	4.076	0.000	39
0	1	2	22.125	22.152	-0.027	4.010	4.005	0.005	33
1	2	1	22.665	22.629	0.036	3.916	3.922	-0.006	3
1	2	-2	23.235	23.250	-0.015	3.821	3.819	0.002	4
0	3	2	25.065	25.061	0.004	3.546	3.547	-0.001	19
0	6	1	27.045	27.032	0.013	3.291	3.293	-0.002	55
1	4	-2	27.315	27.341	-0.026	3.259	3.256	0.003	4
0	5	2	30.105	30.096	0.009	2.964	2.964	-0.001	16
2	3	-1	30.496	30.494	0.002	2.927	2.927	0.000	1
2	2	0	30.736	30.754	-0.019	2.904	2.903	0.002	1
0	7	1	30.916	30.941	-0.025	2.888	2.885	0.002	1
1	2	2	31.401	31.455	-0.053	2.844	2.840	0.005	2
2	3	0	32.176	32.150	0.026	2.778	2.780	-0.002	2
2	4	-1	32.446	32.447	-0.001	2.755	2.755	0.000	2
0	0	3	32.926	32.911	0.015	2.716	2.717	-0.001	6
0	8	0	33.124	33.132	-0.008	2.700	2.700	0.001	6
1	3	-3	33.556	33.561	-0.005	2.667	2.666	0.000	4
2	4	-2	34.606	34.591	0.015	2.588	2.589	-0.001	1
2	5	-1	34.816	34.815	0.001	2.573	2.573	0.000	1
2	1	1	35.266	35.286	-0.020	2.541	2.540	0.001	73
1	5	2	36.886	36.902	-0.016	2.433	2.432	0.001	6
2	2	-3	37.126	37.120	0.006	2.418	2.418	0.000	3
1	7	1	36.016	36.039	-0.023	2.490	2.488	0.002	3
0	7	2	36.496	36.471	0.025	2.458	2.460	-0.002	2
0	8	2	40.037	40.002	0.034	2.249	2.251	-0.002	2
1	2	3	41.627	41.636	-0.010	2.167	2.166	0.000	2
0	10	0	41.717	41.766	-0.049	2.162	2.160	0.002	1
2	7	1	46.068	46.035	0.032	1.968	1.969	-0.001	2
0	11	1	47.628	47.576	0.052	1.907	1.909	-0.002	11
0	5	4	49.428	49.384	0.045	1.842	1.843	-0.002	2

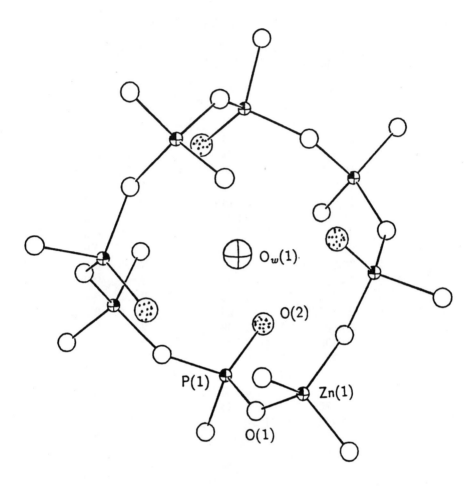

FIGURE 28-4: ORTEP diagram of the 8-ring unit in the semi-condensed phase $Na_6Zn_3(PO_4)_4 \cdot 3H_2O$, surrounding $O_w(1)$, an oxygen atom belonging to an extra-framework water molecule: $Zn(1)$ and $P(1)$ (shaded circles) alternate in the 8-ring; $O(1)$ (plain circles) makes Zn–O–P bridges; $O(2)$ (speckled circles) are 'hanging' P–O vertices.

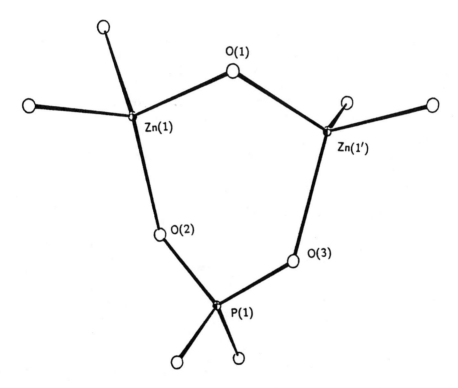

Figure 28-5: ORTEP view of the Zn/P/O 3-ring configuration in the tetrahedral-column containing phase $Na_2ZnPO_4(OH)\cdot 7H_2O$. The polyhedral view shows the tetrahedral-atom connectivity in the 1-dimensional chain.

as the charge-balancing counter-cation, and systematically varying the pH and concentrations of the crystallizing media. The second is the relative ease of forming large (up to 0.5 mm) crystals under mild conditions, suitable for structure studies. This latter fact is probably accounted for by the greater solubility of the zincophosphates in these aqueous systems as compared to the aluminosilicates.

The phases reported above are arranged in Table 28-7 in order of increasing pH. Their thermal stabilities are considerably lower than those of typical aluminosilicate zeolites and it is usual that structure loss accompanies water loss on heating.

The study of non-aluminosilicate microporous framework systems is continuing, using other templating cations (including organic cations such as amine derivatives) and multi-component mixtures. Analogues of zeolites such as zeolite-X (Harrison *et al.* 1991a), RHO (Gier and Stucky 1991), Li-ABW (Harrison *et al.* 1991b), and sodalite (Nenoff *et al.* 1991b) have already been prepared, as well as several new structures that are currently undergoing full structural characterization.

Acknowledgments We thank Nancy Keder for crystallographic assistance, and the National Science Foundation (Division of Materials Research) for partial funding.

TABLE 28-7: Sodium/Zinc/Phosphate Phase-Field Sequence

Phase	Ratio†	pH	Structure
$Zn_3(PO_4)_2 \cdot 4H_2O$	0:3:2	< 2	hopeite
$NaH(ZnPO_4)_2$	1:2:2	2–3	triclinic; corrugated layers
$Na_6(ZnPO_4)_6 \cdot 8H_2O$	1:1:1	6–8	sodalite type
$NaZnPO_4$	1:1:1	*	hexagonal $P6_3(?)$,
$NaZnPO_4 \cdot H_2O$	1:1:1	11.5	hexagonal $P6_1$,
$NaZnPO_4 \cdot nH_2O$	1:1:1	12–13	cubic, $a = 15.20$ Å (pseudo fcc)
$Na_6Zn_3(PO_4)_4 \cdot 3H_2O$	6:3:4	13	cubic, $a = 11.94$ Å (pseudo bcc)
$Na_2ZnPO_4(OH) \cdot 7H_2O$	2:1:1	13	monoclinic, 3-rings
ZnO	0:1:0	> 13	zinc oxide

† ratio of Na:Zn:P respectively.
* reversible sodalite dehydration product.

Note added in press: Since this paper was submitted we have prepared and characterized two further phases in this family: $Zn_3(PO_4)_2 \cdot H_2O$ (monoclinic, space group $P2_1/c$, $a = 8.7018(8)$, $b = 4.8765(5)$, $c = 16.676(2)$ Å, $\beta = 94.972(5)°$) and $Na_2Zn(HPO_4)_2 \cdot 4H_2O$ (monoclinic, space group $P2_1/c$, $a = 8.947(2)$, $b = 13.254(2)$, $c = 10.0977(2)$ Å, $\beta = 116.358(6)°$). The former material is a condensed structure containing 4- and 5-coordinate zinc atoms and 4-coordinate phosphorus atoms. The latter structure contains vertex-sharing ZnO_4 and PO_4 tetrahdra surrounding 10-ring channels, which contain the sodium cations and water molecules. This is the largest channel system observed in these phases so far.

REFERENCES

Bennett, J. M., and Marcus, B. K., 1988. In *Innovations in Zeolite Material Science: Proceedings of the International Symposium, Nieuwport, Belgium*, ed. P. J. Grobet *et al.*, pp. 269–279. Amsterdam: Elsevier.

Bennett, J. M., *et al.*, 1986. *Zeolites* 6: 349.

Breck, D. W., 1984. *Zeolite Molecular Sieves*. Malabar, Florida: Krieger Publishing Co.

Cronstedt, A., 1756. *Akad. Handl. Stockholm* 18: 120.

Dempier, W., and Bührer W., 1991. *Acta Crystallogr.* B47: 197.

Elammari, L., *et al.*, 1987. *Zeit. Kristallogr.* 180: 137.

Felsche, J., Luger, S., and Baerlocher, Ch., 1986. *Zeolites* 6: 367.

Fischer, R. X., 1985. *J. Appl. Cryst.* 18: 258.

Flanigen, E. M., 1991. In *Introduction to Zeolite Science and Practice*, ed. H. Van Bekkum, E. M. Flanigen and J. C. Jansen, pp 13–34. Amsterdam: Elsevier.

Flanigen, E. M., Patton, R. L., and Wilson, S. T., 1988. In *Innovation in Zeolite Materials Science*, ed. P. J. Grobet *et al.*, pp 13–27. Amsterdam: Elsevier.

Frazier, M., *et al.*, 1966. *J. Agric. Food Chem.* 14: 522.

Gier, T. E., and Stucky, G. D., 1991. *Nature* 349: 508.

Grey, I. E., *et al.*, 1989. *J. Sol. St. Chem.* 82: 52.

Harrison, W. T. A., Gier, T. E., and Stucky, G. D., 1991a. In preparation.

Harrison, W. T. A., Gier, T. E., and Stucky, G. D., 1991b. In preparation.

Harrison, W. T. A., *et al.*, 1991a. *Chem. Mater.* 3: 28.

Harrison, W. T. A., *et al.*, 1991b. In preparation.

Harvey, G., and Meier, W. M., 1989. Synthesis of Beryllophosphate Zeolites. In *Zeolites: Facts, Figures, Future*, ed. P. A. Jacobs and R. A. van Santen, pp. 411–420. Amsterdam: Elsevier.

Henderson, C. M. B., and Taylor, D., 1978. *Phys. Chem. Miner.* 2: 337.

Johnson, C. K., 1976. *Oak Ridge National Laboratory Report ORNL-5138*, Oak Ridge, TN 37830, with local modifications.

Kabalov, K., *et al.*, 1972. *Sov. Phys. Dokl.* 17: 72.

Kabalov, K., *et al.*, 1975. *Sov. Phys. Crystallogr. (Eng.)* 20: 91.

Meier, W. M., and Olson, D. H., 1987. *Atlas of Zeolite Structure Types.* London, Butterworths.

Nenoff, T. M., *et al.*, 1991a. In preparation.

Nenoff, T. M., *et al.*, 1991b. *J. Amer. Chem. Soc.* 113: 378.

Newsam, J. M., and Vaughan, D. E. W., 1986. In *New Developments in Zeolite Science and Technology*, ed. Y. Murakami, A. Iijima and J. W. Ward, pp 457–464. Amsterdam: Elsevier.

Peacor, D. R., Rouse, R. C., and Ahn, J.-H., 1987. *Amer. Miner.* 72: 816.

Rouse, R. C., Peacor, D. R., and Merlino, S., 1989. *Amer. Mineral.* 74: 1195.

Smith, J. V., 1988. *Chem. Rev.* 88: 149.

Szostak, R., 1989. *Molecular Sieves. Principles of Synthesis and Investigation.* New York: Van Nostrand Reinhold.

Vaughan, D. E., and Strohmaier, K. G., 1990. United States Patent number 4,960,578.

von Koningsveld, H., 1990. *Acta Crystallogr.* B46: 731.

Watkin, D. J., Carruthers, J. R., and Betteridge, P. W., 1990. *CRYSTALS User Guide*, Chemical Crystallography Laboratory, Oxford University, UK.

Whitaker, A., 1975. *Acta Crystallogr.* B31: 2026.

Williams, P. P., 1973. *Acta Crystallogr.* B29: 1550.

Yvon, K., Jeitschko, W., and Parthe, E., 1976. *LAZY-PULVERIX User Guide*, University of Geneva, Switzerland, with local modifications.

29

Synthesis and Adsorptive Properties of Titanium Silicate Molecular Sieves

S. M. Kuznicki[1], K. A. Thrush[1], F. M. Allen[1], S. M. Levine[1], M. M Hamil[1], David T. Hayhurst[2], and Mahmoud Mansour[2]

[1]*Engelhard Corporation*
[2]*Department of Chemical Engineering, Cleveland State University*

ETS-4 and ETS-10 are two new microporous crystalline solids with uniform pores similar in dimension to classical small- and large-pored zeolites. Like aluminosilicate zeolites, these new titanium-silicates are synthesized under hydrothermal conditions in alkaline environments. Structural and compositional analyses indicate that the titanium is octahedrally coordinated in the framework. This requires two counterbalancing cations per titanium. These cations are readily exchanged by alkaline, alkaline-earth, and multivalent metal cations to a level exceeding 4 meq/g. Adsorption isotherms were measured at 295 K for water, benzene, o-xylene, and triethylamine for Na^+, K^+, Mg^{2+}, Ca^{2+}, Sr^{2+}, and Ba^{2+} exchanged ETS-10 and for water, oxygen, and sulfur dioxide on as-synthesized and Ca^{2+} exchanged ETS-4. ETS-10, in many cases, adsorbed large amounts of these probe molecules, demonstrating capacities often exceeding 20 cc/100 g. ETS-10 demonstrates molecular sieving properties that are similar to traditional large-pore zeolites while ETS-4 behaves as a small-pore molecular sieve. The effective pore sizes of these materials may be modified by ion-exchange. Adsorbate-adsorbent interactions are substantially different when compared to classical aluminosilicate zeolites. ETS-4 and ETS-10 are unique new small- and large-pore molecular sieve titanium silicates constructed from non-traditional primary building units. Their syntheses may represent the discovery of a new class of molecular sieves.

INTRODUCTION

Milton (1959) in the late 1950s discovered that aluminosilicate gels could be crystallized into molecular sieve zeolites containing uniform pores. Since then, numerous commercially important catalytic, adsorptive and ion-exchange applications have been developed based on the unique structure of these materials. This high degree of utility is the result of the zeolite's unique combination of high surface area, its uniform porosity, and most importantly, the electrostatically charged sites within the structure induced by cations located near tetrahedrally coordinated Al^{3+}. Breck (1974) hypothesized that 1,000 aluminosilicate zeolite frameworks are theoretically possible. To date, approximately 150 have been identified. In recent years the rate of discovery of new framework topologies has slowed and much research in the field has focused on varying the nature of the charged sites within the framework by substituting elements other than silicon or aluminum into the tetrahedrally coordinated sites within these known framework topologies. This approach has been particularly successful in the substitution of P^{5+} for Si^{4+} in the aluminophosphate sieves $AlPO_4$ (Flanigen et al., 1982) and for the other metals in the mixed aluminophosphate-metaloxide systems, SAPO's and MeAPO's (Flanigen et al., 1986). Although many SAPO's and MeAPO's have been reported, their utility in commercial applications appears to have been quite limited to date. Reports on the incorporation of other elements into known zeolite structures have also appeared in the literature. For example, Notari et al. (1980) and Szostak and Thomas (1986) have reported the incorporation of small amounts of boron and iron into ZSM-5 analogs. These authors have reported that even at these low levels of substitution, heating these materials to typical catalytic reaction temperatures generally results in amigration of these elements from tetrahedral framework sites into the sieve's pore system as an occluded oxide. The difficulty in incorporating elements other than silicon and aluminum into a zeolite's structure results from the fact that larger cations prefer to coordinate with oxygen ions in an octahedral configuration rather than tetrahedral coordination as is found with Al^{3+} and Si^{4+}. It is not surprising, therefore, that metal ions with atomic radii much greater than that of Si^{4+} and Al^{3+} will not exist stably in the relatively small void formed by tetrahedral oxygen anions. In this research, a novel synthetic approach for creating new molecular sieve structures is described. This approach combines the use of two different structural sub-units, namely tetrahedral silicate units and octahedral titanate units, to produce unique framework structures having uniform pores of molecular dimensions. Like zeolites, these new titanium silicates demonstrate distinct molecular sieving properties. Like zeolites, they also represent framework structures and have exchangeable cations within their pores. The first report of a synthesis of a molecular sieve containing titanium was by Young (1967) in which a titanium silicate was prepared under conditions similar to those used to prepare aluminosilicate zeolites. While these materials were called "titanium zeolites," later reports indicate that the phases reported are too dense to be molecular sieves (Breck, 1974; Barrer, 1982; Bellussi et al., 1986). A naturally occurring titanosilicate, "Zorite," was reported (Bussen et al. 1973) to occur in trace quantities in Russia on the Kola Peninsula. The structure of zorite was reported by Belov and Sandomirskii (1979). Other reports of titanium-bearing molecular sieves include the incorporation of Ti into silicalite (Notari et al., 1983) and into ZSM-5 (Bellussi et al., 1986). The first reports of true molecular sieves incorporating a substantial fraction of titaniumin the structure were presented by Kuznicki (1989, 1990) wherein he described the synthesis of both large- and small-pore materials. The objective of this report is to begin to described the chemistry, properties, and adsorptive behavior of these materials, denoted ETS-4 and ETS-10. Since these first reports, Chapman and Roe (1990, 1991) have also described the synthesis and characterization of additional microporous titanium silicate materials. These materials appear to be of the small-pored variety, although molecular sieving properties have yet to be reported. In this chapter, the general approach to synthesis is outlined. The characterization of these materials with respect to their chemical and physical properties is discussed along with a description of the proposed structures for these phases. Adsorption data confirm the large- and small-pore molecular sieving properties of these novel materials.

EXPERIMENTAL

Synthesis

Samples of ETS-4 and ETS-10 tested in this study were prepared using hydrothermal synthesis procedures outlined inrecent patents (Kuznicki, 1989, 1990). A typical ETS-4 preparation may be represented by combining an alkaline silicate solution of 1507 g N-Brand[R] sodium silicate (28.8 wt% SiO_2, 9.14 wt% Na_2O), with 459.8 g NaOH, 132.8 g anhydrous KF, and 200.4 g deionized water. To this 979.2 g Fisher $TiCl_3$ solution (20 wt% $TiCl_3$, 20 wt% HCl) is added. The resulting gel is autoclaved at $200^{\circ}C$ to 30 hours. The product is vacuum filtered, washed with deionized water, and dried at 100°C. Samples of ETS-10 may be prepared by combining an alkaline silicate solution of 3487.5 g sodium di-silicate solution (25.6 wt% SiO_2, 14.5 wt% Na_2O) with 556.3 g NaOH and 261.0 g anhydrous KF. To this 2160.0 g of $TiCl_4$ solution (24.7 wt% $TiCl_4$, 20 wt% HCl) is added. After mixing the resulting gel into a uniform mixture, 33.8 g of seed, as detailed in Kuznicki (1989), is added. The gel was autoclaved at 200°C for 18 hours. The product was vacuum filtered, washed with deionized water, and dried at 100°C.

Elemental Analysis

The chemical composition of the ETS-mixtures were determined by x-ray fluorescence. Samples were analyzed for Na_2O, K_2O, CaO, MgO, Al_2O_3, SiO_2, Fe_2O_3, and TiO_2 by alithium tetraborate fusion followed by the X-ray fluorescence analysis. The spectrometer used was a Siemens SRS-200X-ray fluorescence spectrometer with a Criss fundamental parameters program.

XRD Spectra

Synchrotron powder XRD scans of as-synthesized ETS-4 and ETS-10 were obtained at the National Synchrotron Light Source at Brookhaven National Laboratory on Long Island on beam line X7A.

TGA/DTA

TGA/DTA analysis was performed on the as-synthesized ETS-4 and ETS-10 using a PL Thermal Sciences STA 1500 instrument. Samples were run in flowing air, using a heating rate of 5°C/min.

Crystal and Framework Density

Crystal and framework densities were measured by pynchnometric methods using molecules of various sizes.

Adsorption

Adsorptive capacities were measured for the as-synthesized and several ion-exchange forms of ETS-4 and ETS-10. The exchanged forms of the

ETS materials were prepared using an exchange protocol similar to that outlined by Sherry (1968). To exchange a given sample, 1.0 g of material was placed in a 250 ml Erlenmeyer flask with the appropriate volume of a 1.0N solution of the exchanging cation nitrate salt. This solution volume represents a five-fold excess of the exchange ion. The flask was agitated for a minimum of 20 hours on a wrist shaker. The spent solution was then decanted from the flask and a second volume of exchange solution was added. The exchange was repeated three times. Upon completion, the exchanged ETS was washed with a minimum of five 100 ml aliquots of distilled, deionized water and dried overnight at 110°C.

Adsorption capacities were measured isothermally in a McBain-Bakr balance. Select isotherms were replicated using a Cahn 1000 vacuum microbalance to insure accuracy and reproducibility. To initiate a run, approximately 300 mg of a test ETS sample was placed in a quartz sample pan. The sample was activated by heating at a rate of 1°C/min to the final activation temperature (200°C for ETS-4 and 350°C for ETS-10) while maintaining a vacuum of $<10^{-6}$ torr. The sample was held at the final activation temperature for a minimum of 12 hours. After activation, the sample was cooled to the run temperature under vacuum. For the oxygen adsorption experiments at 77 K, the adsorption chamber was surrounded with a dewar of liquid nitrogen. The height of the liquid nitrogen was maintained at a minimum of 10 cm above the sample height. Sample weight was monitored until the weight varied less than $\pm$ 0.05 mg in 30 min. Gas was dosed incrementally up to 90% of the adsorbate's saturation vapor pressure for condensable gases and to atmospheric pressure for the remaining adsorbates. Desorption was also measured for all samples to insure the lack of hysteresis. Benzene, o-xylene, and triethylamine adsorption data and oxygen data at 77 K were fitted to a Langmuir model from which the maximum pore volumes were calculated.

RESULTS AND DISCUSSION

Synthesis

The synthesis of ETS-4 and ETS-10 is easily accomplished via hydrothermal techniques utilizing alkaline titanium silicate gels as described in recent patents (Kuznicki 1989, 1990). These gels are quite similar to gels yielding classical aluminosilicate zeolites, with aluminum being replaced by titanium. The primary difference between and ETS-4 preparation and an ETS-10 preparation is the alkalinity. ETS-10 is grown at lower alkalinities, where increased silica chain length is anticipated. It should be noted that a wide variety of titanium and silica sources may be employed and that these materials need not be synthesized in a mixed sodium/potassium system.

Characterization

Due to the complexity of determining the structure of these novel molecular sieves, several complementary techniques were employed. A few of these methods are discussed in this report. A more complete discussion will be published at a later date.

X-Ray Diffraction. Powder X-ray diffraction is a standard method used to characterize crystalline molecular sieves. Figure 29-1 shows synchrotron powder X-ray diffraction spectra of ETS-4 and ETS-10. It is interesting to note that each of these patterns contains both broad and narrow peaks, implying extensive disorder in the structure of these materials. The spectrum for ETS-4 closely resembles that of the rare titanium silicate mineral zorite (Belov and Sandomirskii, 1979).

Elemental Analysis. Bulk chemical analysis of ETS-4 and ETS-10, as shown in Table 29-1, indicates Si/Ti molar ratios of approximately two and one-half and four and one-half to one, respectively. For both materials, the ratio of the cationic charge (Na + K) to framework titanium is two. This indicates that the coordination of the titanium in both molecular sieves is octahedral. These cations are also readily exchangeableusing the standard methods for classical aluminosilicate zeolites. Standardized electron microprobe data for individual crystals of these molecular sieves confirms the bulk elemental composition. The microprobe experiments further indicate that the compositions of these samples are uniform from crystal to crystal.

TGA/DTA. As shown in Figure 29-2, thermal gravimetric analysis of as-synthesized ETS-4 and ETS-10 shows substantial, systematic water losses. These losses occur from ambient temperature to 200 to 300°C in both material. The ETS-4 structure collapses between 200°C and 350°C, depending upon cationic form. As-synthesized ETS-10 maintains structural integrity until approximately 650°C. Both ETS-4 and ETS-10 are found to undergo a phase transformation between 500 and 700°C.

Structure

Structural models of ETS-4 and ETS-10 are depicted in Figures 29-3

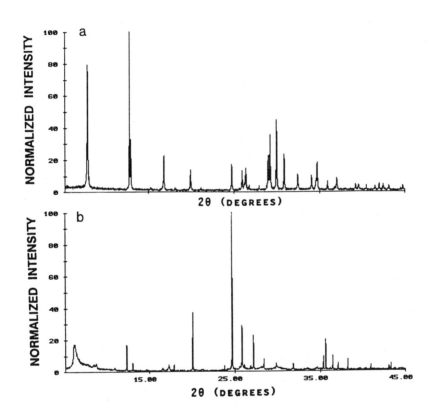

Figure 1. Synchrotron powder x-ray diffraction spectra of (a)
ETS-4 and (b) ETS-10.

Table 29-1

Elemental Analysis of ETS-4 and ETS-10

(wt% of Oxides)

	ETS-4	ETS-10
SiO_2	49.9	66.1
TiO_2	24.9	17.4
Na_2O	16.5	10.0
K_2O	6.8	4.8
Al_2O_3	0.29	0.29
LOI	14.2	15.0
Si/Ti	2.7	5.0
(Na+K)/Ti	2.1	1.9
Ion Exchange	6.0	3.7

Capacity (meq/g, hydrous basis)

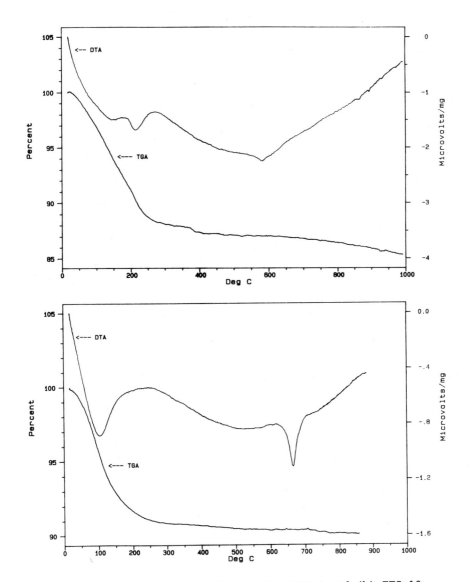

Figure 2.   TGA/DTA spectra of (a) ETS-4 and (b) ETS-10.

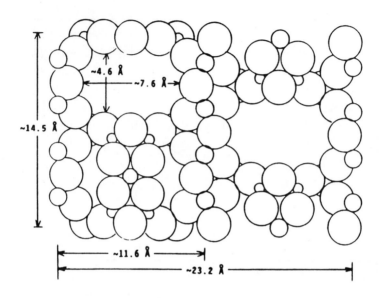

Figure 3.   Structural model of ETS-4.

ETS-10 Structure: Two Projections

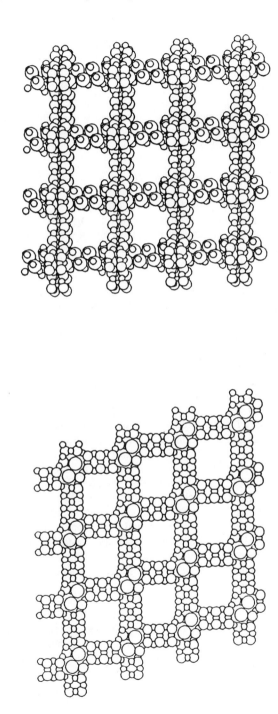

Figure 4. Structural model of ETS-10.

and 29-4, respectively. The structures of both ETS-4 and ETS-10 are markedly different from classical aluminosilicate zeolites. These materials are composed of chains of octahedrally coordinated titanium atoms linked with classical, tetrahedrally coordinated zeolite-type rings. ETS-4 behaves as a small-pored material, with a pore size comparable to zeolite A. ETS-10 is a large-pored material whose structure is a three-dimensional network of interconnecting channels that intersect at a central pore. The effective pore size of the as-synthesized material is approximately 8 A.

Adsorption

Pore size Determination. The effective pore diameters of ETS-4 and ETS-10 were determined by measuring the adsorption capacity of these materials for several plug-gauge molecules. As shown if Figure 29-5, ETS-4 was found of readily adsorb water (2.8 A) and oxygen (3.6 A), while sulfur dioxide (3.8 A) was completely excluded. The critical diameter of the probe molecules used in this report are those listed by Breck (1974). Upon exchange to the Ca-form, ETS-4 was found to readily adsorb the 3.8 A sulfur dioxide molecule (Figure 29-5). These results indicate that, as with aluminosilicate zeolites, the effective pore diameter of the ETS materials can be altered by ion exchange. None of the ETS-4 samples were observed to adsorb methane (4.3 A) from which it is concluded that the as-synthesized ETS-4 has a maximum pore diameter of 3.7 A which can be increased to at least 4.0 A with appropriate ion-exchange. For ETS-10, adsorption capacities were measured using n-hexane (4.3 A), cyclohexane (6.0 A), triethylamine (7.8 A),and 1,3,5 trimethylbenzene (8.1 A) as probe molecules. As shown in Figure 29-6, the as-synthesized form of ETS-10 adsorbs each probe molecule except the trimethylbenzene, which is completely excluded from the structure, indicating an effective pore diameter of approximately 8 A for ETS-10. This effective diameter is typical of large-pore zeolites, such as X and Y. The adsorption and desorption of these probe molecules was observed to be rapid and reversible. This is not unexpected since the proposed structure for this material indicates access to internal pore volume from all crystallographic directions. It can therefore be concluded that the limiting pore sizes of these new titanium silicates are comparable to zeolites and can be easily varied by ion-exchange.

Pore Volume Determination. Pore volumes for various exchanged forms of ETS-10 were estimated by measuring the adsorption capacity for oxygen at liquid nitrogen temperatures (77 K). The condensation of oxygen in a molecular sieve pore structure is generally regarded as a very accurate method to estimate the true free pore volume. Adsorption isotherms for O_2 for the various ion-exchange forms of ETS-10 are shown in Figure 29-7. These data were fit to a Langmuir

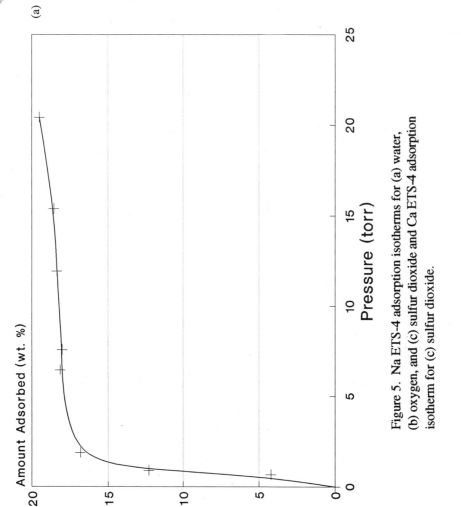

Figure 5. Na ETS-4 adsorption isotherms for (a) water, (b) oxygen, and (c) sulfur dioxide and Ca ETS-4 adsorption isotherm for (c) sulfur dioxide.

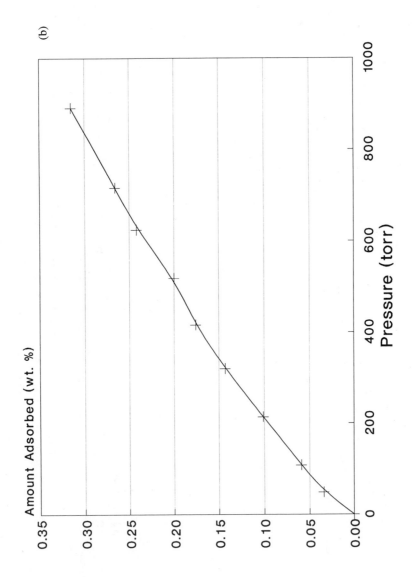

Figure 5. *(continued)*

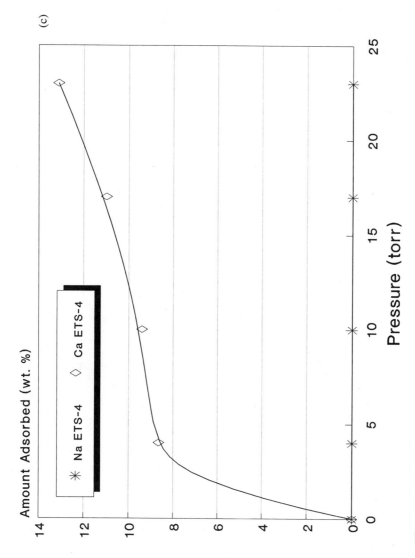

Figure 5. (continued)

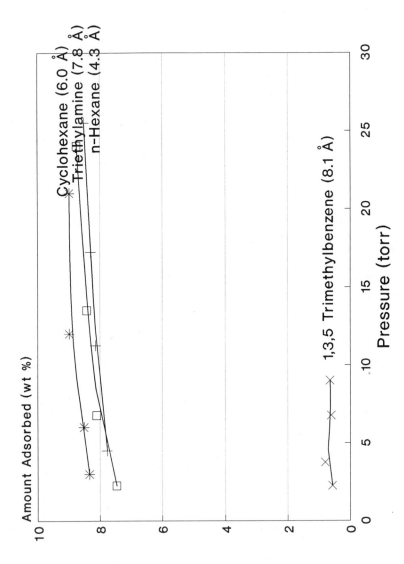

Figure 6. ETS-10 adsorption isotherms for n-hexane, cyclohexane, triethylamine, and 1,3,5 trimethylbenzene.

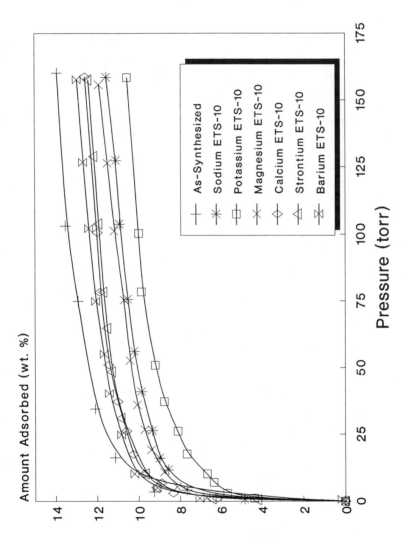

Figure 7. Adsorption isotherms for O_2 for the various ion-exchange forms of ETS-10 at 77 K.

model. The Langmuir maximum volumes corresponded to a complete filling of the ETS-10's pores. The calculated values are listed for each material in Table 29-2. Pore volumes for ETS-10 are found to range from 8.39 to 11.95 cc/100 g for the K- and as-synthesized forms, respectively. As expected, the large monovalent potassium cation-exchanged from of ETS-10 has the lowest pore volume, as potassium occupies a large fraction of the internal pore structure. This O_2 adsorption capacity for as-synthesized ETS-10 is lower than the value of 28.0 c/100 g reported by Breck (1974) for a NaY. If, however, the pore volumes of the ETS-10 and the Y are calculated on a volume basis, the free volume of ETS-10 (30.6 cc O_2/100 cc) is found to be quite close to that of the Y (35.6 cc O_2/100 cc). This is due to the significantly higher framework density of the titanium silicate ETS-10 (2.56 g/cc) when compared to the aluminosilicate Y. Water adsorption isotherms were measured for ETS-4 and ETS-10 at room temperature (298 K). These data are shown in Figure 29-8. For these titanium silicates, adsorption isotherms are observed to be of the Langmuir type, with a steep rise in the isotherms at low relative water pressures. Adsorption capacities are comparable to aluminosilicate molecular sieves, indicating a complete void filling of the ETS's pore structure by water.

Cation Effects. The effect of cation exchange on adsorption capacities of ETS-10 was evaluated. Adsorption isotherms were measured at 298 K for the Na-, K-, Mg-, Ca-, Sr-, and Ba-forms of ETS-10 using benzene, o-xylene, and triethylamine as test adsorbates. Adsorption isotherms were correlated with the Langmuir model. The Langmuir maximum adsorption capacities for these gases are listed in Table 29-3. Figure 29-9 represents a plot of the maximum adsorption capacity for each of these organics as a function of the diameter of the charge-compensating cation. It is observed that the adsorption capacity varies inversely with cation size; that is, the greatest adsorption capacities are observed for the smallest cation. As cation size increases, pore capacity decreases in a linear fashion. This indicates that the cations are most likely located near the walls of the main channels and not isolated in any cage-type structure, which is consistent with the structure proposed for ETS-10, as shown in Figure 29-4. Furthermore, this linear behavior is observed both for aromatic molecules and for an amine, suggesting that for these cation forms of ETS-10, the cation-adsorbate interactions are secondary, and that simply the pore-filling effects dominate at high adsorbate loading. The strong rectangular shape of the isotherms for all three organics, however, indicates that strong cation-adsorbate interactions exist at low adsorbate loadings.

Table 29-2

Langmiur Oxygen Maximum Pore Volume of

Cation Exchange ETS-10's

Cation	<u>Langmiur Maximum Pore Volume</u> (cc/100g)
As synthesized	11.95
Na	9.91
K	8.89
Mg	9.27
Ca	10.25
Sr	9.70
Ba	10.05

Table 29-3

Maximum Capacity Calculated from
Langmuir Isotherm of ETS-10's
(moles/kg)

Zeolite	Benzene	O-Xylene	Triethylamine
As synthesized	1.09	0.72	0.78
Na	1.06	0.84	0.70
K	0.86	0.69	0.68
Mg	1.07	0.88	0.85
Ca	0.98	0.70	0.79
Ba	0.94	0.71	0.75

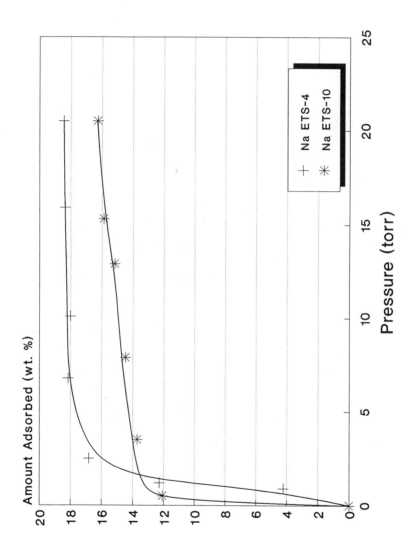

Figure 8.   Water adsorption isotherms for ETS-4 and ETS-10 at room temperature (298 K).

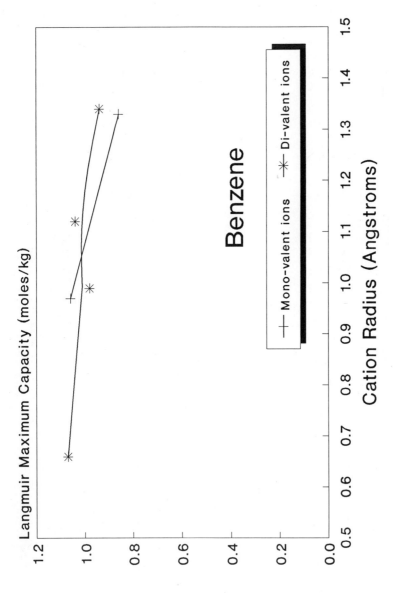

Figure 9. Langmuir maximum adsorption capacity for benzene, ortho-xylene, and triethylamine as a function of the diameter of the charge-compensating cation.

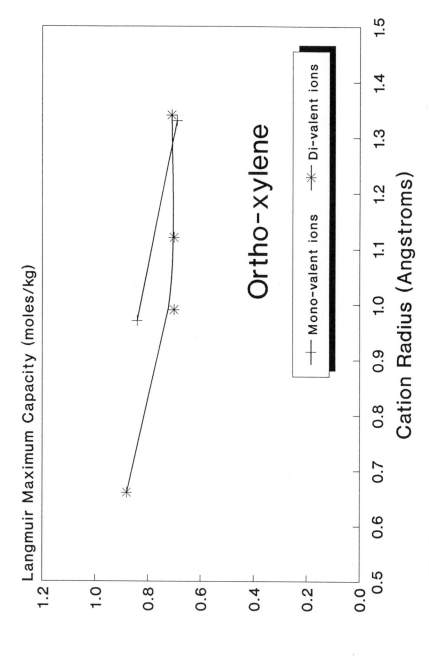

Figure 9. *(continued)*

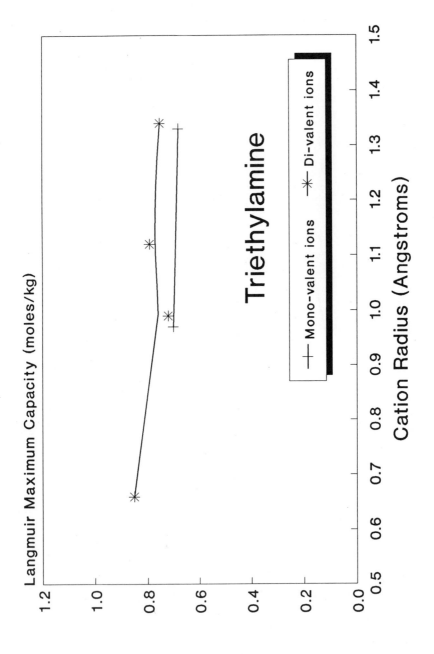

Figure 9. *(continued)*

CONCLUSIONS

ETS-4 and ETS-10 represent small- and large-pored molecular sieves. However, these sieves are constructed using the non-traditional octahedral titanate unit with classical zeolitic tetrahedral silicate units. Chemical analysis shows that each octahedral titanium imparts a net negative divalent charge to the framework requiring either two monovalent ions or none divalent ion to maintain charge neutrality in the structure. A combination of X-ray diffraction analysis, high-resolution transmission electron microscopy, and nuclear magnetic resonance spectroscopy have been used to postulate structures for these titanium silicates. These techniques show that each material has a rigid three-dimensional framework structure. Adsorption data indicate small-pore behavior for ETS-4 and large-pore behavior for ETS-10. As with aluminosilicate zeolites, the effective pore diameters of ETS materials can be modified in a predictable fashion by ion-exchange. The pore volume of ETS-10 was found to be comparable to zeolite Y on a volume basis. This is partly due to the much greater framework density of these titanium silicates when compared to aluminosilicate zeolites. Both ETS materials were found to undergo rapid and reversible adsorption of both polar and non-polar gases. Additionally, adsorption was found to be of the Langmuir type, with a rapid uptake at low adsorbate pressures and complete void filling as pressures approach saturation. The degree of void filling varies inversely with the radius of the charge-compensation cations.

The discovery of these two new titanium silicates which combine both octahedral and tetrahedral primary building units indicates a new approach in the preparation of novel molecular sieves. Adsorptive analyses show that these materials have open structures and their adsorptive properties can be readily modified by ion-exchange. These materials may well provide the basis for the development of many new adsorptive, ion-exchange, and catalytic processes which are fundamentally different from those that use classical aluminosilicate zeolites.

ACKNOWLEDGEMENTS
The authors wish to gratefully acknowledge the help provided by Dr. David Cox of the Brookhaven National Laboratory in acquiring the X-ray diffraction data reported. The assistance of Raj Bhasin and Earl Waterman of Engelhard Corp. in measuring X-ray fluorescence and TGA/DTA data, respectively, is also gratefully acknowledged.

REFERENCES

Barrer, R. M., Hydrothermal Chemistry of Zeolites, Wiley, New York (1982), 293.

Bellussi, G., Buonamo, F., Curmo, C., Esposito, A., Perego, G., and Tarramasso, M., New Devel. Zeol. Sci. and Tech., Murakami, Y., Iijima, A., and Ward, J. W. (ed) (1986), 129-136.

Belov, N. V., and Sandomirskii, P. A., Sov. Phys. Crystallorg., 24 (1979), 686.

Breck, D. W., Zeolite Molecular Sieves, Wiley, New York (1974).

Bussen, I. V., Goika, E. A., Kul'chitskaya, E. A., Men'shikov, Y. P., Mer'kov, A. N., and Nedorezova, A. P., Zapiski Vses. Mineralog. Obshch., 102 (1973), 54-62.

Chapman, D. M., and Roe, A. L., Zeolite, 10 (1990), 730-737.

Chapman, D. M., U.S. Patent 5,015,453 (1991).

Flanigen, E. M., Lok, B. M., Patton, R. L., and Wilson, S. T., New Devel. Zeol. Sci. and Tech., Murakami, Y., Iijima, A., and Ward, J. W. (ed.) (1986), 103-112.

Kuznicki, S. M., "Large-pored crystalline titanium moleculer sieve zeolites," U.S. Patent 4,853,202 (1989).

Kuznicki, S. M., "Preparation of small-pored crystalline titanium molecular sieve zeolites," U.S. Patent 4,938,989 (1990).

Milton, R. M., U.S. Patents 2,882,243 and 2,882,244 (1959).

Notari, B., Perego, G., and Tarramasso, M., Proc. Fifth Intl. Conf. on Zeol., Rees, L.V.C, (ed.) (1980), 40-48.

Notari, B., Perego, G., and Tarramasso, M., U.S. Patent 4,410,501 (1983).

Sherry, H. S., J. Phys. Chem., (1968), 72,4086.

Szostak, R., and Thomas, T. L., J. Catalysis, 100 (1986), 555.

Young, D. A., U.S. Patents 3,329,480 and 3,329,481 (1967).

30

Intergrowths In Zeolite Structures

J. M. Newsam[1], M. M. J. Treacy[2], D. E. W. Vaughan[3], K. G. Strohmaier[3] and M. T. Melchior[3]

[1]Biosym Technologies Inc., San Diego, CA 92121

[2]NEC Research Institute Inc., Princeton NJ 08540

[3]Exxon Research and Engineering Company, Annandale NJ 08801

Intergrowths or stacking disorder, incorporated during hydrothermal synthesis or generated during reconstructive phase transformations, are quite common in zeolite materials. It is argued that faulting patterns reflect the conditions of synthesis and that analyses of faulting patterns can therefore yield insight into synthesis phenomena. Even small numbers of planar faults can have a dramatic effect on a zeolite's macroscopic properties, such as sorption capacity. Monte Carlo simulations are used to quantify the reduction of sorption capacity in a zeolite with a unidimensional channel system as a function both of faulting probability and crystallite length. Several analytical methods, including catalysis or sorption measurements, nmr, electron microscopy and diffraction, can be used to quantify intergrowth occurrence. The combination of powder diffraction data measured under high instrumental resolution and diffraction pattern simulations that take an explicit and accurate account of the effects of stacking disorder is a particularly effective means of studying zeolite intergrowth phenomena.

INTRODUCTION

The regularity of zeolite pore systems is usually a key feature in technological applications. Structural irregularities or imperfections can then detract from a zeolite's performance, providing an incentive for understanding their occurrence. Defect sites may also be active catalytically, contributing in a positive sense to, for example, acid catalysis or perhaps providing an undesirable conversion pathway in sorption applications. A planar defect is an irregularity or fault in the stacking of successive sheets or blocks along the direction perpendicular to the fault plane (Figure 30-1). Planar defects may involve a mismatch in the chemical connectivity across the fault giving what, depending on its characteristics, is more properly termed an interface, a grain boundary, or an overlayer. They may also allow proper chemical connectivity to be maintained. In this case the alternative mode of sheet connection or stacking will be characteristic of a structure type different from that of the unfaulted sequence. When the alternative stacking mode recurs, the region can be described as an intergrowth between the two structure types. This description should, it must be noted, be used cautiously, for the presence of a single planar fault in a given structure does not strictly qualify as an instance of the different structure type.

Although, as discussed further below, intergrowths can be formed during thermally induced framework transformations, we are most interested in the factors that lead to their incorporation under hydrothermal synthesis conditions. Just as the aluminum distributions measured in a fully crystallized faujasite zeolite can convey insight into subtle atomic-scale factors in the process of framework crystallization, so can the manner and extent of planar faulting. Although we are generally not able at present to infer such information, substantial strides have been made in the characterization of intergrowth structures and in the reproducible and controlled synthesis of desired intergrowth materials.

FORMATION OF PLANAR DEFECTS AND INTERGROWTH STRUCTURES

Incorporation of intergrowths under hydrothermal synthesis conditions

A number of defects are expected in zeolite materials produced under the typical hydrothermal conditions of nucleation and rapid crystal growth (1). A key question (2) is "What promotes a particular mode of stacking during synthesis ?" Strictly energetic arguments can be based on the known end-member crystal

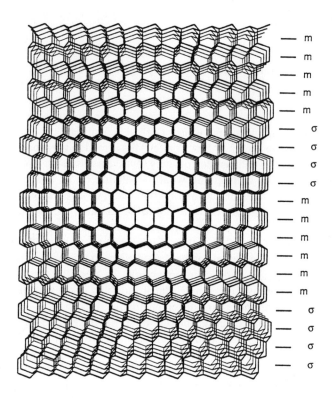

— m
— m
— m
— m
— m
— σ
— σ
— σ
— σ
— m
— m
— m
— m
— m
— σ
— σ
— σ
— σ

FIGURE 30-1. Planar faults in the structure of diamond, drawn as straight lines connecting carbon atom sites.   *The diamond structure is comprised of sheets related by the inversion centers, σ; in the hexagonal analog, lonsdaleite, the sheets are related by mirror planes, m. Diamond and faujasite share the same space group $F d\overline{3}m$, and this model is also a simple schematic for the m/σ-faulting in FAU-EMT framework materials.*

structures. Interatomic potentials are used as a basis for computing the crystal energy of a given configuration of the atoms within the unit cell. The atomic coordinates are then adjusted so as to minimize the computed energy. When potentials are validated, and there remains substantial room for improvement here, the relative optimized energies of different framework models can be compared. For example, the two polymorphs of zeolite beta, the *BEA-framework with ABAB. . . .   pore stackings and its counterpart with ABCABC. . . . pore stackings (3), were reported to have closely similar lattice energies after crystal energy minimizations (4).

Grossly disparate crystal energies might exclude possible formation of some structure types. As a rough guide, for example, the probabilities of intergrowth or fault occurrence could be computed assuming Boltzmann statistics. However, such calculations involve no entropic contribution and relate to energies at 0 K. Further, hydrothermal syntheses of zeolites proceed under kinetic control in complicated synthesis media, preventing such arguments from being used quantitatively unless the complete details of the phase fields are known. Techniques are being developed for performing free energy calculations in which the entropic contribution from the lattice vibrations is included (5), but such computations of course still rely critically on the chosen interatomic potentials, and relate to an infinite, perfect crystal in isolation.

Planar defects likely grow from a single definition point. For example, in a developing ZSM-20 crystallite, the observed morphology demonstrates that lateral growth perpendicular to $[111]_{cubic}$ is much faster than that along $[111]_{cubic}$. For a single faujasite sheet (that appears in the complete structure of faujasite (FAU) as the section xxx, $0 \leq x \leq 1/3$, and in that of ECR-30 or EMC-2 (EMT) as the section $1/4 \leq z \leq 3/4$) to which the next layer is being added, the development of approximately one half of a sodalite cage in the mirror-related configuration will be sufficient, if growth continues laterally from that point, to give a full sheet in the mirror-related configuration. Addition of a sodalite cage in the alternate orientation at some other point on the same plane will inevitably lead to a local defect at the point(s) of contact of the two growing layers, as they cannot then interconnect properly. The high hydroxyl content of beta zeolites may be indicative of mismatches of this type (3). However, no evidence for such defects has yet been reported in the FAU-EMT system, implying that the rate of lateral growth is greater than that of layer addition by a substantial margin, as indeed evidenced by the observed crystallite morphologies. A prerequisite for formation of the EMT-framework may be that the synthesis composition and conditions act so as to define such disparate growth rates.

We know very little about the sizes or the size distributions of those pieces that add to a growing zeolite crystal. Reasonable physical intuition dictates that they are small, probably less that some 30 atoms. Further, giving the rate at which small silicate species equilibrate (1), it is unlikely that nucleation and growth occur through only one type of silicate conformer. For these reasons, although secondary building units (SBU's) are useful in rationalizing interrelationships in some families of zeolite crystal structures, they probably have no general significance at the synthesis stage (6,7).

In the absence of both a strong orienting influence and a low activation barrier to sheet alignment, the fusion or "zippering" of fully formed sheets is an unlikely route to either intergrowth or near-perfect crystal structures. The chance of two sheets that are formed independently happening both to be of identical dimensions and to meet in fully correct registry is remote. For example, even if the origins of two similar 1000 x 1000 Å sheets are coincident, there would be a probability of only ~0.0001 that they would be in orientational registry. The unlikelihood of chance registry is often evidenced directly in lamellar systems such as clays. Solvent-induced swelling and exfoliation forms a suspension of extended sheets. These can be flocculated to produce disordered structures, typically with sheets intertwined and joined in rather random fashions. Even when mechanical means are used to promote parallel layering, the products usually have extensive layer-rotational or turbostratic disorder. That is, there is no long-range coherence to the relative orientations of successive sheets along the unique direction. We observe, in diffraction, peaks arising only from the unidimensionally coherent structure.

This physical tendency toward orientational disorder can be counteracted by the interlayer cations that promote ordering in certain clay systems. The lack of strong covalent bonding between the clay layers apparently allows at least partial registry to be developed. However, neither the degree of 3-dimensional order nor the crystallite morphologies approach the level of perfection usually displayed by zeolite synthesis products.

In some dispersed lamellar systems, the presence of sheets is evidenced directly by the particular pattern of light scattering that can be seen by the eye directly. If extended sheets occur in substantial concentration in zeolite synthesis media, they should also be detectable by light or, depending on their extents, by small-angle X-ray or neutron-scattering experiments. Without more experimental data we cannot yet draw definitive conclusions about the importance of these mechanical or physical factors during zeolite synthesis. The same arguments, however, do also apply to other types of extended structure such as chains (particularly when they have conformational flexibility), or even large cage units found in zeolite crystal structures such as α-, β- or ε-cages.

The incorporation of intergrowths during synthesis probably relates to particular local patterns of preorganization in the synthesis medium at the nucleation stage, and similar organization on the surface of the developing crystal during growth. This local organization may, during nucleation, reflect the local coordination requirements of the cations in the system, a role previously used to rationalize the dependence of the crystallized structure type on the nature of the inorganic base present   (1). The character and influence of this preorganization may be different during nucleation and during growth, depending on the nature of the developing crystal surface. There is perhaps evidence for this distinction in some faulted materials. CSZ-1 crystallites have typically one or two planar faults (the hexagonal rather than cubic mode of stacking – see below) close to the centers of the crystallites (8,9), consistent with the hexagonal mode being promoted on nucleation, but superseded during growth by the cubic mode of stacking.

Post-synthesis generation of planar defects

An open zeolite structure is metastable and invariably transforms to a more condensed phase at elevated temperatures. The temperature of the transition(s) and the structure of the final phase depend on the initial zeolite composition. The phase transition(s) may entail complete framework restructuring. Thus zeolites A and X evolve through an amorphous phase at <~600°C to a carnegeite phase, the exact temperature depending on the chemical composition. Simpler topotactic transitions are observed in the conversion of (Na, TMA)-E(AB) to sodalite at above 633 K (10), sodium ZK-14 to sodalite at above 610 K (11), $AlPO_4$-C to $AlPO_4$-D at ~633 K(12) and, perhaps, $AlPO_4$-21 to $AlPO_4$-25 at ~530 K (13). Hydroxylated structures such as that of the mineral sepiolite are particularly susceptible to such restructuring. Reconstructive transformations may also be induced by cation rearrangement on dehydration, such as in barrerite and heulandite B.

The conversion of VPI-5 or $AlPO_4$-54 (VFI) to $AlPO_4$-8 (AET) proceeds by a solid state transformation (14), although this transformation apparently requires the presence of water (15). The bonding in successive blocks along $[100]_{hex}$ is reconfigured, giving a new framework connectivity which has orthorhombic symmetry and in which the 18-ring apertures are reduced to 14-rings (16,17). As the solid state transformation does not require a complete restructuring of the bonding, it can occur progressively through a crystallite. The hexagonal symmetry of the VFI-framework parent implies that there are three equivalent possible directions for that which becomes [001] in the AET-framework Transformation may occur simultaneously along different directions in different regions of a given crystallite. $AlPO_4$-8 materials are therefore frequently disordered, with stacking faults common. There is evidence that the $AlPO_4$-54 – $AlPO_4$-8 transformation is reversible (14), consistent with the proposed transformation mechanism. This reversibility implies possibilities for synthesizing other new framework types by transformations of known systems.

METHODS OF OBSERVATION AND MEASUREMENT

Sorption and catalysis

From an applications standpoint, the effect of intergrowths on macroscopic sorption or catalytic performance is a key issue. The alteration of sorption

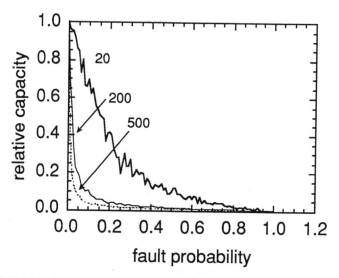

FIGURE 30-2. Results of Monte Carlo simulations of the reduced sorption capacity that accompanies planar faulting in a unidimensional pore system.   *A single fault is assumed to block the channels. The effect of faulting depends on the channel lengths, results for 20, 200, and 500 unit cell lengths being shown.*

capacity from that expected based on known crystallographic data (18) can be one indication of intergrowth occurrence. For example, the low sorption capacities of typical natural and synthetic gmelinites have been attributed to planar faults or limited intergrowths of other ABC-6 structures such as that of cancrinite. The sorption capacity of a zeolite with unidimensional channels like gmelinite is significantly reduced even for a low fault probability, particularly when the channels are long (Figure 30-2). Partly because of a dependence on size and degree of fault clustering, a structural interpretation of sorption capacity reduction is difficult without considerable corroborating evidence also being available. Thus, the structural basis for the different sorptive properties of small-port and large-port mordenites is still debated and may, in different samples, be related to detritus in the channels, faulting, or intergrowth of related structures (19).

The product distributions obtained from certain catalytic test reactions, such as hydrocracking, also depend on the geometries of the pore spaces within which they occur. Such reaction studies have been made on materials with planar faults (20,21) and catalytic data have also been related quantitiatively to intergrowth occurrence in the ZSM-5 – ZSM-11 pentasil family (22). As with sorption experiments, however, interpretations of these catalytic data from materials with intergrowth structures are rarely simple. Indeed, now that diffraction methods for quantifying intergrowth occurrence have been developed (see below) it may be informative to reexamine many of the classical intergrowth problems.

Nuclear Magnetic Resonance

Although nmr techniques sample the local environments of the target nuclei, they can yield quantitative data on intergrowth characteristics. The chemical shift of a ^{29}Si nucleus in a zeolite depends on the composition and geometry of its local coordination environment (23). Were, say, the geometries of the ^{29}Si nuclei in the the parent and intergrowth structures to differ significantly, resolvable resonances would be observed, allowing measurement of the relative amounts of the two components. The ^{29}Si spectra are quite sensitive to changes in local environment (23), and intergrowth structures often give poorly resolved ^{29}Si spectra, consistent with there being a distribution of local environments. There are exceptions, notably dealuminated zeolite beta, which gives a well-resolved spectrum (24) (that is consistent in form with the structural and stacking model (3,25)) despite the presence of near-random stacking disorder. The character of the intergrowths may be such that little change in local coordination occurs. Thus dealuminated ZSM-20 (26) and EMC-2 both show single resonances, consistent with the four crystallographically equivalent T-sites (T = tetrahederal species, Si, or Al, etc.) in the EMT-framework components of these materials having geometries that are similar both to each other and to that in the FAU-framework component. In contrast, four separate peaks are observed for dealuminated CSZ-1 (27), demonstrating that compositional or strain effects can remove the degeneracy of the T-sites in these frameworks.

The chemical shifts of ^{129}Xe in sorbed atomic xenon (28,29) and ^{13}C in occluded tetramethylammonium cations (30,31) depend on the accessible extents of the pores within which they reside. The sizes of the cages in many intergrowth materials, for example intergrowths of the OFF- and ERI-frameworks in zeolite T, depend on the form and separation between the intergrowth boundaries. Although the ^{129}Xe or ^{13}C spectra measured from such materials can be complex, interpretations in terms of intergrowth occurrence are in some cases possible (32).

Microscopy

Crystallite morphology often conveys information about the symmetry of the crystallographic unit cell. In CSZ-1, for example, the observed platelet morphology is consistent with a structural model that has a unique crystallographic direction. Once a single mirror-plane fault occurs in the cubic FAU-framework, the normal to the plane becomes a unique axis, as there is no simple way of forming a similar defect in other than a parallel plane, as such faults cannot interpenetrate coherently.

The structure in the edges of zeolite crystallites seen in the optical or electron microscopes can also indicate when planar faulting is present. The relative face angles at discontinuities provide clues as to the nature of intergrowths at a unit cell level. Similarly, the morphology of zeolite beta crystallites conveys that extensive planar faulting is present and, perhaps with a large measure of hindsight, suggests the 4-fold character of the faulting operations (3,25)

Considerable detail is extractable from electron lattice images if, as is the case with numerous minerals (33), the material is stable under electron irradiation. As zeolites degrade rapidly in the electron beam it is difficult, except in special cases (34), to exploit in zeolite studies the atomic resolution now available in high voltage instruments. However, zeolite pore systems provide sufficient scattering potential contrast that stacking arrangements can be inferred even under moderate-resolution conditions (35,36). Even when only relatively subtle changes in projected structure are involved, such as in the ZSM-5 – ZSM-11 pentasil family, transmission electron microscopy has been used to measure intergrowth occurrence (37). Direct inspection of lattice images provides one means of quantifying stacking fault populations, albeit a tedious one, and these data are a valuable corroboration of bulk diffraction results.

It is sometimes argued that what is actually observed in the electron microscope is not representative of the bulk material from which minute amounts are extracted for examination. Some specimen preparation procedures, such as suspension in solvent, may preselect or preconcentrate particular particle sizes or shapes. High-resolution images can be obtained only from crystallites or portions of crystallites somewhere between 50 and 200-300 Å in thickness, perhaps biasing against observation of certain particle types, for example, those larger crystallites that do not have tapered edges. However, if an examined crystallite is truly a sport then the chance of it being observed is correspondingly slim. Sampling a reasonable number of particles then reduces the probability of repeated observations of atypical crystallites to a vanishingly small level.

The key test for an image interpretation is whether the image simulated based on a structural model and known microscope parameters matches that observed. An image match is usually not unique, for the measured image represents the convolution of the effects of several factors such as defocus, crystallite thickness, etc. Different sets of parameters might yield similar images, or, indeed, different structures might, by suitable choice of the other adjustable parameters, also give rise to similar computed images, particularly when the structures are closely related (35). Computational tools are being

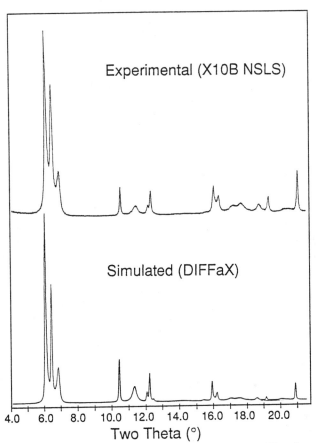

FIGURE 30-3. Comparison of synchrotron powder X-ray diffraction data (X10B, NSLS) for an ECR-30 material (upper) compared with that simulated using the DIFFaX code (41).

developed to facilitate both construction of appropriate intergrowth models and simulation of the corresponding lattice images.

Lattice images also provide data on another important component of planar defects, that is the degree of strain present at the defect or intergrowth boundary and its propagation throughout the crystallite (9). Observations of strain can, in principle, be related to the interatomic interactions across the boundary plane(s). Additionally, the STEM microscope allows elemental compositions or composition gradients to be measured with good spatial resolution, potentially revealing compositional changes in the vicinity of planar defects. Where such compositional changes are present, they may provide insight into the reasons for defect occurrence at synthesis.

Diffraction

Diffraction, being the premier method for studying structures with translational periodicity, can also be the most informative measure of planar faulting and intergrowth behavior. Intergrowths that are regularly recurrent are fully ordered and then correspond to a superlattice, or more accurately to a new framework topology. The corresponding diffraction patterns consist of sharp spots or peaks, occurring at positions dictated by the changed periodicity. The AABBCCAACCBB.... stacking sequence of 6-rings in the AFT-framework of $AlPO_4$-52 with a 28.95 Å c-axis repeat distance is one such example. ZSM-3 has been reported (38) to have a c-axis repeat distance of ≤ 129 Å based on intergrown ABC... and ABAB... stacking of faujasite sheets, but the stacking is highly disordered.

Simple intepretations of the relative amounts of two intergrown structures present can be made based on incoherent, addititive contributions by the different structure types to the diffraction pattern (39,40). However, when planar faults or intergrowths occur in a disordered fashion, this disorder is manifested in the diffraction data as broadening or streaking of particular classes of reflections. In single-crystal diffraction measurements in the electron microscope or using X-ray Weissenberg, Precession or area-detector techniques, streaking is seen along the reciprocal lattice direction perpendicular to the fault plane. In a powder diffraction scan (that involves sampling reciprocal space radially, hence intersecting the streaks at successively more obtuse angles the higher the index (typically l) associated with the faulted direction), the peak widths are index-dependent.

The pattern of streaking or powder peak broadening depends on the character of the intergrowths and a proper analysis can then provide quantitative data on the character and order of the intergrowth sequence. Although this simulation was traditionally difficult for complex structures such as those of zeolites, a new method for simulating the effects of planar faults or intergrowths on the corresponding diffraction patterns has recently been developed (41). Embodied in the program DIFFaX (41), this method has already been used to analyze the patterns of faulting that occur in the structural families based on the framework types MFI-MEL (ZSM-5 - ZSM-11), FAU–EMT (faujasite - EMC-2), *BEA (zeolite beta), FER (ferrierite), etc. For zeolite beta, still known only in a highly faulted form, the reproduction of measured diffraction patterns based on a structural model and the associated pattern of planar faulting was a key step in proving the correctness of that framework model (3).

The effect of planar faulting on the diffraction pattern of a ZSM-20 material (40,42) is shown in Figure 30-3. The pattern of measured peak widths in the powder diffraction pattern is the convolution of the sample contributions (arising from stacking disorder, finite particle size and shape, strain, etc.) and the intrinsic

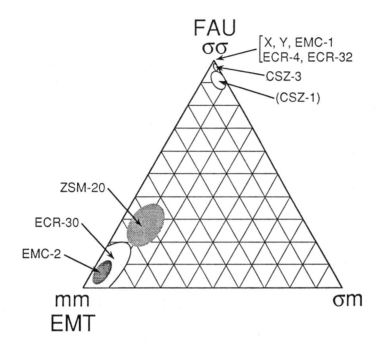

FIGURE 30-4. A phase diagram for materials related to faujasite that comprise intergrowths between the cubic and hexagonal modes of stacking of sheets. *The fault-clustering tendency necessitates use of a triangular phase diagram, with the apices represented by pure ABC... (FAU - cubic), AB... (EMT - hexagonal) and ABCB... (hexagonal) stackings.*

instrumental resolution function. The most informative data thus come from diffraction patterns in which the latter, instrumental contribution is minimized. The diffraction pattern shown in Figure 30-3 was measured using synchrotron X-radiation (43) on the Exxon X10B beam line at the NSLS at Brookhaven. The diffraction pattern simulated using DIFFaX based on models for faujasite sheets clipped from the zeolite Y crystal structure, and with an intergrowth pattern adjusted so as to best reproduce the experimental pattern, is also given in Figure 30-3. This type of fitting enables the identification, classification, and differentiation of intergrowth materials (44)

Most materials so far studied in the FAU–EMT structural family are appropriately termed intergrowth structures as there is a pronounced tendency for faults in this system to be clustered. The successive faujasite sheets can be interlinked related by inversion centers (σ as occurs in the FAU-framework) or mirror planes (m – EMT). There are materials, such as CSZ-1 and CSZ-3, in which only a small number of m-stackings occur in an otherwise perfect σσσσ.... stacking sequence. Likewise, EMC-2, produced using 18-crown-6 ether as a template in synthesis, contains only a small number of σ-stackings in an almost pure mmmmm... sequence. However, when both stacking modes are present in the same crystallite, the clustering tendency is manifested in higher proportions of a given stacking operation (σ or m) being followed by the same operation (rather than the alternative, faulting mode) than would be expected for a random fault distribution. To quantify the stacking characteristics in this system then requires a 3-component structural phase diagram (Figure 30-4), with apices corresponding to pure σσ... (FAU), mm... (EMT) and recurrent σm... (not yet observed in pure form) stacking sequences. Atomic coordinates for these three structure types, optimized by distance least-squares are given in Tables 30-1, 30-2 and 30-3. Measurement of the powder X-ray diffraction (PXD) patterns of materials in this structural family under conditions of high instrumental resolution, coupled with simulations of these PXD patterns using DIFFaX (Figure 30-3) then enables them to be placed on the structural phase diagram (45). A concerted synthesis effort over the past several years has yielded materials than span almost half of the full phase field (Figure 30-4). The production of materials that have higher proportions of recurrent σm... stackings remains a synthetic challenge.

CONCLUSION

Planar faults, even in low concentrations, can have a dramatic effect on macroscopic properties. Such faults are incorporated during synthesis or, in some systems, they can be introduced during phase transitions at higher temperatures. It is anticipated that the pattern of incorporation of faults can convey insight into the atomic-level factors responsible for their occurrence. A variety of methods are now available for both indirect and direct measurement of faulting and intergrowth populations. In particular, quantitative interpretations of diffraction data measured from zeolites subject to planar faulting now enable patterns of faulting as a function of synthesis conditions to be studied, and materials to be defined based on their intergrowth characteristics.

ACKNOWLEDGEMENTS

The Biosym Catalysis and Sorption project is supported by a consortium of member companies and universities.

TABLE 30-1   Framework atomic coordinates for the ABC sequence structure
(FAU-framework, space group Fd3̄m (No. 227), lattice constant a = 24.639 Å)

Atom	x	y	z
T1	0.1250	0.9470	0.0365
O1	0.1041	0.8959	0.0000
O2	0.2534	0.2534	0.1433
O3	0.1740	0.1740	0.9675
O4	0.1766	0.1766	0.3218

Coordinates and unit cell constants optimized by distance least squares for a
composition of Si:Al = 4.6, with targets of T-O 1.631 Å and T-O-T 144°

TABLE 30-2   Framework atomic coordinates for the AB sequence structure
(EMT-framework, space group P63/mmc (No. 194), lattice constants a =
17.425 Å, c = 28.423 Å)

Atom	x	y	z
T1	0.0368	0.4300	0.1074
T2	0.0965	0.3702	0.0179
T3	0.1552	0.4887	0.1957
T4	0.1558	0.4885	-0.0706
O1	0.1299	0.4564	-0.0162
O2	0.0582	0.5291	0.0924
O3	0.1879	0.5939	0.1908
O4	0.1678	0.3356	0.0159
O5	0.0705	0.4319	0.1611
O6	0.2365	0.4729	-0.0874
O7	0.2359	0.4718	0.1807
O8	0.0882	0.3973	0.0718
O9	0.1870	0.5935	-0.0748
O10	-0.0697	0.3616	0.1042
O11	0.0000	0.2914	0.0000
O12	0.1265	0.4574	0.2500

Coordinates and unit cell constants optimized by distance least squares for a
composition of Si:Al = 4.6, with targets of T-O 1.631 Å and T-O-T 144°

TABLE 30-3   Framework atomic coordinates for the ABCB sequence structure
(Space group P63/mmc (No. 194), lattice constants a = 17.426 Å, c = 56.856 Å)

Atom	x	y	z
T1	0.3700	0.0966	0.6787
T2	0.4297	0.0368	0.6339
T3	0.4885	0.1554	0.7228
T4	0.4890	0.1552	0.5897
T5	0.1776	0.1781	0.6603
T6	0.0599	0.2966	0.0713
T7	0.9404	0.2370	0.1161
T8	0.0002	0.1782	0.0272
O1	0.4627	0.1229	0.6169
O2	0.3914	0.1957	0.6713
O3	0.5211	0.2606	0.7204
O4	0.5012	0.0026	0.6330
O5	0.4038	0.0984	0.7056
O6	0.5698	0.1396	0.5814
O7	0.5692	0.1385	0.7154
O8	0.4213	0.0639	0.6609
O9	0.5205	0.2602	0.5877
O10	0.2635	0.0282	0.6772
O11	0.3333	0.9578	0.6250
O12	0.4598	0.1240	0.7500
O13	0.0981	0.4031	0.0728
O14	0.2038	0.2104	0.6331
O15	0.0969	0.1937	0.6686
O16	0.1463	0.0731	0.6624
O17	0.1376	0.2752	0.0787
O18	0.0728	0.1455	0.0297
O19	0.8346	0.1654	0.1170
O20	0.0279	0.2628	0.0444
O21	0.9025	0.0974	0.0346
O22	0.9759	0.2454	0.0891
O23	-0.0022	0.2069	0.0000

Coordinates and unit cell constants optimized by distance least squares for a
composition of Si:Al = 4.6, with targets of T-O 1.631 Å and T-O-T 144°

REFERENCES

1. Barrer, R. M. 1982 Hydrothermal Chemistry of Zeolites. London: Academic Press.
2. Vaughan, D. E. W. 1991. The roles of metal and organic cations in zeolite synthesis. In Catalysis and Adsorption by Zeolites (Stud. Surf. Sci. Catal. No. 65), ed. G. Öhlmann, H. Pfiefer, R. Fricke, pp. 275-286. Amsterdam: Elsevier.
3. Newsam, J. M., Treacy, M. M. J., Koetsier, W. T., and deGruyter, C. B. 1988. Structural characterization of zeolite beta. Proc. Roy. Soc. (London) A420: 375-405.
4. Tomlinson, S. M., Jackson, R. A., and Catlow, C. R. A. 1990. A Computational study of zeolite beta. J. Chem. Soc. Chem. Comm. : 813.
5. Parker, S. C. 1991. In Modeling Zeolite Structure and Chemistry, ed. C. R. A. Catlow, in press.
6. McCormick, A. V., and Bell, A. T. 1989. The solution chemistry of zeolite precursors. Catal. Rev. Sci. Eng. 31: 97-127.
7. Knight, C. T. G. 1990. Are zeolite secondary building units really red herrings ? Zeolites 10: 140-144.
8. Treacy, M. M. J., Newsam, J. M., Beyerlein, R. A., Leonowicz, M. E., and Vaughan, D. E. W. 1986. The structure of zeolite CSZ-1 interpreted as a rhombohedrally distorted variant of the faujasite framework. J. Chem. Soc. Chem. Comm. 1211-1214.
9. Treacy, M. M. J., Newsam, J. M., Vaughan, D. E. W., Beyerlein, R. A., Rice, S. B., and DeGruyter, C. B. 1988. On the propagation of twin-fault induced stress in platelet FAU-framework zeolites. In Microstructure and Properties of Catalysts (MRS Symp. Proc. Vol. 111), ed. M. M. J. Treacy, J. M. White, J. M. Thomas, pp. 177-190. Pittsburgh, PA: Materials Research Society.
10. Meier, W. M., and Groner, M. 1981. Zeolite structure type EAB: crystal structure and mechanism for the topotactic transformation of the Na,TMA form. J. Solid State Chem. 37: 204-218.
11. Cartlidge, S., and Meier, W. M. 1984. Solid state transformations of synthetic CHA- and EAB-type zeolites in the sodium form. Zeolites 4: 218-225.
12. Keller, E. B., Meier, W. M., and Kirchner, R. M. 1990. Synthesis, Structures of $AlPO_4$-C and $AlPO_4$-D and their topotactic transformation. Solid State Ionics 43: 93-102.
13. Richardson, J. W., Smith, J. V., and Pluth, J. J. 1990. $ALPO_4$-25: Framework topology, topotactic transformation from $ALPO_4$-21, and high-low displacive transition. J. Phys. Chem. 94: 3365-3367.

14. Vogt, E. T. C., and Richardson, J. W. 1990. The reversible transition of the molecular sieve VPI-5 into AlPO4-8 and the structure of AlPO4-8. J. Solid State Chem. 87: 469-471.

15. Annen, M. J., Young, D., Davis, M. E., Cavin, O. B., and Hubbard, G. R. 1991. J. Phys. Chem. 95: 1380.

16. Richardson, J. W., and Vogt, E. T. C. 1990. The framework structure of AlPO4-8. Zeolites 12: 13.

17. Dessau, R. M., Schlenker, J. L., and Higgins, J. B. 1990. Framework topology of AlPO4-8: The first 14-ring molecular sieve. Zeolites 10: 522-524.

18. Breck, D. W. 1973. Zeolite Molecular Sieves: Structure, Chemistry and Use. London: Wiley and Sons (reprinted R. E. Krieger, Malabar FL, 1984).

19. Sherman, J. D., and Bennett, J. M. 1973. Proc 3rd IZC Zurich 1973. In Molecular Sieves (ACS Adv. Chem. Ser. No. 121), ed. W. M. Meier, J. B. Uytterhoeven, p. 52. Washington, DC: American Chemical Society.

20. Martens, J. A., Perez-Pariente, J., and Jacobs, P. A. 1986. Isomerization and cracking of the n C10 - n C17 alkanes on Pt/H-beta. Acta Phys. Chem. 31: 487-495.

21. Martens, J. A., Jacobs, P. A., and Cartlidge, S. 1989. Investigation of the pore architecture of CSZ-1 zeolites with the decane test reaction. Zeolites 9: 423-427.

22. Jacobs, P. A., and Martens, J. A. 1987 Synthesis of High-Silica Aluminosilicate Zeolites (Stud. Surf. Sci. Catal. No. 33). Amsterdam: Elsevier.

23. Engelhardt, G., and Michel, D. 1987 High-Resolution Solid-State NMR of Silicates and Zeolites. New York: John Wiley.

24. Fyfe, C. A., Strobl, H., Kokotailo, G. T., Pasztor, C. T., Barlow, G. E., and Bradley, S. 1988. Correlations between lattice structures of zeolites and their ^{29}Si nmr spectra: zeolites KZ-2, ZSM-12, and Beta. Zeolites 8: 132-136.

25. Higgins, J. B., LaPierre, R. B., Schlenker, J. L., Rohrman, A. C., Wood, J. D., Kerr, G. T., and Rohrbaugh, W. J. 1988. The framework topology of zeolite beta. Zeolites 8: 446-452.

26. Stöcker, M., Ernst, S., Karge, H. G., and Weitkamp, J. 1990. ^{29}Si-MAS NMR studies of hydrothermal dealumination of zeolite ZSM-20. Acta Chem. Scand. 44: 519-521.

27. Beyerlein, R. A., Newsam, J. M., Melchior, M. T., and Malone, H. 1988. Framework compositions in as-synthesized and dealuminated CSZ-1 zeolites by ^{29}Si MASNMR. J. Phys. Chem. in preparation.

28. Fraissard, J., Ito, T., Springuel-Huet, M., and Demarquay, J. 1986. In New Developments in Zeolite Science and Technology, ed. Y. Murakami, A. Iijima, J. W. Ward, p. 393. Tokyo and Amsterdam: Kodansha and Elsevier.

29. Benslama, R., Fraissard, J., Albizane, A., Fajula, F., and Figueras, F. 1988. An example of the technique of studying adsorbed xenon by ^{129}Xe nmr: Approximate determination of the internal void space of zeolite beta. Zeolites 8: 196-198.

30. Jarman, R. H., and Melchior, M. T. 1984. J. Chem. Soc. Chem. Commun. 414-415.

31. Hayashi, S., Suzuki, K., Shin, S., Hayamizu, K., and Yamamoto, O. 1985. Chem. Phys. Lett. 113: 368-371.

32. Melchior, M. T., Vaughan, D. E. W., Jarman, R. H., and Jacobson, A. J. 1984. presented at Rocky Mtn. Conf. Applied Spectroscopy, Denver CO, Aug. 8, 1984, 393.

33. Buseck, P. R., and Iijima, S. 1974. Amer. Mineral. 59: 1-21.

34. Terasaki, O., Thomas, J. M., Millward, G. R., and Watanabe, D. 1989. Chem. Mater. 1: 158.

35. Millward, G. R., Ramdas, S., and Thomas, J. M. 1985. On the direct imaging of offretite, cancrinite, chabazite and related other ABC-6 zeolites and their intergrowths. Proc. Roy. Soc. (London) 399: 73-91.

36. Vaughan, D. E. W., Treacy, M. M. J., and Newsam, J. M. 1989. Recent advances in techniques for characterizing zeolite structures. In Guidelines for Mastering the Properties of Molecular Sieves (NATO ASI Series), ed. D. Barthomeuf, E. G. Derouane, W. Hölderich, pp. 99-120. New York: Plenum Press.

37. Thomas, J. M., and Millward, G. R. 1982. Direct, real-space determination of intergrowths in ZSM-5/ZSM-11 catalysts. J. Chem. Soc. Chem. Commun. 1380-1383.

38. Kokotailo, G. T., and Ciric, J. 1971. Synthesis and structural features of zeolite ZSM-3. In Molecular Sieve Zeolites – I (ACS Adv. Chem. Ser. No. 101), ed. E. M. Flanigen, L. B. Sand, pp. 109-121. Washington, DC: American Chemical Society.

39. Gard, J. A., and Tait, J. M. 1972. Crystal structure of the zeolite offretite, K1.1Ca1.1Mg0.7[Si12.8Al5.2O36]15.2H2O. Acta Crystallogr. B28: 825-834.

40. Vaughan, D. E. W., Treacy, M. M. J., Newsam, J. M., Strohmaier, K. G., and Mortier, W. J. 1989. Syntheses and characterization of zeolite ZSM-20. In Zeolite Synthesis (ACS Symp. Ser. 398), ed. M. Occelli, H. E. Robson, pp. 544-559. Washington, DC: American Chemical Society.

41. Treacy, M. M. J., Newsam, J. M., and Deem, M. W. 1991. A general recursion algorithm for calculating diffracted intensities from crystals containing planar faults. Proc. Roy. Soc. (London) 433: 499-520.

42. Newsam, J. M., Treacy, M. M. J., Vaughan, D. E. W., Strohmaier, K. G., and Mortier, W. J. 1989. The structure of zeolite ZSM-20 - Mixed cubic and hexagonal stackings of faujasite sheets. J. Chem. Soc. Chem. Comm. 493-495.

43. Newsam, J. M., and Liang, K. S. 1989. Synchrotron X-ray diffraction studies of inorganic materials and heterogeneous catalysts. Int. Rev. Phys. Chem. 8: 289-338.

44. Vaughan, D. E. W., Strohmaier, K. G., Treacy, M. M. J., and Newsam, J. M. 1990. ECR-35 patent. US Patent Application 686,451.

45. Treacy, M. M. J., Newsam, J. M., Vaughan, D. E. W., Yang, C. Z., and Strohmaier, K. G. 1991. in preparation.

31

Two-Dimensional NMR Studies of the Structure and Reactivity of Molecular Sieve Catalysts

Waclaw Kolodziejski and Jacek Klinowski

Department of Chemistry, University of Cambridge, Lensfield Road, Cambridge CB2 1EW, U.K.

Two-dimensional solid-state NMR techniques, such as spin diffusion and homonuclear correlation spectroscopy (COSY), are shown to provide important information about molecular sieves. The details of these methods are given and their performance discussed. Thus ^{31}P spin diffusion in hydrated aluminophosphate molecular sieve VPI-5 shows that spin diffusion can be suppressed by fast magic-angle spinning but not by high-power decoupling, so that the widely accepted "extraneous spins" mechanism is not valid in this case. The complete understanding of the spectra requires the knowledge of zero-quantum lineshape functions for the spins involved. A ^{13}C NMR spin-diffusion experiment *in situ* on the products of methanol conversion into gasoline over zeolite H-ZSM-5 substantially aids the spectral assignment. Combined with model studies, it is likely to provide new information concerning the distribution of organic species in the intracrystalline space of the catalyst. A two-dimensional ^{29}Si solid-state J-scaled COSY spectrum of highly siliceous mordenite reveals the connectivities of tetrahedral sites, permitting an unambiguous assignment of all signals.

INTRODUCTION

We describe and critically review the use of two two-dimensional (2D) NMR methods[1,2] for the study of structure and reactivity of molecular sieves, something with which our laboratory has been involved for a number of years. Spin-diffusion NMR is usually used in solids; COSY is well established in solution NMR, but has been adapted for use with solid samples. The common feature of the two methods is that both rely on homonuclear correlation of chemical shifts, and conclusions are drawn from the so-called cross-peaks in the spectra. However, the methods differ with respect to the type of spin-spin interactions which give rise to the cross-peaks. Spin diffusion relies on through-space dipolar interactions; COSY on electron-mediated interactions via the chemical bonds.

Spectral Spin Diffusion in VPI-5

Spectral spin diffusion in the solid state involves simultaneous flip-flop transitions of dipolar-coupled spins with *different* resonance frequences,[1-16] while spatial spin diffusion transports spin polarization between spatially separated *equivalent* spins. In this paper we shall only be concerned with the former. The interaction of the X nuclei undergoing spin diffusion with the proton reservoir provides compensation for the energy imbalance (the "extraneous spins mechanism").[8,10,13,14] Spin diffusion results in an exchange of magnetization between nuclei responsible for resolved NMR signals, which can be conveniently detected by observing the relevant cross-peaks in the 2D spin-diffusion spectrum.[3-5] The technique, formally analogous to the NOESY experiment in liquids, is already well established for solids. The few published ^{31}P solid-state spin-diffusion studies deal mostly with model cases of magnetization transfer between two sites.[13,15,16] We undertook a 2D ^{31}P NMR study of spin diffusion in aluminophosphates, and chose VPI-5 as the model sample.

VPI-5 is a crystalline aluminophosphate molecular sieve containing 18-membered rings of tetrahedral atoms.[17,18] The large channel diameter (ca. 12 Å) gives this remarkable material a considerable potential for the separation of large molecules and for catalytic cracking of heavy fractions of petroleum which at present are discarded as bottom-of-the-barrel residue. According to the X-ray powder diffraction study with Rietveld refinement reported by Rudolf and Crowder,[19] the hydrated material has a space group $P6_3cm$ and two kinds of crystallographically inequivalent T-sites (T = P or Al). The model provides for two kinds of T-sites: those linking 6-membered and 4-membered rings (6-4 sites) and those linking two 4-membered rings (4-4 sites), in the population ratio of

2 : 1. However, this structure does not agree with ^{31}P NMR spectra of VPI-5. Accordingly, McCusker et al.[20] refined the structure in a lower symmetry P63 space group. This model calls for two crystallographically inequivalent kinds of 6-4 sites, so that in total there are three kinds of equally populated T-sites (see Fig. 31-1).

The ^{31}P magic-angle-spinning (MAS) NMR spectrum of hydrated VPI-5 (Fig. 31-2) shows three signals in an intensity ratio of 1 : 1 : 1. It has been shown that the structure is considerably affected by adsorbed water, which is reflected in the ^{31}P MAS NMR spectra.[21-23] During dehydration the intensity of signal **2** increases at the expense of signal **1**, while the intensity of signal **3** remains virtually unchanged. The same is observed on heating, as discussed in Ch. 22. On the basis of variable-temperature ^{31}P MAS NMR,[24] the signals at -23 ppm (**1**) and -27 ppm (**2**) have been assigned to phosphorus atoms in 6-4 sites and the signal at -33 ppm (**3**) to phosphorus atoms in 4-4 sites.

VPI-5 was synthesized according to ref. 25 using n-dipropylamine as the organic template. Two-dimensional ^{31}P MAS NMR spectra were recorded at 162.0 MHz on a Bruker MSL-400 spectrometer using a double-bearing probehead and zirconia rotors. The experiments were performed with a standard NOESY pulse program for liquids without cross-polarization and proton decoupling. In some experiments high-power proton decoupling was used during the mixing period. Sixty-four experiments were carried out for the F_1 dimension with 8 scans in each. The [31]P $\pi/2$ pulse was typically 3 μs and the relaxation delay was 80 s in all cases. To prevent possible coherent magnetization transfer by scalar or dipolar couplings, the mixing time was randomly varied by ±20 %.

2D NMR reveals the presence of spin diffusion in a wide range of mixing times (from 0.1 s to 10 s), with all diagonal and cross-peaks of almost the same intensity in the latter case. The spectra are of excellent quality with very high signal-to-noise ratio (Fig. 31-3). There are cross-peaks for each pair of signals, even when high-speed MAS is applied [Fig. 31-3(c)] and even for short mixing times (Fig. 31-4), with the intensities decreasing in the order: (**1-2**) > (**2-3**) > (**1-3**). Under such conditions the cross-peaks must originate from the closest P atoms which are located in the second tetrahedral coordination sphere of one another in the framework. We have calculated the corresponding dipolar interactions from the X-ray structure of VPI-5[20] and found that they decrease in the order P1-P3 > P2-P3 > P1-P2. Considering this order and the intensity of the cross-peaks one would assign the diagonal peaks, or the peaks in 1D spectrum, at -23 (**1**), -27 (**2**), and -33 ppm (**3**) to P1, P3, and P2, respectively. However, such interpretation is unacceptable, since measurements on dehydrated samples and variable-temperature studies[21-24] unequivocally assign signals **1** and **2** to the 6-4 sites. Further, we note that the greater the distance between the diagonal signals, the weaker their cross-peaks, suggesting that the energy imbalance during spin diffusion has the predominant influence on the cross-peaks

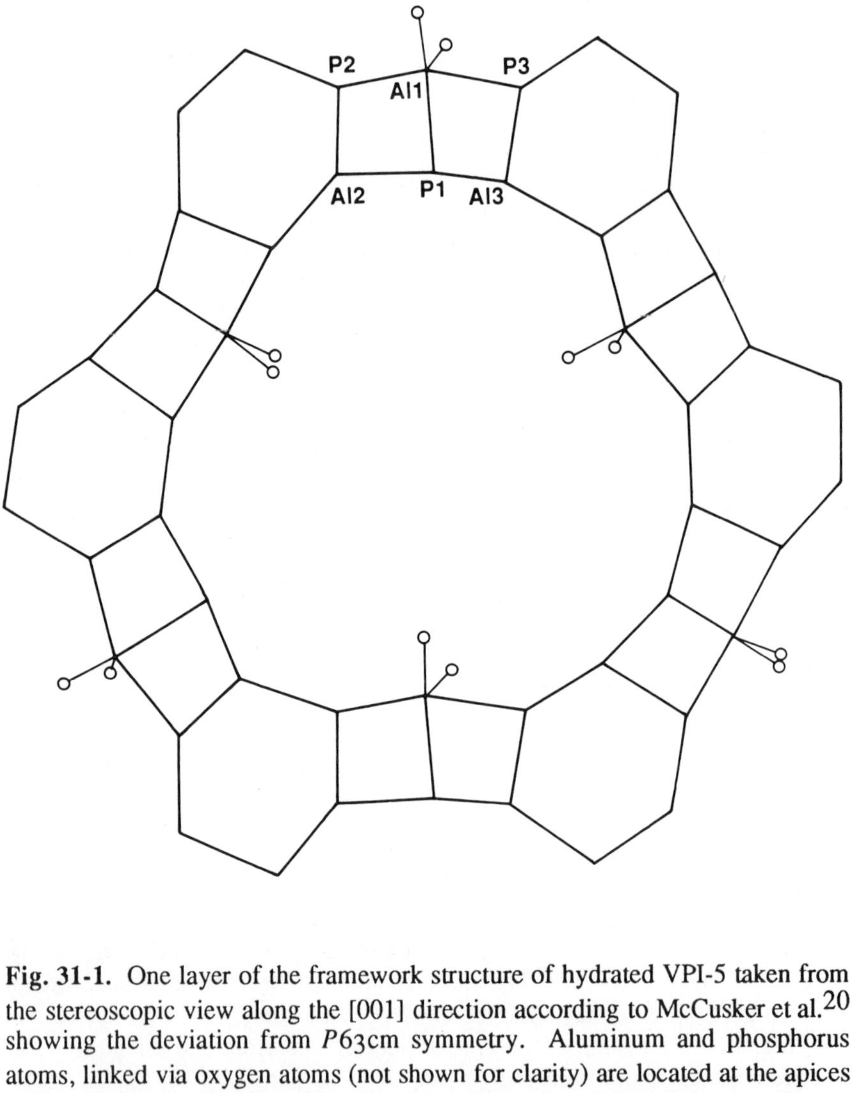

Fig. 31-1. One layer of the framework structure of hydrated VPI-5 taken from the stereoscopic view along the [001] direction according to McCusker et al.[20] showing the deviation from $P6_3cm$ symmetry. Aluminum and phosphorus atoms, linked via oxygen atoms (not shown for clarity) are located at the apices of the polygons. Sites located between two fused 4-membered rings are known as 4-4 sites; those located between 6-membered and 4-membered rings are known as 6-4 sites. P2 and P3, and Al2 and Al3 sites are inequivalent as a result of the distortion. The Al1 site is 6-coordinated as a result of bonding to four bridging oxygens and two "framework" water molecules. Other intracrystalline water is not shown.

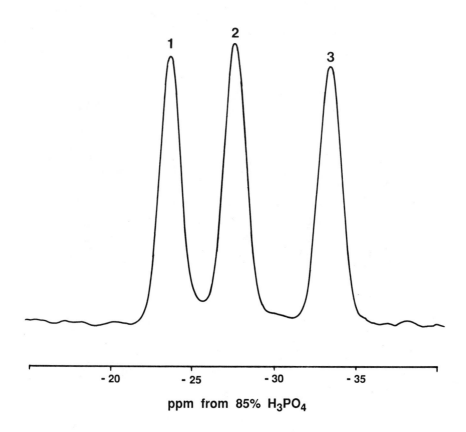

Fig. 31-2. ^{31}P NMR spectrum of hydrated VPI-5 (MAS at 12.2 kHz).

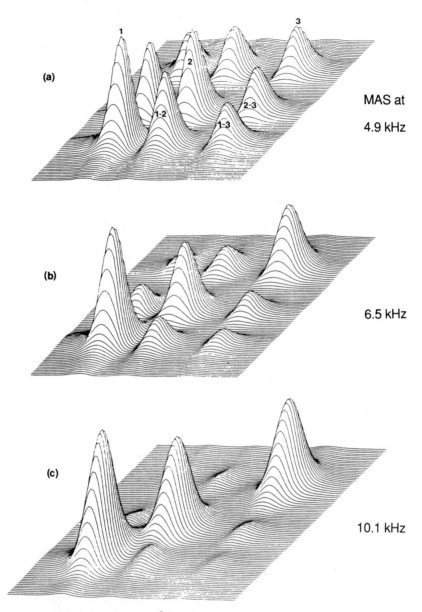

Fig. 31-3. Experimental 2D ^{31}P NMR spin-diffusion spectra of hydrated VPI-5 recorded at various MAS speeds: **(a)** 4.9 kHz; **(b)** 6.5 kHz; and **(c)** 10.1 kHz. The three spectra are not on the same intensity scale, so that only the relative intensities within each can be compared.

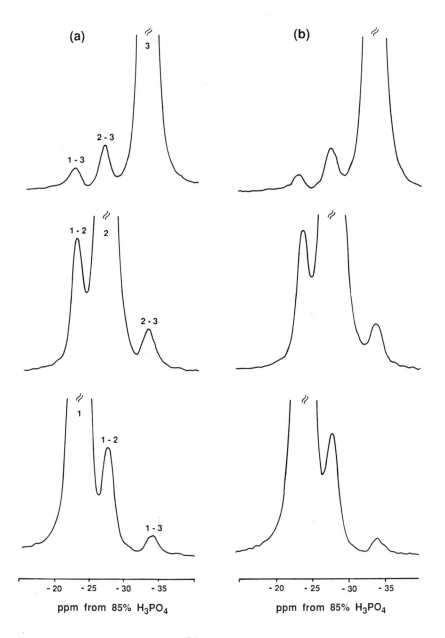

Fig. 31-4. Cross-sections of 2D ^{31}P NMR spin-diffusion spectra chosen at the positions of diagonal peaks. The spectra were acquired (a) without and (b) with high-power proton decoupling (200 ms mixing time, MAS at 5.3 kHz).

intensities. Therefore, the assessment of dipolar interactions between the three P sites in VPI-5 is not straightforward. First, we must calculate spin-diffusion rates from the cross-peak intensities in the complicated spin-diffusion network.[9] Second, we must relate the rates of spin diffusion to dipolar interactions using existing theories,[8,10,13,14] but all of them require the knowledge of the zero-quantum (ZQ) lineshape function at the position of the ZQ resonance frequency for the spins involved. The ZQ coherence is unobservable in 1D pulsed NMR, so one has to resort to yet another 2D solid-state experiment. However, this is difficult in practice, because phase cycling cannot distinguish between longitudinal magnetization and ZQ coherence, and therefore the ZQ signals, which we wish to record, are obscured by strong spurious axial peaks. Even if we were successful in evaluating the ZQ lineshapes, the available theories[8,10,13,14] are not necessarily valid in our case. The "extraneous spins" mechanism of spin diffusion[8,10,13,14] implies that protons under decoupling cannot compensate for the energy imbalance during flip-flops of X spins and therefore that spin diffusion of X spins must be quenched. Our results show that this is not the case in VPI-5. It is not clear what mechanism may satisfy the energy conservation requirement during ^{31}P spin diffusion in VPI-5. We suggest that this is achieved by ^{31}P - ^{31}P dipolar interactions among the participant spins themselves,[8] especially if crystal lattice vibrations allow motion-induced flip-flops.[26] Finally, we note that our spectra demonstrate nicely how spin diffusion can be slowed down by fast MAS. ^{31}P - ^{31}P dipolar interactions in VPI-5 are weak (below 1 kHz for either site) and can be averaged out, thus reducing the intensity of cross-peaks.

Spin Diffusion in the Products of Methanol Conversion on Zeolite ZSM-5

A 2D solid-state spin-diffusion ^{13}C NMR experiment *in situ* on the products of methanol conversion into gasoline over zeolite H-ZSM-5 substantially aids the spectral assignment. Combined with model studies, the experiment is likely to provide new information concerning the distribution of organic species in the intracrystalline space of the catalyst.

Zeolite H-ZSM-5 is a powerful heterogeneous catalyst. The high silica content of the zeolite gives it high thermal stability, while the crystal structure, involving two intersecting intracrystalline channel systems, is responsible for the striking shape selectivity. Channels 5.6×5.3 Å in diameter run in the [010] direction, and 5.5×5.1 Å channels run in the [100] direction. H-ZSM-5 is capable of converting methanol into gasoline (a mixture of hydrocarbons up to C_{10}). Three of our recent papers[27-29] reported ^{13}C NMR spectra of the products of the methanol-to-gasoline reaction *in situ* using samples contained in specially designed capsules[30] spun at the magic angle (MAS). It has been possible to identify a number of different organic species *in the adsorbed phase* and to monitor their fate during the course of the reaction.[27,28] In addition to conventional spectral assignment based on chemical shifts and signal intensities, two-dimensional ^{13}C NMR has been used[29] to determine the number of protons coupled to each carbon in the various organics. The corresponding ^{13}C - ^{1}H J-couplings have also been measured. We now describe a two-dimensional solid-state spin-diffusion ^{13}C NMR experiment in this system.

The rate of spin diffusion is very strongly dependent on the internuclear separation r, being proportional to $1/r^3$ for a rigid crystal lattice and to $1/r^6$ for species undergoing rapid isotropic motion.[5] As a result, spin diffusion occurs only between nuclei in adjacent functional groups within the same molecule (the intramolecular case) or between nuclei in neighbouring molecules mixed on a microscopic level (the intermolecular case). Both cases are observed in our system.

Zeolite H-ZSM-5 (Si/Al = 30) was prepared by ammonium exchanging the Na^+ form followed by calcination in air at 550°C; 99.9% ^{13}C-enriched methanol, supplied by Aldrich, was diluted to 50% isotopic abundance (w/w) with ordinary methanol. The NMR samples were prepared in specially designed Pyrex capsules.[30] After activating the zeolite under vacuum at 400°C, 50 torr of enriched methanol, previously degassed by the freeze-pump-thaw method, was adsorbed at room temperature. The capsule was sealed with a microtorch while the sample was maintained at liquid nitrogen temperature so as to prevent any reaction occurring. The capsule was then heated at 300°C for 45 min and quenched to room temperature.

2D ^{13}C NMR spectra were recorded at 100.61 MHz using the spin-diffusion pulse program of Szeverenyi et al.[3] (see Fig. 31-5). The magic angle was set precisely by observing the ^{79}Br resonance of KBr.[31] The spinning rate was ca. 2.8 kHz, and it was not found necessary to synchronize the rotation period with the parameters of the pulse program. ^{13}C spin-lattice relaxation times of all species observed are very similar and of the order of 2.5 s.[28] The mixing time $\Delta = 1.5$ s was found to yield the best signal-to-noise ratio. We confirmed that high-power proton decoupling suppresses the cross-peaks, so that chemical exchange is not involved in their formation.[32] The lengths of $\pi/2$ pulses for proton and carbon channels were found to be 4.8 μs using the 2D method described by Nielsen et al.[33] A 5 ms contact time (optimized) and 2 s recycle delay were used. For the F_2 dimension, corresponding to the spectrum width of 12.2 kHz, 512 real data points were collected. For the F_1 dimension, corresponding to the spectrum width of ±6100 Hz, 96 experiments were carried out with 192 real and 8 dummy scans in each. The total time of our 2D experiment was ca. 19 hours. Zero-filling to 1K and 512 points, and Gaussian and trapezoidal apodizations were applied in the F_2 and F_1 dimensions,

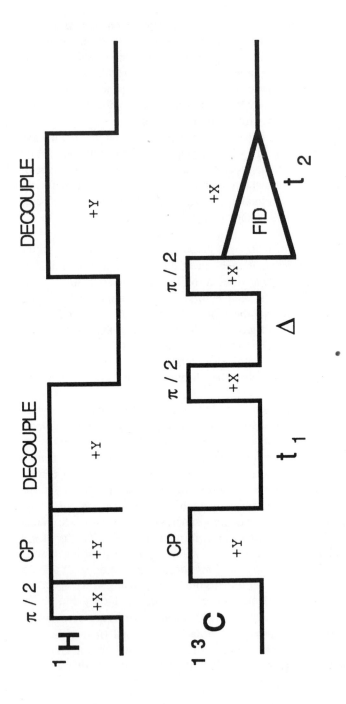

Fig. 31-5. The spin-diffusion pulse sequence.

respectively. The spectrum in Fig. 31-6 was obtained in the power mode and was symmetrized.

Fig. 31-6 shows the 2D spin-diffusion spectrum of aliphatic hydrocarbons trapped in the zeolite. The 1D signals, corresponding to the diagonal peaks in the 2D spectrum, have been assigned previously[27-29] (see Tab. 31-1), but the assigment of signal b[27,28] has subsequently been questioned.[29] It is clear that n-hexane and n-heptane are present, since signal e comes exclusively from their CH_3 groups. Therefore, considering the chemical shift of CH_2 groups of n-hexane and n-heptane,[34] both hydrocarbons must contribute to signal b. In the J-resolved experiment[29] this signal was not split into a triplet, because no homonuclear proton decoupling was applied during the first half of the evolution period, so that the splitting was obscured by substantial dipolar broadening. By contrast, CH_3 groups of isobutane undergo free rotation, which reduces the dipolar interaction and allows the quartet splitting of signal b to be observed. We note that signals of n-butane in the spin-diffusion spectrum are missing, since cross-polarization has a tendency to underestimate signals of mobile products, so that only those which are as abundant as propane are able to appear.

In order to interpret the cross-peaks correctly one must consider the features of the system studied and the nature of the spin-diffusion phenomenon. Thus zeolite ZSM-5 contains no cages and its channel diameter only allows the hydrocarbon species in the channels to be lined up sequentially. For any two molecules to exchange their positions, access to an unocuppied channel crossing is required, which is difficult to satisfy at high adsorbate loadings (30% w/w). Hydrocarbon molecules are capable of limited motion along the channels, which disfavours *intermolecular* spin diffusion. Free isotropic molecular rotation cannot occur, so that *intramolecular* dipolar interactions are present even for relatively mobile functional groups, and make the intramolecular spin diffusion possible. Thus intramolecular spin diffusion in our system is preferred to intermolecular spin diffusion. We note that the optimal mixing time (one which produces the most intensive cross-peaks) is generally different for each pair of the participating diagonal peaks, and in unfavourable cases short longitudinal relaxation times can hinder or even prevent the detection of cross-peaks. This means that no conclusions can be drawn from the absence of some cross-peaks and the presence of others, unless dipolar interactions underlying the corresponding spin-diffusion cases are closely similar (i.e., involve the same functional groups in similar molecules), and unless the relevant longitudinal relaxation times T_1 are alike.

Signal a (see Tab. 31-1) produces only one cross-peak (with signal d) and its assignment is obvious, since a and d each belong to a single compound. The b - d cross-peak must be classified as intermolecular and may be assigned to isopentane - propane, n-hexane (CH_2) - propane or n-heptane (CH_2) - propane

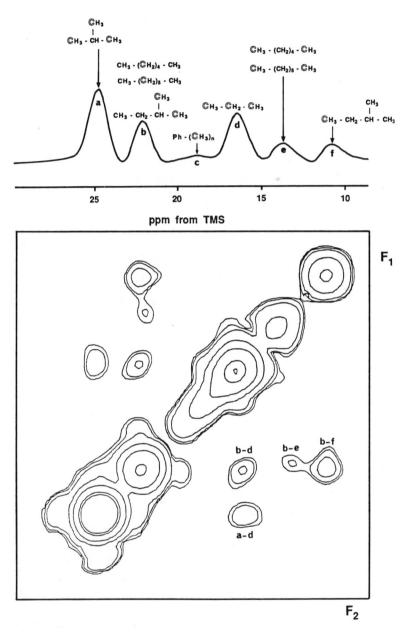

Fig. 31-6. ^{13}C NMR spin-diffusion spectrum of products of methanol conversion into gasoline over zeolite H-ZSM-5 with the projection at the top. Carbon atoms to which individual resonances are assigned are highlighted.

Tab. 31-1. Assignment of the 2D ^{13}C NMR spin-diffusion spectrum shown in Fig. 31-6.

Diagonal peaks[27-29]			
Signal	Chemical shift (ppm)	Group	Assignment
a	24.7	CH$_3$	isobutane
b	22.3	CH$_3$CH$_2$CH(C̲H$_3$)$_2$ CH$_2$	isopentane n-hexane + n-heptane
c	18.7	CH$_3$	methyl-substituted benzenes
d	16.7	CH$_3$ + CH$_2$	propane
e	14.3	CH$_3$	n-hexane + n-heptane
f	11.2	C̲H$_3$CH$_2$CH(CH$_3$)$_2$	isopentane

Cross-peaks		
Signal	Assignment	Type of spin diffusion
a - d	isobutane - propane	intermolecular
b - d	isopentane - propane	intermolecular
b - e	n-hexane	intramolecular
	n-heptane	intramolecular
b - f	isopentane	intramolecular

spin diffusion. Thus its assignment poses severe problems. We note that spin diffusion between propane and n-hexane or n-heptane would have to involve first of all their CH$_3$ groups, but the **d - e** cross-peak is missing. This indication, while not conclusive (see above), inclines us to suggest that the **b - d** cross-peak be assigned to isopentane - propane. The **b - e** and **b - f** cross-peaks can come from inter- and intramolecular spin diffusion, the latter being more likely.

Further studies, especially model studies, are needed, if spin diffusion is to be a useful tool for the investigation of molecular catalysis on zeolites. By adsorbing various compounds and their mixtures, typical mixing times for the communication between various functional groups in various molecules under inter- and intramolecular spin diffusion can be established. This would provide an insight into the redistribution of the intermediates and reaction products on the catalyst and allow us to address the problems which in this work cannot be explained. For example, we note that prominent intermolecular **a** - **d** and **b** - **d** cross-peaks are present, but there is no intermolecular **a** - **b** cross-peak. It would be interesting to learn whether branched hydrocarbons such as isobutane and isopentane occupy channel crossings, and whether their consequent remoteness prevents the spin diffusion. Further, we could inquire whether propane molecules mostly occupy zeolite channels, so that the efficient spin diffusion between propane and adjacent isobutane or isopentane may take place. Such model studies are now in progress.

J-Scaled ^{29}Si NMR COSY of Highly Siliceous Mordenite

Applications of two-dimensional NMR techniques to the study of solids are still in their infancy. There have been, however, several successful investigations into the connectivities of tetrahedral sites in highly siliceous zeolites[35-41] using either COSY or INADEQUATE pulse sequences. However, use of conventional COSY is limited by the need to resolve weak cross-peaks which are very close to the intense signals on the main diagonal of the 2D spectrum. This is not possible except for extremely narrow signals. By contrast, the INADEQUATE experiment gives no diagonal peaks, but suffers from low sensitivity. We report here the application of the J-scaled COSY pulse sequence,[42] for the first time in the solid state, to the study of highly siliceous mordenite. The technique *scales up* the scalar couplings involved in the COSY experiment, thereby enhancing cross-peak intensities and consequently improving spectral resolution between adjacent diagonal and cross-peaks. Mordenite is a zeolite of considerable commercial interest.[43] Our highly siliceous sample was prepared from a synthetic mordenite with natural ^{29}Si isotopic abundance. It was calcined at 600°C for 12 h and then twice ion-exchanged in 1 M NH_4Cl aqueous solution at 80°C using a 1 : 10 w/w solid-to-liquid ratio. The NH_4^+-mordenite was then hydrothermally dealuminated at 600°C for 12 h, exchanged with NH_4^+ again, and then dealuminated at 800°C. The hydrothermal dealuminations were carried out using a vertical tube-furnace with a water injection rate of about 1 ml/min. The final product was highly crystalline (by XRD) and had a Si/Al ratio of several hundred.[44]

The J-scaled COSY experiment (Fig. 31-7) was performed using a scaling factor of 5. The 2D ^{29}Si NMR spectrum (Fig. 31-8) was recorded at 79.5 MHz with MAS at 4.2 kHz using the following parameters: ^{29}Si $\pi/2$ pulse of 4.8 μs, 33 ms acquisition time, 6 s recycle delay, 256 real data points in the F_2 dimension, 96 experiments for the second spectral dimension with 16 dummy and 288 real scans in each. Zero-filling to 2K and to 1K was used in F_2 and F_1 dimensions, respectively, and sine bell squared apodization with power calculation followed by symmetrization was applied for data processing.

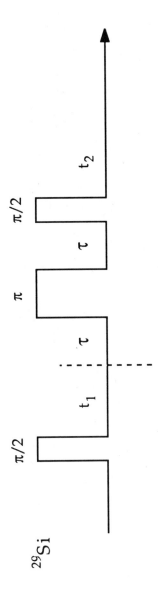

Fig. 31-7. J-scaled COSY ^{29}Si MAS NMR pulse sequence. The relationship between τ and t_1 is chosen so that J-couplings are scaled up by, in our case, a factor of five.

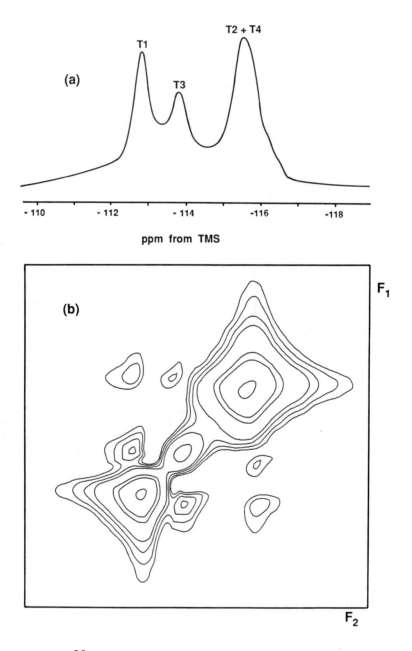

Fig. 31-8. (a) ^{29}Si MAS NMR spectrum of highly siliceous mordenite; (b) J-scaled COSY spectrum.

The conventional ^{29}Si MAS NMR spectrum of highly siliceous mordenite consists of three peaks in the intensity ratio of 2 : 1 : 3.[44-46] This may be explained on the basis of the known structure of mordenite[47,48] which contains four distinct tetrahedral crystallographic sites in the intensity ratio T1 : T2 : T3 : T4 = 2 : 2 : 1 : 1 (Fig. 31-9) with two of the peaks overlapping. Assignment of the signals has, in the past, relied on the correlation between ^{29}Si chemical shifts and mean Si-O-Si bond angle, α. This is theoretically a $<\cos \alpha/(\cos \alpha - 1)>$ dependence, but follows an approximately linear relationship in the regime under investigation.[49,50] The published structure solutions for various forms of mordenite[47,51,52] show that the mean T-O-T bond angles vary slightly with the degree of dealumination, cation type and water content, but the relative values remain approximately constant. The values for a siliceous mordenite prepared by acid leaching[51] given in Tab. 31-2 permit the immediate assignment of the downfield peak to the T1 site and shows that the T2 site is a component of the strongest peak. However, the $<$T3-O-T$>$ and $<$T4-O-T$>$ bond angles are fairly similar, and so it is not possible to assign confidently the spectrum completely on the basis of bond angles alone. The two possible assignments of the three peaks in the spectrum are to T1 : T4 : T2+T3 (as in refs. 44 and 46), or to T1 : T3 : T2+T4 crystallographic sites.

The 2D J-scaled COSY spectrum of highly siliceous mordenite (Fig. 31-9) reveals three cross-peaks. We were unable to resolve these by the COSY experiment with two extra delays, as described by Fyfe et al.[35,36] Attempts to do so yielded a spectrum with the cross-peaks obscured by the intense peaks on the main diagonal. The couplings between the tetrahedral sites are expected to be in the range of 10-15 Hz,[38] too small to give rise to prominent cross-peaks for this sample, unless they are scaled up by the particular pulse sequence used as is the case in J-scaled COSY. On the basis of the known connectivities of the mordenite structure (Tab. 31-2) only two cross-peaks are expected for the T1 : T4 : T2+T3 assignment, while the T1 : T3 : T2+T4 assignment implies that three cross-peaks should be observed. Thus the presence of three cross-peaks in the two dimensional J-scaled COSY experiment shows that the correct interpretation is T1 : T3 : T2+T4. Such unambiguous assignment is not possible by one-dimensional NMR or conventional COSY techniques.

CONCLUSIONS

Spin diffusion allows good quality spectra to be obtained relatively easily, but the interpretation of the results is often difficult and requires additional information on the shape of the zero-quantum signal (ZQ lineshape). In COSY, it is more difficult to obtain cross-peaks, because NMR signals from solids are

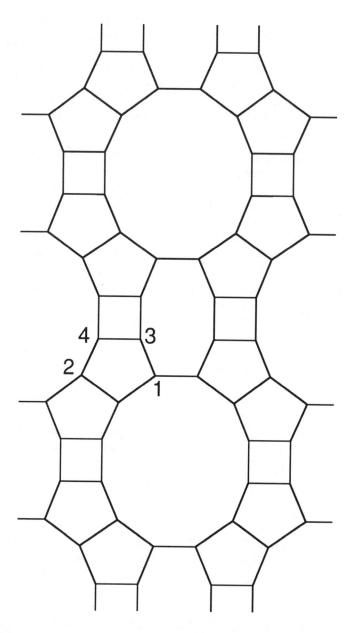

Fig. 31-9. The structure of mordenite viewed along [001]. The four kinds of crystallographic sites are indicated. Their relative populations (16 : 16 : 8 : 8 per unit cell) are not reflected in this projection.

Tab. 31-2.   The connectivities and typical mean T-O-T bond angles of the mordenite structure.

T-site	No. of sites per unit cell	Neighbouring sites	Mean T-O-T bond angle[51]
T1	16	T1, T1, T2, T3	150.4°
T2	16	T1, T2, T2, T4	158.1°
T3	8	T1, T1, T3, T4	153.9°
T4	8	T2, T2, T3, T4	152.3°

broad in comparison with the J-coupling constants, so that the cross-peaks are very weak because of the destructive interference of their broad antiphase components, and are obscured by the intense diagonal peaks. However, when a COSY spectrum can be obtained, its interpretation is much easier than that of a spin-diffusion spectrum.

Acknowledgments. We are grateful to Unilever Research, Port Sunlight, for support.

References

1.   Abragam, A. *The Principles of Nuclear Magnetism*, Oxford University Press, 1961.

2.   Ernst, R. R.; Bodenhausen, G.; Wokaun, A. *Principles of Nuclear Magnetic Resonance in One and Two Dimensions*, Oxford University Press, 1987.

3.   Szeverenyi, N. M.; Sullivan, M. J.; Maciel, G. E. *J. Magn. Reson.* **1982,** *47,* 462.

4.   Bronnimann, C. E.; Szeverenyi, N. M.; Maciel, G. E. *J. Chem. Phys.* **1983,** *79,* 3694.

5.   Caravatti, P.; Deli, J. A.; Bodenhausen, G.; Ernst, R. R. *J. Am. Chem. Soc.* **1982,** *104,* 5506.

6.   Caravatti, P.; Bodenhausen, G.; Ernst, R. R. *J. Magn. Reson.* **1983,** *55,* 88.

7.   Frey, M. H.; Opella, S. J. *J. Am. Chem. Soc.* **1984,** *106,* 4942.

8.   Suter, D.; Ernst, R. R. *Phys. Rev.* **1985,** *B32,* 5608.

9.   Linder, M.; Henrichs, P. M.; Hewitt, J. M.; Massa, D. J. *J. Chem. Phys.* **1985,** *82,* 1585.

10.   Henrichs, P. M.; Linder, M.; Hewitt, J. M. *J. Chem. Phys.* **1986,** *85,* 7077.

11.   Takegoshi, K.; McDowell, C. A. *J. Chem. Phys.* **1986,** *84,* 2084.

12.   VanderHart, D. L. *J. Magn. Reson.* **1987,** *72,* 13.

13. Kubo, A.; McDowell, C. A. *J. Chem. Phys.* **1988,** *89,* 63.
14. Kubo, A.; McDowell, C. A. *J. Chem. Soc., Faraday Trans. I* **1988,** *84,* 3713.
15. Connor, C.; Naito, A.; Takegoshi, K.; McDowell, C. A. *Chem. Phys. Lett.* **1985,** *113,* 123.
16. Clayden, N. J. *J. Magn. Reson.* **1986,** *68,* 360.
17. Davis, M. E.; Saldarriaga, C.; Montes, C.; Garces, J.; Crowder, C. *Nature* **1988,** *331,* 698.
18. Davis, M. E.; Saldarriaga, C.; Montes, C.; Garces, J.; Crowder, C. *Zeolites* **1988,** *8,* 362.
19. Rudolf, P. R.; Crowder, C. E. *Zeolites* **1990,** *10,* 163.
20. McCusker, L. B.; Baerlocher, Ch.; Jahn, E.; Bülow, M. *Zeolites* **1991,** *11,* 308.
21. Davis, M. E.; Montes, C.; Hathaway, P. E.; Arhancet, J. P.; Hasha, D. L.; Garces, J. M. *J. Am. Chem. Soc.* **1989,** *111,* 3919.
22. Grobet, P. E.; Martens, J. A.; Balakrishnan, I.; Mertens, M.; Jacobs, P. A. *Appl. Catal.* **1989,** *56,* L21.
23. Stocker, M.; Akporiaye, D.; Lillerud, K. P. *Appl. Catal.* **1991,** *69,* L7.
24. van Braam Houckgeest, J. P.; Kraushaar-Czarnetzki, B.; Dogterom, R. J.; de Groot, A. *J. Chem. Soc., Chem. Comm.* **1991,** 666.
25. Davis, M. E.; Montes, C.; Hathaway, P. E.; Garces, J. M. in *Zeolites: Facts, Figures, Future,* Jacobs, P. A.; van Santen, R. A., Eds. Elsevier, Amsterdam, 1989, p. 199.
26. Virlet, J.; Ghesquieres, D. *Chem. Phys. Lett.* **1980,** *73,* 323.
27. Anderson, M. W.; Klinowski, J. *Nature* **1989,** *339,* 200.
28. Anderson, M. W.; Klinowski, J. *J. Am. Chem. Soc.* **1990,** *112,* 10.
29. Anderson, M. W.; Klinowski, J. *Chem. Phys. Lett.* **1990,** *172,* 275.
30. Carpenter, T. A.; Klinowski, J.; Tennakoon, D. T. B.; Smith, C. J.; Edwards, D. C. *J. Magn. Reson.* **1986,** *68,* 561.
31. Frye, J. S.; Maciel, G. E. *J. Magn. Reson.* **1982,** *48,* 125.
32. Limbach, H. H.; Wehrle, B.; Schlabach, M.; Kendrick, R.; Yannoni, C. S. *J. Magn. Reson.* **1988,** *77,* 84.
33. Nielsen, N. C.; Bildsoe, H.; Jakobsen, H. J.; Sørensen, O. W. *J. Magn. Reson.* **1988,** *79,* 554.
34. Kalinowski, H. O.; Berger, S.; Braun, S. *^{13}C-NMR-Spektroskopie,* Georg Thieme, Stuttgart, 1987.
35. Fyfe, C. A.; Gies, H.; Feng, Y. *J. Chem. Soc., Chem. Commun.* **1989,** 1240.
36. Fyfe, C. A.; Gies, H.; Feng, Y. *J. Am. Chem. Soc.* **1989,** *111,* 7702.
37. Fyfe, C. A.; Feng, Y.; Kokotailo, G. T. *Nature* **1989,** *341,* 223.
38. Fyfe, C. A.; Feng, Y.; Gies, H.; Grondey, H.; Kokotailo, G. T. *J. Am. Chem. Soc.* **1990,** *112,* 3264.

39. Fyfe, C. A.; Grondey, H.; Feng, Y.; Kokotailo, G. T. *Chem. Phys. Lett.* **1990,** *173,* 211.

40. Fyfe, C. A.; Grondey, H.; Feng, Y.; Kokotailo, G. T. *J. Am. Chem. Soc.* **1990,** *112,* 8812.

41. Fyfe, C. A.; Gies, H.; Feng, Y.; Grondey, H. *Zeolites* **1990,** *10,* 278.

42. Hosur, R. V.; Chary, K. V. R.; Ravi Kumar, M. *Chem. Phys. Lett.* **1985,** *116,* 105.

43. Breck, D. W. *Zeolite Molecular Sieves,* John Wiley, New York, 1974.

44. Thomas, J. M.; Klinowski, J.; Ramdas, S.; Hunter, B. K.; Tennakoon, D. T. B. *Chem. Phys. Lett.* **1983,** *102,* 158.

45. Fyfe, C. A.; Gobbi, G. C.; Murphy, W. J.; Ozubko, R. S.; Slack, D. A. *J. Am. Chem. Soc.* **1984,** *106,* 4435.

46. Bodart, P.; Nagy, J. B.; Debras, G.; Gabelica, Z.; Jacobs, P. A. *J. Phys. Chem.* **1986,** *90,* 5183.

47. Meier, W. M. *Z. Kristallogr.* **1961,** *115,* 439.

48. Meier, W. M.; Olson, D. H. *Atlas of Zeolite Structure Types,* Butterworths, London, 1987.

49. Engelhardt, G.; Michel, D. *High-Resolution Solid-State NMR of Silicates and Zeolites,* John Wiley, New York, 1987.

50. Engelhardt, G. in *Recent Advances in Zeolite Science,* Klinowski, J.; Barrie, P. J., Eds. *Stud. Surf. Sci. Catal.* **1989,** *52,* 151.

51. Schlenker, J. L.; Pluth, J. J.; Smith, J. V. *Mat. Res. Bull.* **1979,** *14,* 849.

52. Mortier, W. J.; Pluth, J. J.; Smith, J. V. in *Natural Zeolites, Occurrence, Properties, Use,* Sand, L. B.; Mumpton, F. A., Eds. Pergamon Press, Oxford, 1978, p. 53.

32

Model Considerations for the Crystallization of Zeolites

H. Lechert, H. Kacirek, *Institute of Physical Chemistry University of Hamburg, Bundesstr. 45, D-2000 Hamburg 13, Germany*
and

H. Weyda, *Fa. Süd - Chemie, Katalyse-Labor, Waldheimer Str. 13, D8206 Heufeld - Bruckmühl, Germany*

In a large number of crystallization experiments with different zeolites the Si/Al ratio of the crystals has been studied in terms of dependence on the excess alkalinity $m = (Na-Al)/Si$ in the batch. The results are explained by a model assuming that at the surface of the growing crystal

- an equilibrium of $\equiv Si-OH$ and $\equiv SiO^-$ is established with the alkali in the solution
- aluminate from the solution is preferably attached to charged silicate surface species
- silicate from the solution is attached to aluminate and in a condensation reaction to the $\equiv Si-OH$ groups.

The model explains the dependence of the Si/Al ratio in the crystal on the alkalinity avoiding the assumption of the incorporation of silica oligomers which should be unfavourable from kinetic reasons .

1. INTRODUCTION

In several preceding papers the crystallization of X and Y zeolites has been thoroughly studied (Kacirek and Lechert 1975, 1976; Lechert 1983; Lechert and Kacirek 1991; Kostinco 1983). Another recent paper on this problem has been published by Robson (1989). From these and other investigations summarized, e.g., by Barrer (1983) and Occelli and Robson (1989) the conditions of the crystallization of most of the zeolites are known in great detail.

Problems arise, however, in the explanation of of the mechanism of the crystallization process. Generally, zeolites crystallize in the presence of a gel phase which is dissolved during the crystallization, supporting the growing crystal with the material necessary for the growth process. The crystallization via the solution phase has been proved in a large number of investigations.

In many of the papers dealing with zeolite crystallization, silicate oligomers have been discussed as precursors for the nucleation and building units for the growth. The role of special oligomers as so called secondary building units as precursors in the nucleation process can be easily imagined.

Difficult to understand, however, is that these comparatively large molecules, which are present in the solutions in very small concentrations, are expected to be incorporated as building blocks at the surface of the growing crystals.

A recent critical review of these ideas has been given by Keijsper and Post (1989).

Attempts have been made, therefore, to find a mechanism for the crystal growth based on the incorporation of monomer building units, which are present in the solution in great excess even in the batches of silica rich zeolites (Lechert and Kacirek 1991).

A model describing the crystallization of aluminosilicate zeolites has to explain some common observations. First the strict obedience to the Loewenstein rule has to be explained. This rule describes the experimental observation that never are two adjecent Al-ions found in zeolite structures. A further observation is that the Si/Al ratio in the final product generally increases with decreasing alkali content in the solution phase.

This is demonstrated by a large number of samples of faujasite and P zeolite which occur frequently together (Kacirek and Lechert 1975; Kacirek and Lechert 1976; Robson 1989).

In addition a series of experiments with the zeolites L, omega, offretite, and mordenite will be presented. In contrast to the other zeolites the structure of mordenite contains preferably five-membered rings.

2 MODEL CONSIDERATIONS FOR THE CRYSTAL GROWTH

For the exact dependence of the composition of the crystal on the alkalinity of the solution a model has been developed by Lechert and Kacirek (1991) taking into account the processes which may occur at the surface of the growing crystal.

Generally, it can be imagined that at the surface of a growing zeolite crystal three different species are present which may be represented schematically in the following way

1. $[Zeo\equiv Al-OH]^- Na^+$
2. $Zeo\equiv Si-OH$
3. $Zeo\equiv Si-O^- Na^+$

The species 2 and 3 are related by the equilibria

$$Zeo\equiv Si-OH + Na^+ + OH^+ \longleftrightarrow Zeo\equiv Si-O^- Na^+ + H_2O \qquad (1)$$
$$Zeo\equiv Si-OH + {}^-O-Si\equiv + Na^+ \longleftrightarrow Zeo\equiv Si-O^- Na^+ + HO-Si\equiv \qquad (1a)$$
$$Zeo\equiv Si-OH + {}^{2-}O_2\!=\!Si\equiv + Na^+ \longleftrightarrow Zeo\equiv Si-O^- Na^+ + {}^-O-Si\equiv \qquad (1b)$$

where ${}^{2-}O_2\!=\!Si\equiv$, ${}^-O-Si\equiv$ and $HO-Si\equiv$ are the silicate species in the solution phase.

The aluminium should be present primarily as aluminate (Aveston 1965; Barrer 1989). Some of these aluminate ions form aluminosilicate complexes (Thangaraj and Kumar 1990). The Al-OH groups have a very low acidity. Therefore, in the reactions with the surface preferably substitutions of the OH groups should be taken into account.

For the hydrolysis reaction of the silicate ions according to Lagerström (1959) and Caullet and Guth (1989) the following equilibria are established:

$$[SiO(OH)_3^{1-}][H_2O]/[Si(OH)_4][OH^-] = 6.918 \ 10^3 \ \text{for} \ 50°C$$
$$= 1.862 \ 10^4 \ \text{for} \ 25°C \quad (1c)$$
$$[SiO_2(OH)_2^{2-}][H_2O]/[Si(OH)_4][OH^-]^2 = 7.943 \ 10^4 \ \text{for} \ 50° \ C$$
$$= 2.754 \ 10^5 \ \text{for} \ 25°C \quad (1d)$$

This means that under the conditions of the zeolite synthesis the silicate is present in the solution mainly as $[SiO_2(OH)_2^{2-}]$ and as $[SiO(OH)_3^-]$. Further, it can be concluded that the amount of $Si(OH)_4$ increases with temperature.

The reaction with the surface can occur via the $HO-Si\equiv$ groups or the ${}^-O-Si\equiv$ groups of the silicate. The $HO-Si\equiv$ groups react preferably acidic. The unspecified bonds may be occupied with -OH or $^-$O- groups. In principle, with these surface groups and ions in the solution nine different reactions of the first step of the attachment can be formulated schematically.

Taking into account the arguments preferring one or the other reaction the following reactions are the most probable

$$[Zeo\equiv Al-OH]^-Na^+ +{}^- O-Si\equiv \longleftrightarrow [Zeo\equiv Al-O-Si\equiv]^-Na^+ + OH^-$$
$$(2)$$

$$[Zeo\equiv Al-OH]^- Na^+ + HO-Si\equiv \longleftrightarrow [Zeo\equiv Al-O-Si\equiv]^- Na^+ + H_2O$$
$$(3)$$

for the attachment of the silicate ions from the solution to the aluminate at the surface.

$$Zeo\equiv Si-O^-Na^+ + Al(OH)_4^- \longleftrightarrow [Zeo\equiv Si-O-Al(OH)_3]^-Na^+ + OH^-$$
$$(4)$$

should be the preferred reaction for the incorporation of aluminate at the silicate sites. Both reactions can be regarded as substitution reactions of an OH group at the aluminate against a silicate species. The dependence of the reactions (2) and (4) on the concentration of OH$^-$ ions explains the strong influence of the alkalinity on the Si/Al-ratio of the products. The attachment of the silicate to the silicate at the surface may occur via a condensation reaction

$$Zeo\equiv Si-OH + HO-Si\equiv \longleftrightarrow Zeo\equiv Si-O-Si\equiv + H_2O \qquad (5)$$

Reactions similar to (3) with aluminate should be excluded because of the avoidance of Al-O-Al bonds in the crystals.

According to these arguments the following assumptions can be made for further discussion

A. Formation of Al-O-Al bonds is not possible
B. Aluminate is attached to the surface preferably on the silicate species Zeo$\equiv$Si-O$^-$Na$^+$ according to (4).
C. Silicate is attached at the aluminate according to (2) or (3).
D. The binding to the silica occurs via the condensation reaction (5).

At the surface, only a limited number of Zeo$\equiv$Si-OH groups is available. These groups react with the OH$^-$ ions in the solution according to the equilibrium reaction (1). The rate of the forward reaction in (1) is proportional to the OH$^-$ concentration in the solution and the amount of unreacted Zeo$\equiv$Si-OH groups at the surface. The backward rate is proportional to the amount of $Zeo \equiv Si-O^-Na^+$ at the surface.

These conditions are similar to the assumptions made for the derivation of the Langmuir equation.

Therefore, a Langmuir function can be derived for the relative amount of $Zeo \equiv Si-O^-Na^+$ in dependence on the NaOH and the $Na^{+-}O-Si \equiv$ concentration in the solution.

Simplifying, it will be assumed that the reaction (1) is the fastest reaction for the attack of the $\equiv Si-OH$ groups of the surface.

Abbreviating SOH = Amount of $Zeo \equiv Si-OH$ at the surface

SO = Amount of $Zeo \equiv Si-O^- Na^+$ at the surface

S = Total amount of silicate at the surface

A = Amount of $[Zeol. \equiv Al-OH]^- Na^+$ at the surface

$$A + SO + SOH = A + S = 1 \qquad (6)$$

$$SO = S \frac{K [OH^-]}{1 + K [OH^-]} \qquad (7)$$

can be obtained, where

$$K = [Zeo \equiv Si-O^-]/[Zeo \equiv SiOH] = SO/SOH$$

To get an impression of the influence of the OH concentration on the Si/Al ratio of the crystalline product the following model has been discussed (Lechert and Kacirek 1991).

On a layer n-1 of the growing surface an arbitrary layer n is crystallized. The relative amount of Si at a layer n-1 is denoted as S_{n-1} the relative amount of Al as A_{n-1}. In the layer n A_n and S_n are defined in the same way. According to the reactions discussed above it follows, that

$$SO_{n-1} = S_{n-1} \frac{K[OH^-]}{1 + K[OH^-]} \qquad (8)$$

According to assumption B it follows that for the amount of aluminum A_n in the layer n

$$A_n = S_{n-1} \frac{K[OH^-]}{1 + K[OH^-]} \qquad (9)$$

This includes the assumption that the equilibrium of the

reaction (4) lies far at the right side. Then, from (9) can be concluded that the Zeo≡Si-O⁻ sites are saturated with the aluminate inspite of its low concentration in the solution phase.

The residual surface silicate is present as Zeo≡Si-OH which can attach ≡Si(OH) from the solution via condensation.

Then the relative amount of Si in the n-th layer is given by

$$S_n = A_{n-1} + SOH_{n-1} = 1 - S_{n-1} + SO_{n-1} + SO_{n-1}/K[OH^-]$$
$$= 1 - S_{n-1} + S_{n-1}/(1 + K[OH^-]) \qquad (10)$$

$$S_n = \frac{1 + K[OH^-] + K[OH^-] S_{n-1}}{1 + K[OH^-]} = S_{n-1} \frac{K[OH^-]}{1 + K[OH^-]} \qquad (11)$$

This may be taken as a recursion formula for the calculation of S_n. With

$$y = \frac{K[OH^-]}{1 + K[OH^-]} \qquad (12)$$

S_n can be written

$$S_n = 1 - y(1 - y(1 - ... y(1 - y S_0)) = 1 - y + y^2 - y^3 ... - y^{n-1} S_0 = 1/(1+y) \qquad (13)$$

S_0 is the silica content at the nucleus from which the crystal grows.

From (9) it follows in the same way that

$$A_n = y(1 - y(1 - ... y(1 - y S_0)) = y + y^2 - y^3 ... - y^{n-1} S_0 = y/(1+y) \qquad (14)$$

For large n the term $y^{n-1} S_0$ becomes very small because $y < 1$.
Finally, the Si/Al-ratio in the crystal is

$$S_n/A_n = 1/y = \frac{1 + K[OH^-]}{K[OH^-]} = 1 + \frac{1}{K[OH^-]} \qquad (15)$$

It can be seen that the averaging over the whole crystal leads to a rather simple function compared to the relation derived in the preceding paper (Lechert and Kacirek 1991).

This function will now be compared with an extended set of data obtained for different zeolites.

From the papers mentioned above a large number of data of

experiments for the crystallization of Y and P zeolites have been taken (Kacirek and Lechert 1975; 1976; Kacirek 1974) which have been extended recently (Wienecke 1985; Wulff-Döring 1990; Wulff-Döring and Lechert 1991). Further experiments have been carried out higher temperatures and conditions under which mordenite crystallizes.

To study the influence of other cations a series of crystallization experiments has been carried out with the zeolites L, omega and offretite.

3 EXPERIMENTAL

The detailed experimental procedures for the crystallization of the X-zeolites have been discussed in our former paper (Lechert and Kacirek 1991). The procedures for the synthesis of Y zeolites and mixtures of P and Y are discussed in detail by Kacirek and Lechert (1975, 1976) .

The compositions of the batches for the X zeolites are given by

$$NaAlO_2 \ 2(Na_m H_{4-m} SiO_4) \ 195 \ H_2O$$

and

$$NaAlO_2 \ 1.5(Na_m H_{4-m} SiO_4) \ 163 \ H_2O$$

with m = 2.8, 3.0, 3.5, 4.0 and some batches with an excess of NaOH corresponding to m = 4. The quantity m is the so called excess alkalinity is calculated from the batch composition by

$$m = ([NaOH] - [AlO_2^-])/ [SiO_2]$$

In our earlier papers, m has proved to be the most realistic measure for the alkali content of the batch, which can be obtained directly from the batch composition. It has been shown that m can be very well related to the Si/Al ratio of the final products.

For the Y zeolites generally batches with

$$NaAlO_2 \ 2.0 (Na_m H_{4-m} SiO_4) \ 195 \ H_2O$$

are used. As described by Kacirek (1974) and by Kacirek and Lechert (1975, 1976) the Y zeolites were grown using X seeds. From unseeded batches products with more or less P zeolite crystallized. Similar experiments have been repeated more recently using nucleation gels (Wienecke 1985; Wulff-Döring 1990; Wulff-Döring and Lechert 1991).

The crystallization has usually been carried out at 361 K.

Going to higher temperatures with the batch compositions given above almost only P zeolite crystallizes at about 400 K. Mordenite is obtained at higher temperatures at comparatively high dilutions.

For a more exact study of this phenomenon a series of experiments has been carried out at with different batch compositions

$$NaAlO_2 \ 9.2(Na_m H_{4-m} SiO_4) \ x \ H_2O$$

m = 0.38 and 120 $\leq$ x $\leq$ 656 at 420 K.

Further, a series of kinetic experiments at different temperatures and experiments to study the dependence of the composition of the final products on the batch compositions have been done. Using K^+ or tetramethylammonium ions the zeolites L, omega and offretite can be obtained under the conditions where mordenite crystallizes.

For the crystallization of L zeolite the crystallization field of Breck (1974) has been taken as a basis from which optimization experiments for the L crystallization have been carried out. Pure L-samples were crystallized at 400 K from batches around the composition

$$(Na_{0.2}K_{0.8}) \ AlO_2 \ 14.5 (Na_{0.16}K_{0.64}H_{3.2}SiO_4) \ 200 \ H_2O$$

within 18 hours.

The experiments for the crystallization of omega started from batches

$$NaAlO_2 \ 4.35(Na_{0.28}H_{3.72}SiO_4) \ n \ TMA\text{-}Cl \ 80 \ H_2O$$

where the influence of the content of TMA-Cl on the kinetics of crystallization and the kind of the crystallizing product was studied systematically.

The starting point of the synthesis of pure offretite without template was a batch composition

$$(Na_{0.34} \ K_{0.6}) AlO_2 \ 12,5 (Na_{0.24} \ K_{0.46} H_{3.3} SiO_4) \ 200 \ H_2O$$

from which offretite or a mixture of offretite and erionite could be obtained at 373 K in about 6 days. Without template the range where pure offretite is obtained is rather narrow under the given conditions. The range of crystallization of offretite can be largely extended by adding TMA ions to the batch.

From the preceding data it can be seen that there exists a temperature range and a range of Si/Al-ratios and water contents of the batches where the crystallizing structure type is only dependent on the cations present in the solution.

After a series of preliminary experiments such a range

could be found for

$$(Na,K,TMA)AlO_2 \; 12.0((Na,K,TMA)_{0.8}H_{3.2}SiO_4) \; 200\,H_2O$$

at rather high crystallization temperatures of 460 K. An extended study of a crystallization diagram in dependence on the ratio of the cation concentration is given in a separate paper of this issue .

The samples were washed and dried at 373 K. Before further treatment they were equilibrated over saturated NaCl solution.

The phase composition and the crystallinity were checked by X-ray diffraction. The chemical composition of the samples was analyzed by X-ray fluorescence spectroscopy or by EDAX.

4 RESULTS AND DISCUSSION

In Fig. 1 the data of the different experiments, which are described in the preceding section, are summarized. Further, the curves which can be obtained from the theoretical considerations according to Eq. 15 are demonstrated.

As an approximation $m \approx [OH^-]$ has been taken. For the upper curve is $K = 1$ and for the lower $K = 3$. The data for the faujasites include for $m < 2.5$ experiments which have been carried out without seeds. The resulting zeolites contain, therefore, varying amounts of P zeolite as it was found in the above mentioned papers in great detail (Kacirek 1974; Kacirek and Lechert 1975, 1976).

It can be seen that the samples with P content show no systematic devaition from the data. Our data agree perfectly with data of the dependence of the Si/Al ratio of a large number of zeolites on the excess alkalinity m which have been reported in an early paper by Zdhanov (1968) These data are shown also in the book of Barrer (1982) This shows that our theoretical considerations may be of more general importance.

The comparatively large scattering of the data for mordenite has different origins. At first the value of m may be too rough an approximation for the concentration of the free OH ions in the solution under the conditions of the synthesis. Further, it can be seen from the equations 1c and 1d that the equilibrium constant for the formation of the higher loaded ion decreases with increasing temperature. This means that the OH^- concentration in the solution increases with increasing temperature.

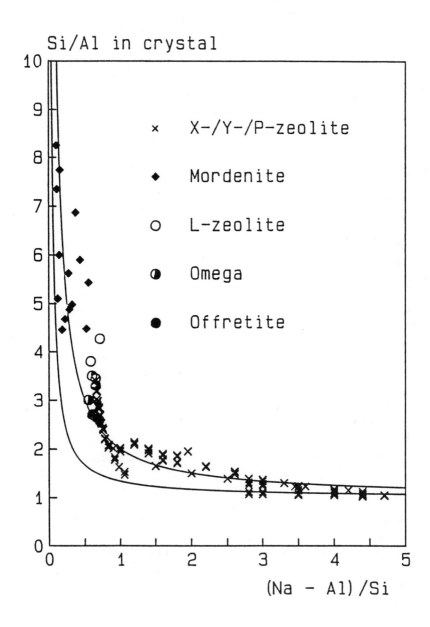

Figure 1. *Dependence of the Si/Al ratio of several zeolites on the excess alkalinity* $m = (Na-Al)/Si$ *in the batch. The curves correspond to the function of Eq.15 with* $K = 1$ *(upper curve) and* $K = 3$ *(lower curve)*

A similar effect may be expected for the formation of the charged species at the surface. This means that lower values of K have to be expected, favoring a higher amount of Si-OH groups the incorporation of silicate.

Another source of error is certainly the amount of gel present in the crystallizing batch. To form the gel processes like

$$kNaAl(OH)_4 + l(Na_2H_2SiO_4) \longleftrightarrow Na_kAl_kSi_lO_{k \cdot l} + 21NaOH$$
$$+ k H_2O$$

may be expected which release NaOH.

Controlled experiments of a template free synthesis of ZSM 5 are in progress.

More detailed information should be obtained from an exact chemical analysis of the solution phase, as it has been carried out in our former paper (Lechert and Kacirek 1991). Experiments of this kind are in progress.

Looking at the data in Fig. 1 it should be pointed out that from a refinement of the theoretical considerations given above and a careful analysis of the solution phases and the crystal composition in the X and Y region a more detailed picture of the crystallization mechanism may be expected. Additional information should be expected from NMR-MAS investigations of the Si-Al distribution.

5 CONCLUSIONS

It can be seen that from a model taking into account the surface composition of the growing crystal and assuming that
- an equilibrium of $\equiv$ Si-OH and $\equiv$ Si-O$^-$Na$^+$ is established at the surface with the alkali in the solution
- aluminate from the solution is preferably attached to charged silicate species at the surface
- silicate from the solution is attached to aluminate
and in a condensation reaction to the Si-OH groups
the Si/Al ratio of a large number of zeolites can be described with good accuracy in dependence on the excess alkalinity in the solution.

This model explains the dependence of the Si/Al ratio in the crystal on the alkalinity avoiding the assumption of the incorporation of silica oligomers which should be unfavorable from kinetic reasons.

6 ACKNOWLEDGMENTS

The authors thank the Deutsche Forschungsgemeinschaft for the support of their work.

7 REFERENCES

Aveston, J., 1965, J. Chem. Soc., 4445.

Barrer, R.M., 1982, *Hydrothermal Chemistry of Zeolites* , Academic : London.

Barrer, R.M.; 1989, in *Zeolite Synthesis* Occelli, M.L., Robson, H.E. Eds., ACS Symp. Ser. 398, Washington, p. 11.

Breck, D.W. 1974, *Zeolite Molecular Sieves*, John Wiley and Sons, New York .

Caullet, P., Guth, J.L.; 1989, in *Zeolite Synthesis*, Occelli, M.L., Robson H.E., Eds., ACS Symp. Ser. 398, Washington, p. 83.

Kacirek H. 1974, *Untersuchungen zur Darstellung und zur Kristalli-sationskinetik von Faujasiten unterschiedlicher Zusammensetzung* , Ph.D. Thesis University of Hamburg, .

Kacirek, H., Lechert, H., 1975, J. Phys. Chem. 79: 1589.

Kacirek, H., Lechert, H. 1976, J. Phys. Chem. 80:   1291.

Keijsper, J.J., Post, M.F.M.; 1989 in *Zeolite Synthesis*, Occelli, M.L., Robson H.E. Eds., ACS Symp. Ser. 398, Washington , p. 28.

Kostinco, J.A.   1983, in *Intrazeolite Chemistry* , Stucky, J.A. Ed., ACS-Symp. Ser.218, Washington, : p.1.

Lagerström, G., 1959, Acta.Chim. Scand., 13 : 722.

Lechert, H. 1984, in *Structure and Reactivity of Modified Zeolites* Jacobs, P.A., Jaeger N.I., Jiru, P., Kazansky, V.B., Schulz-Ekloff, G.,Eds. Stud. in Surf. Sci. and Catal., 18, p. 107.

Lechert, H., Kacirek. H. 1991, Zeolites 11, 720.

Occelli, M.L., Robson,H.E. Eds., 1989, *Zeolite Synthesis* ACS Symp. Ser. 398, Washington .

Robson, H.E. *1989,* in *Zeolite Synthesis*, Occelli, M.L., Robson H.E. Eds., ACS Symp. Ser. 398, Washington, p. 436.

Thangaraj, A., Kumar, R. 1990, Zeolites, 10, 117.

Wienecke J., 1985, *Zur Bildungskinetik der Zeolithe an Beispielen der Faujasitgruppe und Pentasilen ergänzt durch NMR-Untersuchungen zur Beweglichkeit von Molekülen in Pentasilen* , PhD Thesis, University of Hamburg.

Wulff-Döring, J., 1990. *Kristallisationskinetische Untersuchungen zur Synthese faujasitähnlicher Zeolithe mit Keimbildungsgelen.* Ph.D. Thesis, University of Hamburg.

Wulff-Döring, J., Lechert, H., 1991, Catalysis Today, 8, 395 .

Zdhanov, S.P. 1968 in *Molecular Sieves* , Society of Chemical Industry, London.

33

MICROWAVE TECHNIQUES IN ZEOLITE SYNTHESIS

J. C. Jansen[1], A. Arafat[2], A. K. Barakat[2] and H. van Bekkum[1]

1) Laboratory of Organic Chemistry and Catalysis, Delft University of Technology Julianalaan 136, 2628 BL Delft, The Netherlands.
2) Faculty of Science, Helwan University, Cairo, Egypt.

Microwave heating of zeolite synthesis mixtures resulted in simultaneous and abundant nucleation, and compared to conventional, heating the crystal growth was approximately 10 times faster. The short crystallization time in a microwave-heated system is mainly due to the fast dissolution of the gel. Hydroxysodalite could be prepared with sodium as well as with tetramethylammonium ions. The crystallization field of zeolite NaA when applying microwave heating is essentially the same as with conventional heating.

INTRODUCTION

Microwave heating of zeolite synthesis mixtures is a rather unexplored field. Very recently the use of microwave heating in the unseeded preparation of zeolite NaA and in the seeded preparation of ZSM-5 was reported (Chu and Dwyer, 1990). The influences of microwave heating on zeolite preparation, based on the specific properties of microwave radiation, are as yet not well documented. Typical properties include the high transmissibility of autoclave materials such as Teflon for microwaves and the fact that microwave energy is absorbed maximally without a temperature gradient in a solvent such as water. Based on the latter property, microwave heating rates of aqueous zeolite synthesis mixtures can be very high. Furthermore, the homogeneous heating results in a more simultaneous nucleation compared to a conventional heating system.

Moreover, the effects of the meandering electromagnetic wave, causing ion oscillation and water dipole rotation, on the zeolite nucleation and crystallization mechanism might differ from those of a conventional heating system. In this study high heating rates and homogeneous heating by microwave were applied to various zeolite synthesis mixtures and the effects compared with conventional heating systems.

EXPERIMENTAL

Two types of microwave ovens have been used. Figure 33-1a shows a microwave oven with a tubelike cavity of 100 x 5 x 10 cm. A continuous variable power with a maximum of 2500 W can be applied. A 3 ml autoclave (see Fig. 33-1b) of polyether ether ketone (PEEK) with glass fiber reinforced, was used. Work temperatures up to 250°C and corresponding autogenic pressures of water are allowed in this thermostable material. As shown in the schematic drawing in Figure 33-1c, the autoclave is placed in the cavity via a tube perpendicular to the cavity. The temperature was monitored with an infrared thermometer on the outer wall of the autoclave(see Fig. 33-1a and c). Differences in temperature between the aqueous synthesis mixture and the outer wall of the autoclave were estimated to be less than 5°C at constant temperature. A constant temperature was obtained in the tubelike cavity by means of the continuously variable power adjustment.

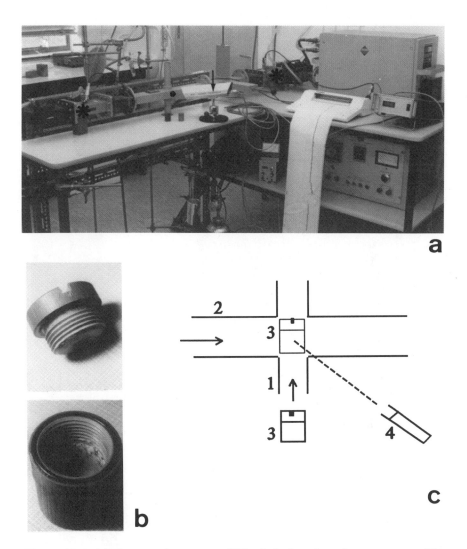

Figure 33-1. (a)The complete setup of The "tube"-cavity microwave oven. The cavity (∗   ∗) contains a hole (•) in one side of the wall. The heat radiated through this hole by the autoclave when positioned in the cavity is sensored with an infrared thermometer (↓). (b) The polyether ether ketone glass fiber reinforced 3 ml autoclave with a viton ring (c). Schematic drawing of the (1) of the cavity (2) for the autoclave (3) together with the infrared thermometer (4). The direction of the microwave radiation is indicated by the horizontal arrow.

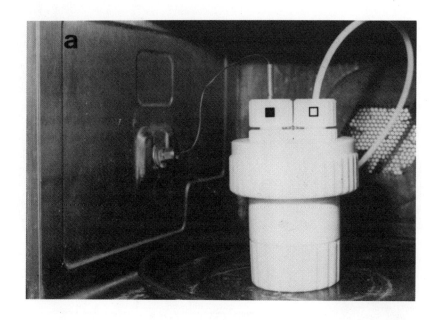

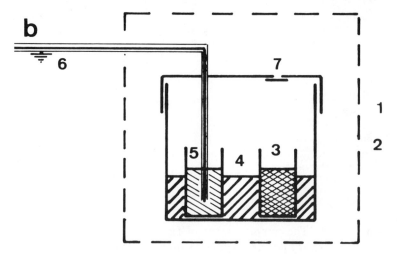

Figure 33-2. (a) General picture of the setup in the large microwave oven. (b) Schematic drawing of the perforated stainless steel cage (1), the Teflon autoclave (2), the vessel containing the synthesis mixture (3),the vessel with and water (5) and water (4) to equilibrate pressure in vessels (3) and (5), the earth-connected gold-plated thermocouple (6) and the safety plate (7).

With this setup, high microwave powers could be dissipated in a small, 2 ml synthesis mixture, resulting in high heating rates (100°C/10 sec) and high temperatures In a second household-type oven (Sharp R-10R 50 (W)) with a large cavity of 35 x 35 x 25 cm the microwave power could be adjusted from 100 to 1000 W in ten steps. A large 200 ml Teflon autoclave was used (see Fig. 33-2a), in which two 30 ml vessels (see Fig. 33-2b) could be inserted. In one bottle containing only water, a gold-plated earth-connected thermocouple was situated to monitor the temperature of water exposed to microwave. The second bottle contained the synthesis mixture. The setup is schematically shown in Figure 33-2b.

Heating rates up to 20°C/10 sec could be achieved in this oven. The temperature of the autoclave was kept at the desired level using a perforated stainless steel cage around the autoclave. Series of experiments were carried out in this oven in new bottles to preclude memory effects, with an identical temperature profile in the different experiments. Products were characterized by XRD, FTIR and SEM.

RESULTS AND DISCUSSION

The synthesis mixtures applied are given in Table 33-1 in molar oxide ratios. The amount of water in each synthesis mixture was 90-93 mole %. Whereas aerosil 200 (Degussa, Germany) was used as a silicon source and sodium aluminate (Riedel-De Haen, Germany) as aluminum source. Synthesis mixtures were carefully prepared as the homogeneous dense gel and high heating rates easily lead to for example hydroxysodalite formation in products via pockets which are not adequately mixed or aged. (Thompson, 1990).

Table 33-1. Zeolite synthesis formulations used: (I) A synthesis mixture to study the crystallization rate of zeolite NaA, (II) to compare crystallization fields in the preparation of three zeolites and (III) to study the effect of TMA.

Zeolite type prepared	Molar oxide ratio			Ref.
	Na_2O	SiO_2	Al_2O_3	
(I) NaA	3.1	2.1	1	Breck, 1974
(II) NaA	1.3-3.5	1.2-5.2	1	Breck, 1974
(II) NaX	1.3-1.5	5.3-7.0	1	Breck, 1974
(II) HS[a]	1.0-1.2	0.6-0.7	1	Breck, 1974
(III)TMA-HS[a] 0.5TMA$_2$O	0.5	0.6	1	Aeillo, Barrer, 1970

a: hydroxysodalite

In Table 33-2, the percentage crystallinity and the average particle size are given for the products of a zeolite NaA synthesis mixture after subsequent various heating times in the large microwave oven. For this series of experiments the temperature was increased to 120°C in 40 sec using a power of 800 W. Furthermore reduction of the microwave was obtained by using a perforated stainless steel cage covering the autoclave. Subsequently, the power was switched off and the mixtures were, after cooling, kept at 95°C with a power adjustment of 100 W by which the microwave energy was further reduced using perforated stainless steel cage. This temperature profile was chosen mainly to initiate nucleation at 120°C followed by crystal growth at 95°C. Based on reference experiments it can be concluded that, during the nucleation step, essentially 100% of the 800 W microwave energy was absorbed whereas in the crystal growth step about 50% of the heating energy was dissipated by the heat capacity of the total system and 50% by the reduced microwave energy. As given in Table 33-2 and shown in Fig. 33-3 complete crystallization of pure zeolite NaA is obtained after 10 minutes which is fast compared to a conventional heating method *vide infra*. From the crystal size distribution (see Fig. 33-4) which is rather small, it is concluded that the main part of the nucleation took place in the high temperature (120-95°C) region and thus the crystal growth at 95°C. Although it can be concluded that the overall crystallization time of the zeolite synthesis mixture is much shorter with microwave heating than with a conventional heating method, it is not clear if this is due to the nucleation step, the crystal growth step or both steps. For this reason experiments were carried out using the same system as described above, i.e., initial microwave heating to 120°C followed by conventional heating at 95°C. As shown in Table 33-3 and in Fig. 33-5 at least 90 min are required at 95°C to achieve complete crystallization of pure zeolite NaA when applying conventional heating in the crystal growth step. In Fig. 33-5 the SEM photographs show that only gel seems to be present in the first 90 min of the synthesis time. The XRD characterization (see Table 33-3) indicates, however, a crystallinity of 30% after 60 min. Apparently the dissolution and crystallization of zeolite takes about a factor 10 more in time when the conventional heating method is applied. Moreover Fig. 33-4 and the SEM photographs (see Fig. 33-3) show that the crystal size distribution is substantially larger than in the microwave experiments, indicating further nucleation or different diffusion rates in the heterogeneous gel during the crystal growth step. The conclusion cannot be drawn that only additional nucleation occurred in the crystal growth step as the microwave experiment indicated that nucleation mainly takes place in the high-temperature region. As part of the gel is still present after 90 min, "cakes" of crystals are often found at the autoclave wall with conventional heating whereas this is not the case in the microwave heating. Apparently heterogeneous nucleation can be avoided using microwave heating, and it seems that the zeolite crystal growth is mainly dependent on the dissolution of the gel.

Although the history and the temperature profile were the same in both crystal growth step experiments, an essential difference between conventional and microwave heating is the enhancement of the Brownian motion and the rotation dynamics of the water molecules, respectively. In the case of the rotational motion far more hydrogen bridges of water molecules are destroyed (Walker, 1987) resulting in so-called active water molecules (Symons, 1981). The active water molecules have a higher potential compared to the hydrogen-bonded water molecules to dissolve the gel because the lone pairs and OH groups of the active water molecules are available to attack the gel bondings.

Table 33-2. Average crystal size and XRD crystallinity of zeolite NaA (gel composition code I in Table 33-1) prepared by microwave heating as a function of time.

Heating time, min.	Average crystal size, μm	XRD crystallinity, %
1	> 0.5	10
2	0.75	18
3	0.85	20
4	0.95	23
6	1.10	64
8	1.25	81
11	1.25	100

Table 33-3 Average crystal size and XRD crystallinity of zeolite NaA (gel composition code I in Table 33-1) prepared conventionally as a function of time after short microwave heating to 120°C.

Heating time, min.	Average crystal size, μm	XRD crystallinity, %
15	> 0.5	10
30	> 0.5	30
60	0.5	60
120	1	100

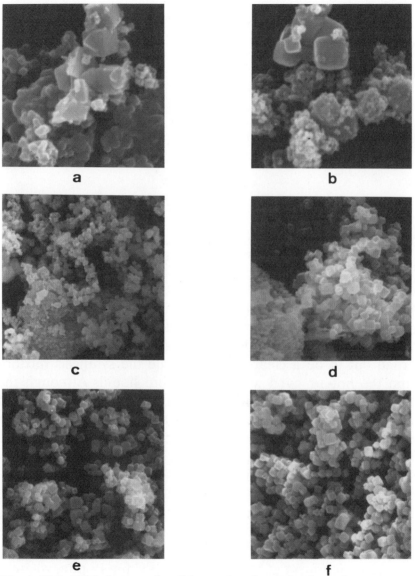

Figure 33-3. SEM photographs of the product formation in a microwave oven
after (a) 2 min (1000 x) (b) 3 min (1000 x) (c) 4 min (2000 x) (d)
6 min (3000 x) (e) 8 min (3000 x) (f) 11 min (3000 x).

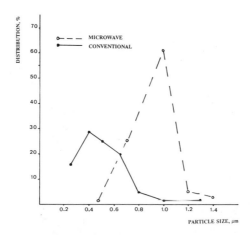

Figure 33-4. Particle size distribution diagrams of zeolite NaA obtained by microwave (120-95°C, 11 min) and by conventional heating (120-95°C, 120 min). (cf. text).

An experiment to support the above idea was carried out with a very diluted but still turbid synthesis mixture (molar ratio **I**). After 15 min the synthesis mixture from the microwave oven turned out to be a clear solution, while the synthesis mixture from the hot-air oven was still a turbid liquid. Although the active water molecules might also affect the dissolution of zeolite crystals formed, it was established that at higher microwave powers the crystallization time decreased under the same conditions. As the zeolite is the more stable phase it can be concluded herein that mainly the gel is dissoluted. Furthermore the interesting observation was made that zeolite template precursors, for example tetraalkylammonium cations and in particular TPA^+, in an aged solution (Arafat et al. to be published, 1992) are not degradated under microwave radiation into the amine form at high pH and 140°C for 15 min, whereas a fresh solution of TPA^+ at high pH undergoes Hofmann degradation in a few minutes even at 75°C. This indicates that the water clathrate formation around the tetrapropylammonium cation is not yet fully developed in a fresh solution and thus the hydroxide anion can attack the TPA$^+$, whereas in an aged system TPA$^+$ is clathrated by water, obstructing the hydroxide from attacking the TPA$^+$. As one of the crystal growth models of zeolites starts with clathrated cations, it looks like these systems are dissipated by microwave conditions. In this respect a relatively low temperature zeolite crystallization experiment with an organic template cation was performed in a microwave oven. The synthesis mixture used is formulated in Table 33-1, type **III**.

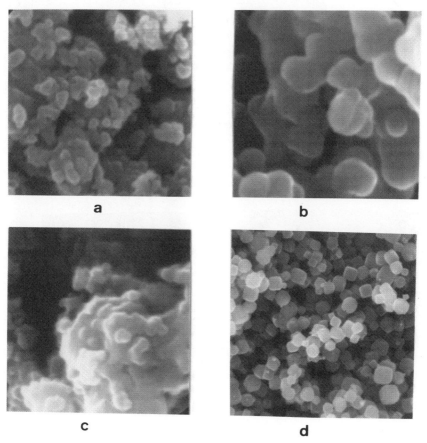

Figure 33-5 SEM photographs of zeolite NaA formation in hot-air oven after (a) 15 min (30000 x) (b) 30 min (30000 x) (c) 60 min (30000 x) (d) 120 min (5400 x).

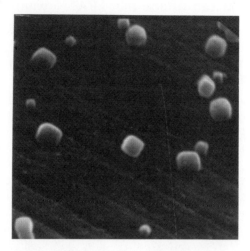

Figure 33-6 SEM photographs of zeolite NaA crystals grown on the surface of a copper plate by microwave heating.

a

b

Figure 33-7 SEM photographs of (a) the cordierite module and (b) the zeolite NaA crystals grown on the inner wall of the cordierite module.

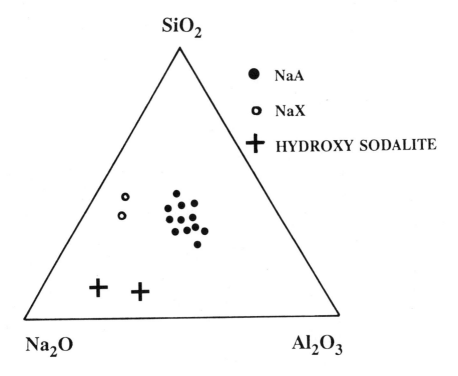

Figure 33-8. Ternary composition diagram of different synthesis mixtures leading to pure zeolite products upon microwave heating (120-95°C) (cf. Table 1).

A completely crystalline product was obtained at 95°C after 15 min identified as hydroxysodalite. TMA^+ was shown to be present in the lattice according to XRD and FTIR. It is therefore concluded that zeolites can be formed by microwave heating with organic templates under certain conditions.

In order to understand the influence of heterogeneous nucleation effects on zeolite synthesis mixtures under microwave irradiation, some experiments were carried out with metal and ceramic supports on which zeolite NaA is formed. Thus an activated Cu metal support platelet was immersed in the synthesis mixture type **I**, Table 1. After fast heating to 120°C followed by 10 min at 95°C in the microwave oven, zeolite NaA crystals, firmly attached to the metal oxide surface but not fully grown, with smoothed edges and corners were observed (see Fig. 33-6). According to recent publications (Ichert et al., 1990) crystals grow more "droplet"-like on a surface when the interaction with the surface is large. When using a ceramic support and under the above described conditions, fully grown cube-type crystals were found on a module of cordierite (see Fig. 33-7 a and b). In conclusion microwaves seem to stimulate mainly the interaction between the growing zeolite crystal formation and a conductor such as copper whereas this is not the case with a nonconductor such as cordierite as shown by the crystal form. To explore somewhat further the scope of microwave zeolite synthesis, a series of formulations within the crystallization field of zeolite NaA were subjected to our standard conditions (large oven). At the same time, synthesis mixtures neighboring the crystallization field of zeolite A were heated using microwave energy. The synthesis mixtures used are given in Table 1, code **II**. As shown in Fig. 33-8 the ternary composition diagram for the different synthesis mixtures shows almost the same feature as the original one (Breck, 1974) using a conventional heating method. It can be concluded that microwave heating of especially zeolite NaA but also of other low Si/Al synthesis mixtures results in same zeolites as found with conventional heating methods.

CONCLUSIONS

Microwave energy can be applied to achieve high heating rates and relatively homogeneous heating to increase a variety of reaction rates. The temperature can be excellently monitored with a gold-plated earth-connected thermocouple (Kingston and Jassie, 1986) resulting in adequate adjustment of required microwave energy in the synthesis system. The fast crystallization of zeolite A with microwave energy compared to conventional heating methods is ascribed to fast dissolution of the gel. Heterogeneous nucleation and crystal growth of zeolite A on microwave energy conductors such as metals leads to a strong interaction between the metal surface and the zeolite crystal in contrast to nonconducting material.Zeolite NaA, NaX, Na hydroxysodalite and TMA$^+$-containing hydroxysodalite were prepared with microwave energy.

Acknowledgment The authors would like to thank from our laboratory G. C. A. Luijkx for constructing the tube microwave oven and his guidance during the preliminary steps of this work and R. de Ruiter for valuable discussions. Also they would like to acknowledge prof. A. R. Ebaid for his encouragement during all the steps of this work.

REFERENCES

- Aiello, R. and Barrer, R.M. 1970. J. Chem. Soc. A: 1470.
- Arafat, A., Jansen, J.C., de Ruiter, R. and van Bekkum, H. To be published.
- Breck, D.W. 1974. Zeolite Molecular Sieves. New York: John Wiley & Sons.
- Chu, P. and Dwyer, F.G. 1990 Eur.Pat. 0 358 827.
- Ichert, L. and Schneider, H.G. 1990. In Advances in Epitaxy and Endotaxy. eds. Schneider, H.G., Ruth, V. and Kormany, T. Amsterdam: Elsevier: P. 53.
- Kingston, H.M. and Jassie, L.B. 1986. Anal. Chem. 58: 2534.
- Symons, M.C.R. 1981. Acc. Chem. res. 14: 179.
- Thompson, B. 1990. Private communication.
- Walker, J. 1987. Sc. Amer. 256: 98

34

Dynamic Features of the Synthesis of Zeolites in a Mixed Alkali Ion System

Prabir K. Dutta and Reza Asiaie <u>The Ohio State University</u>, Department of Chemistry, The Ohio State University, Columbus, OH 43210

Starting with a sodium aluminosilicate composition that results in zeolite Y synthesis, the influence of $[Li^+]$ and $[K^+]$ on nucleation of this system has been examined. With increasing $[K^+]$, zeolite Y is replaced by zeolite D and finally chabazite is formed. The process essentially consists of competitive nucleation between Na^+ and K^+ directed zeolites. In the Li^+ system, it was discovered that at high $[Li^+]$, zeolite Z (Li,Na-E) is formed. However, at intermediate $[Li^+]$, the system remains amorphous for extended periods of time. Raman spectra of the Na^+, Li^+ exchanged forms of both zeolite Z and X are different, indicating that the framework structure is sensitive to the cation. It is prpoposed that Na^+ and Li^+ by distorting the framework of zeolites Z and X, respectively, destabilizes the nuclei and thereby disrupts the crystal growth process. The influence of various solvents, complexing agents, and other cations on the Na^+-Li^+ system has also been examined.

522

Introduction

Zeolite synthesis is a dynamic process, with various polymerization, depolymerization, and nucleation events occurring during crystal growth.[1] The complexity of the gel chemistry allows for the formation of a variety of frameworks by changing the starting composition, nature of reactants, and effects such as stirring and temperature.[2] The ways in which these variables influence the crystallization pathway has been an area of research for many decades. However, the level of understanding of these phenomena at a molecular level is still quite primitive.

The important role that a co-cation can play in influencing zeolite synthesis has been recognized since the early work of Barrer and co-workers.[3] For example, in the Na^+-Li^+ system, ZSM-3 crystallizes, whereas in the Na^+-K^+ a chabazite type or an offretite-erionite type of crystal can be formed.[4] The crystallization fields of Li_2O: Na_2O: Al_2O_3: SiO_2: H_2O, and the corresponding K_2O system have been mapped.[5,6] The roles of SiO_2/Al_2O_3, OH^-/Al_2O_3, and temperature have been found to be very critical in influencing the formation of different zeolitic species.

In this study, we focus on a narrow aspect of zeolite synthesis in a mixed cation system. Keeping the composition fixed (i.e., concentrations of total monovalent cations, hydroxide, silicate and aluminate) for a zeolite Y synthesis, we have explored the role of K^+ and Li^+ in this system. Co-cations have been thought to be primarily involved in establishing the nucleation of potentially competing species, and this concept has been explored. A particular Li-Na aluminosilicate system that exhibits interesting nucleation behavior has been used as starting material for zeolite synthesis in the presence of solvents, other cations, and complexing agents.

Experimental

NaY (Si/Al = 2.6) and NaX (13X, Si/Al = 1.3) zeolites were obtained from Union Carbide Co.. The ion-exchanged samples were made by stirring the zeolites in 1M solution of the corresponding chloride solution for 24 hours. Dimethyl sulfoxide (DMSO), hexamethylphosphoramide (HMPA), TritonX-100, 18-Crown-6, cesium chloride, tetramethylammonium chloride (TMACl), and strontium chloride were purchased from Aldrich Chemical Company. Barium chloride and potassium chloride were provided by MCB Reagents and Allied Chemicals, respectively. All reagents were used as received. All experiments were performed in 250 ml capacity Teflon bottles.

In the synthesis experiments, the aluminum and silicon sources were sodium aluminate (Chem Services) and Ludox (DuPont). Gels were formed by the addition of appropriate amounts of Ludox to aqueous solutions composed of NaOH and Na_2AlO_4. Exact compositions are presented in the next section.The mixtures were aged for 12 hours at room temperature, mixed with the corresponding amounts of lithium or potassium hydroxides and then placed in a 95°C oven. The perturbation experiments were performed on the composition corresponding to r_{Li+} = 0.25. The perturbing cations (in this case as chloride salts), cosolvents, and complexing agents were added to the reaction mixtures after 48 hours of heating. At this time no crystals were yet formed and the reaction mixture was susceptible to influence. The solid samples after various times of heating were recovered by filtration or centrifugation and the crystallinity examined by diffraction measurements. The amorphous gels used for the Raman spectroscopic study were recovered from the reactor and examined without any pretreatment. The crystals for the Raman experiments were extensively washed with water.

Powder X-ray diffraction patterns were obtained on a Rigaku Geigerflex D/Max-2b diffractometer with nickel

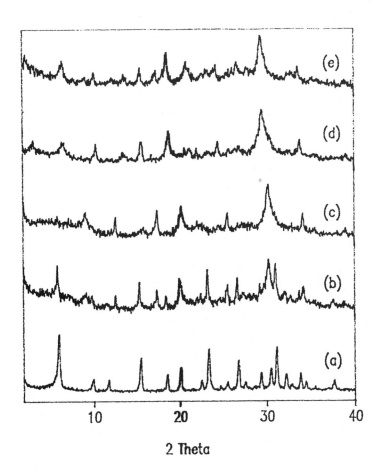

2 Theta

Fig 34-1. X-ray diffraction powder patterns of the influence of K^+ on $13(K^+ + Na^+)_2O$: Al_2O_3: $28\ SiO_2$: $502\ H_2O$. The mole fractions of K^+ $(K^+/(K^+ + Na^+))$ correspond to (a) 0.05 (b) 0.07 (c) 0.1 (d) 0.15 and (e) 0.3. (All patterns recorded after 60 hours of heating).

filtered CuK_α (1.5405 Å) source. The Raman spectra were obtained using a Spectra-Physics Model 171 Argon laser using 514.5 nm excitation. A Spex 1403 double monochromator and a RCA C131034 GaAs photon counting photomultiplier tube was used to filter and detect the scattered light. Slit widths typically of ~6 cm^{-1} and laser power of ~100 mW were used for all samples. The data was collected with a Spex Datamate computer. Spectra Calc programs were used to correct for any sloping baselines. The surface area was measured with a Micromeritics Pulse-Chemisorb 2700. The sample was dehydrated at 400°C for four hours prior to the analysis.

Results and Discussion

The reactant composition used in this study corresponds to $xNa_2O:yM_2O:Al_2O_3:28\ SiO_2:502H_2O$, where $M^+ = Li^+, K^+$, and $x + y = 13$. In the absence of M^+, this composition after aging for 24 hours leads to the formation of zeolite Y (Si/Al = 2.5). The role of M^+ in influencing the nucleation of this system has been explored for varying M^+, but with $[M^+ + Na^+]$ being held constant. Also, prior to addition of MOH, all samples were aged with the Na^+ system for 12 hours in order to examine the influence of added M^+ on zeolite Y nucleation.

K^+-Na^+ system

Figure 34-1 shows the XRD patterns obtained at various r_{K^+} (moles of K^+/moles of $(K^+ + Na^+)$) values after 50-60 hours of heating at 95°C. Zeolite Y crystallizes at $r_{K^+} < 0.05$ and chabazite for $r_{K^+} > 0.15$. In the intermediate range a structure related to chabazite is formed, and referred to in the literature as zeolite D.[7] Though the powder patterns are distinct between these two forms of chabazite, the Raman spectra are similar, as shown in Figure 34-2. In the case of (Na,K)F and edingtonite, unit cell size has been shown to vary with cation fraction of Na^+.[8] A similar effect is occurring here since the intermediate zeolite D can incorporate both Na^+ and K^+. At the two extreme r_K+ values, Na^+ helps nucleate zeolite Y and K^+ does so for chabazite. At the intermediate r_K+ values, both zeolite Y and zeolite D crystals form and the amounts of each structure is controlled by kinetic factors affecting nucleation. Zeolite Y disappears between r_K+ of 0.07 and 0.1. The overall process indicates a competing nucleation process, in which Na^+ favors zeolite Y nucleation and K^+ does so for chabazite. We have proposed in an earlier study that both these frameworks can be built from similar, yet unique building blocks, with the electrostatics of the cation-water complex controlling the pathway.[10a]

Li$^+$-Na$^+$ system

The comparable results for the Li^+ system were quite different, especially in the intermediate $r_{Li}+$ range. These data are shown in Figure 34-3 for samples heated for 60 hours. Zeolite Y (Figure 34-3a) is formed within 60 hours for $r_{Li}+ \leq$ 0.2. Within the same time frame, a zeolite resembling Li A(BW) and similar to (Li, Na)-E as reported by Borer and Meier and zeolite Z as reported by Barrer is formed for $r_{Li}+$ $\geq$ 0.3 [6-9] (Figure 34-3c). The structure of this material is unknown. We will refer to it as zeolite Z. The surprising result is that within the range of $r_{Li}+$ = 0.2 - 0.3, the gel remains amorphous for extended periods, and the time before crystals appear gradually increases as $r_{Li}+$ does. At $r_{Li}+$ = 0.25, the gel is amorphous for 180 hours followed by the formation of zeolite Z (Figure 3b). At $r_{Li}+$ of 0.215 and 0.235, zeolite Y is formed after 100 and 120 hours, followed by zeolite Z at 140 and 160 hours, respectively. Thus, it appears that as the fraction of Li^+ increases, zeolite Y nucleation is disrupted but zeolite Z also does not form readily until $r_{Li}+$ exceeds 0.3. Therefore, both Li^+ and Na^+ cannot successfully nucleate zeolite Z or zeolite Y in these $r_{Li}+$ ranges. For $r_{Li}+$ = 0.25, this process continues for 180 hours. If the

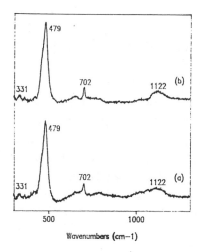

Fig 34-2. Raman spectra of (a) zeolite D and (b) chabazite.

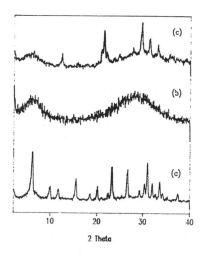

Fig 34-3. X-ray diffraction powder patterns of the influence of Li^+ on $13(Li^+ + Na^+)_2O$: Al_2O_3: $28\ SiO_2$: $502\ H_2O$. The mole fraction of Li^+ corresponds to (a) 0.2 (b) 0.25 (c) 0.3. (Patterns recorded after 60 hours of heating.)

phenomenon controlling this process was competitive nucleation as in the K^+ - Na^+ case discussed earlier, then it is expected that crystals of either zeolite Y or Z should be formed. Rather, it appears that Li^+ and Na^+, besides nucleating zeolite Z and zeolite Y respectively, are also interfering with the nucleation of each other's systems, thus leaving the system amorphous. In order to examine this process, various spectroscopic studies were carried out. The Raman spectra of the gel phase after heating for 45 hours for $r_{Li}+$ = 0.15, 0.25, and 0.3, are shown in Figure 34-4. The gel structure in all three cases is very similar, though for each of these compositions, the gel evolves to a different species, zeolite Y for $r_{Li}+$ = 0.15, zeolite Z for $r_{Li}+$ = 0.3 and amorphous for $r_{Li}+$ = 0.25. In all cases, the gel is characterized by dimeric silicate solution species (~ 600 cm^{-1}) trapped in a solid phase made up of disordered four-membered aluminosilicate rings ($\sim$ 495 cm^{-1}) that is also considerably depolymerized (Si-O$^-$ at 1040 cm^{-1}).[10,11,12] Therefore, the systems are all evolving from a similar global aluminosilicate structure.

A more subtle effect would be the possibility that Na^+ destroys the nuclei of zeolite Z and Li^+ destroys that of zeolite Y. This destabilization of the zeolite nuclei could arise from structural changes that occur upon interaction of the cation with the aluminsolicate framework, primarily due to electrostatic effects. Since it is not possible to isolate the zeolite nuclei, we examined the influence of Na^+ and Li^+ on fully formed zeolite Z and Y, respectively. The framework changes were monitored by Raman spectroscopy for completely exchanged Na^+ and Li^+ samples of both zeolite Z and Y, and are shown in Figures 34-5 and 34-6, respectively. In the case of zeolite Z, there are significant structural difference between the Na^+ and Li^+ exchanged forms, as evidenced by the splitting of the bands at 420, 484, and 620 cm^{-1} and the shift of the Si-O stretching mode at 1080 cm^{-1}. Since the exact assignment of the lower frequency bands is unknown, it is not

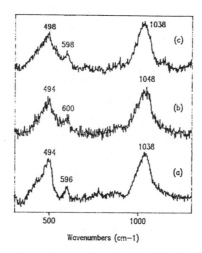

Fig 34-4.   Raman spectra of gel removed after 48 hours of heating from the composition $13(Li^+ + Na^+)_2O$: Al_2O_3: 28 SiO_2: 502 H_2O at Li^+ mole fractions of (a) 0.15 (b) 0.25 and (c) 0.3.

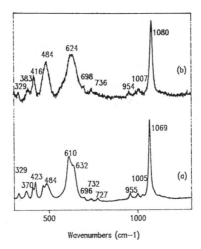

Fig 34-5.   Raman spectra of (a) Li^+ form and (b) Na^+ form of zeolite Z.

possible to deduce the exact nature of the framework difference between the Na^+ and Li^+ ion-exchanged forms. However, it is clear that this difference could lead to the destabilization of zeolite Z nuclei at the lower r_{Li^+} values, due to the higher fraction of Na^+ associated with these nuclei.

From Figure 34-6, we find that the Raman spectra of Na^+ and Li^+ exchanged zeolite Y are identical, thus indicating that Li^+ has no influence on the zeolite Y framework, and should have no influence on its nucleation, contrary to our experimental observation. The mechanism of formation of zeolite Y has been examined and the possibility that zeolite Y actually nucleates from lower Si/Al ratio zeolite X has been proposed.[13,14]   If this is indeed the case, then a more appropriate comparison would be between the Na^+ and Li^+ exchanged form of zeolite X. The Raman spectra of this system are shown in Figure 34-7. Unlike the case of zeolite Y, major differences are observed, including the shift of the ν_s(T-O-T) band from 510 to 516 cm^{-1} in the Li^+-X and the presence of four bands at 953, 995, 1028, and 1084 cm^{-1} in Li^+-X as compared to 1003 and 1085 cm^{-1} in Na^+-X. Also the 368 cm^{-1} band is shifted to 385 cm^{-1} upon Li^+ introduction. Based on our earlier studies of Raman spectra of faujasitic zeolites, the shift of the ν_s(T-O-T) band shows that Li^+ actually brings about structural changes that decrease the average T-O-T angle.[15]   Also, the presence of four bands in the Si-O stretching region indicates that Li^+ has decoupled the SiOSi vibrations.[15] Thus, it appears reasonable to propose that for intermediate ranges of r_{Li^+} (0.2 - 0.3), Li^+ and Na^+ disrupt the nucleation of the competing zeolite framework, while at the same time nucleating zeolites on their own. At an optimum concentration level, r_{Li^+} = 0.25, this leads to the presence of an amorphous phase for 180 hours.

In the second part of this study, we have explored the influence of reagents that bind to Li^+ and/or Na^+ for the composition corresponding to r_{Li^+} = 0.25. All materials were added after the reactant composition was heated for 48 hours,

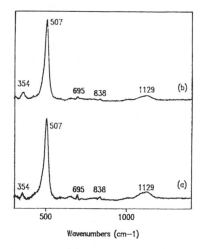

Fig 34-6. Raman spectra of (a) Li$^+$ form and (b) Na$^+$ form of zeolite Y.

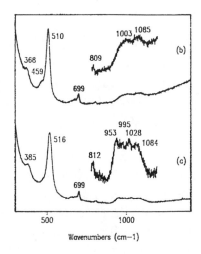

Fig 34-7. Raman spectra of (a) Li$^+$ form and (b) Na$^+$ form of zeolite X.

thus allowing the competitive nucleation to occur. Solvents such as dimethylsulfoxide and hexamethylphosphoramide added at mole fractions corresponding to seven percent of the water content led to rapid crystallization of zeolite Z within 60 hours (Figure 34-8a). In the case of DMSO, NMR studies based on ^{7}Li and ^{23}Na nuclei have shown that DMSO exhibits a preferential solvation for Na$^+$, whereas no particular preference is exhibited for Li$^+$.[16] Also it is known that HMPA binds to Na$^+$ more strongly than DMSO, based on NMR, IR and conductance measurements.[17] The propensity for binding of HMPA to Li$^+$ has not been reported. Therefore, it appears that solvents that tend to bind more strongly with Na$^+$ favor the nucleation and growth of zeolite Z, supporting the proposed hypothesis on competitive inhibition.

However, the process is probably not as simple, since addition of polyethers such as 18-crown-6 and Triton X (15 mole percent of Na$^+$ content) which also show a higher binding constant for Na$^+$ than Li$^+$ results in the rapid formation of zeolite Y (Figure 8b).[18,19] Na$^+$-crown ether complexes have been shown to be effective templates for the formation of cubic and hexagonal forms of faujasite.[20] Thus, in the case of polyethers, binding to Na$^+$ may be producing a more effective structure directing agent that preferentially nucleates zeolite Y.

Monovalent cations K$^+$, Cs$^+$ and tetramethylammonium ion (TMA$^+$) added to the $r_{Li}+$ = 0.25 composition at levels approaching 40-50 mole percent of the Na$^+$ content lead to the formation of zeolite L, pollucite and mazzite respectively, (Figure 34-8c, d, and e). These results are not surprising considering that these zeolites are typically formed in the presence of K$^+$, Cs$^+$ and TMA$^+$ ions. However, there is a difference in the nucleation process if these cations are added at the beginning of the reaction as opposed to the heated gel for 48 hours. Mazzite, zeolite L and pollucite are still formed with these cations present initially in the reactant composition, but they are unstable and disappear, leaving behind zeolite Z.

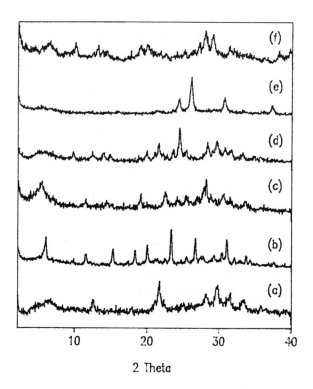

Fig 34-8. X-ray diffraction powder patterns of materials obtained by treatment of the composition $13(0.25 \, Li_2O + 0.75 \, Na_2O): Al_2O_3: 28 \, SiO_2: 502 \, H_2O$ with (a) DMSO - (zeolite Z); (b) 18-crown-6-(zeolite Y); (c) K^+ - (zeolite L); (d) TMA^+ - (mazzite) (e) Cs^+ - (pollucite) and (f) Ba^{+2} - (unidentified). Details of the amounts and procedures can be found in the text. The zeolites formed are indicated in parentheses.

Therefore, the competition between Li^+ and Na^+ in disrupting each other's zeolite nuclei helps in stabilization of zeolite L, mazzite, and pollucite.

Divalent cations Sr^{+2} and Ba^{+2} at the level of 15 mole percent to that of Na^+ also redirect the synthesis to a different product, as shown in Figure 34-8f. We have been unable to relate this pattern to a known zeolite. However, the structure is thermally unstable beyond 300°C and collapses to a dense structure of surface area of 8 m^2/g and resembles the feldspathic species reported in the literature.[21]

In conclusion, there are several features that we have addressed in this study. The most important is that in the presence of co-cations, not only can nucleation to a different zeolite occur, but that cations are able to destroy the nuclei of co-crystallizing zeolites. Thus in the case of Li^+-Na^+ composition, the material remains amorphous for an extended period of time. We have used this composition as a starting material and found that it can be redirected to zeolitic species by the addition of solvents, complexing agents and other cations.

References

1. Szostak, R.M., Molecular Sieves, Principles of Synthesis and Identification, Van Nostrand, Reinhold, NY, 1989.
2. Breck, D.W., Zeolite Molecular Sieves, Wiley, NY, 1976.
3. Barrer, R.M., Hydrothermal Chemistry of Zeolites, Academic Press, Ny. 1982.
4. Colella, C.; Gennaro, M., Ann. di Chimica. 1986, 76, 115.
5. Pereyron, A.; Guth, J.L.; Wey, R., Comp. Rend. 1971, 272, 1331.
6. Borer, H.; Meier, W.M., Advances in Chem. Ser. 1971, 101, 122.
7. Breck, D.W.; Acara, N.A., U.S. Patent 3011869, 1961.
8. Barrer, R.M.; Mainwaring, D.E., J. Chem. Soc. Dalton, 1972, 2534.

9. Barrer, R.M.; Baynham, J.W., J. Chem. Soc. 1956, 2882

10a.Dutta, P.K.; Puri, M.; Shieh, D.C., Materials Research Society Publication, 1988, 111, 101.

10b.Dutta, P.K.; Shieh, D.C.; Puri, M., J. Phys. Chem. 1987, 91, 2332.

11. Dutta, P.K.; Shieh, D.C., Appl. Spect. 1985, 39, 343.

12. Dutta, P.K.; Shieh, D.C., J. Phys. Chem. 1986, 90, 2331.

13. Robson, H., ACS Symp. Ser. 1989, 398, 436.

14. Dutta, P.K.; Twu, J.; Kresge, C.T., Zeolites, 1991, 11, 672.

15. Dutta, P.K.; Twu, J., J. Phys. Chem. 1991, 95, 2498.

16. Johnson, D.A., J. Chem. Soc. Dalton, 1974, 1671.

17. Vasilev, V.P.; Kunin, B.T.; Russ J., Inorg. Chem. 1972, 17, 1129.

18. Frensdorff, H.K., J. Am. Chem. Soc. 1971, 93, 600.

19. Martell, A.; Smith, R.M., Critical Stability Constants. Plenium Press., NY, 1974.

20. Delprato, F.; Delmotte, L.; Guth, J.L.; Huve, L., Zeolites, 1990, 10, 546.

21. Barrer, R.M.; Marshall, D.J., J. Chem. Soc., 1964, 485.

35

EXAFS Studies of Iron-Substituted Zeolites with the ZSM-5 Structure

Sean A. Axon, Katharine K. Fox, Stuart W. Carr, and Jacek Klinowski

Department of Chemistry, University of Cambridge, Lensfield Road, Cambridge CB2 1EW, U.K.
Unilever Research, Port Sunlight Laboratory, Quarry Road East, Bebington, Wirral, Merseyside, L63 3JW, U.K.

Extended X-ray absorption fine structure (EXAFS) used in combination with Mössbauer spectroscopy and X-ray diffraction confirms the substitution of Si by Fe in the framework of zeolite ZSM-5 prepared via the "fluoride route." Each framework Fe atom is coordinated to four oxygens with an Fe-O distance of 1.83 Å, and to four Si atoms in the second coordination shell: two at 3.2 Å and two at 3.38 Å from Fe, compared with an average Si-Si distance in silicalite of 3.1 Å. Calcination causes ca. 40% of the iron to be expelled from the zeolite framework to form a mixture of amorphous Fe oxides and oxyhydroxides in the intracrystalline channel system.

INTRODUCTION

Zeolite ZSM-5[1] is used in many important catalytic processes and has been the subject of intense research. With the Si/Al ratio of above ca. 20, the zeolite is highly siliceous and has two intersecting intracrystalline channel systems: channels 5.6 × 5.3 Å in diameter run in the [010] direction, and 5.5 × 5.1 Å channels in the [100] direction.[2] ZSM-5 is synthesized in the presence of a structure-directing template, typically tetrapropylammonium bromide (TPABr). The catalytically active H^+ form of the zeolite is produced by calcining the as-prepared TPA^+/NH_4^+ cationic form at > 550°C. The purely siliceous end member of the ZSM-5 substitutional series is known as silicalite. Because the catalytic activity and selectivity of zeolites are intimately linked to their elemental composition, there has been considerable interest in introducing heteroatoms such as Fe, B, Ti and Ga into the zeolite framework[3] either directly or by a range of post-synthesis modification methods.

The preparation of zeolite [Si,Fe]-ZSM-5 has been reported[4-6] and several techniques used to confirm the presence of Fe in the framework.[7,8] It is clear that, since for steric reasons the charge-balancing TPA^+ cations can be located only at channel intersections, Fe atoms which make the zeolitic framework electrically negative must also be located at such intersections. Apart from our earlier work[9,10] on [Ga,Si]-zeolites, Extended X-ray Absorption Fine Structure (EXAFS) had not been used to study zeolite *frameworks,* although it had been successful in probing the location and coordination of extra-framework cations and clusters in zeolites.[11-13] Other methods have been employed instead, such as [57]Fe Mössbauer spectroscopy, which is sensitive to the coordination of Fe.[14] Mössbauer results are interpreted by comparison with the spectra of model compounds, and the isomer shift has been used to distinguish between 4- and 6-coordinated Fe.[15] We wish to report that EXAFS can readily establish the isomorphous substitution of Si by Fe in the zeolite framework and monitor the fate of the iron upon thermal treatment of the sample.

The physical basis of EXAFS is the modulation in X-ray absorbance beyond the absorption edge (e.g. K, L$_I$, etc.) of a particular element caused by the back-scattering, by the neighboring atoms, of the outgoing photoelectron wave. The modulation contains information about the number, nature and distance of the neighboring atoms from the absorber species.[16,17] The advantages of EXAFS are that it is element-specific, does not require special sample preparation and allows rapid acquisition of data. EXAFS is extremely sensitive to absorber-nearest neighbor bond distances, while the coordination number of the absorbing element can be inferred by comparison with well-characterized model compounds.

We shall demonstrate that EXAFS can discriminate between 4-coordinated (framework) and 6-coordinated (extra-framework) Fe in silicalite. A comparison with model compounds and an examination of pre-edge features in the EXAFS spectra help to explain the structural changes induced by the thermal treatment of the sample.

EXPERIMENTAL

Zeolite [Si,Fe]-ZSM-5 was synthesized under non-alkaline conditions using the fluoride method.[18] The chemical composition, crystallinity and phase purity of the product were determined by atomic absorption spectroscopy and powder X-ray diffraction.

[Si,Fe]-ZSM-5 was prepared from Ludox AS-40 silica sol, tetrapropylammonium bromide (TPABr), NH_4F, $Fe(NO_3)_3 \cdot 9H_2O$ and distilled water. The mixture of molar composition $SiO_2 : 0.02\ Fe_2O_3 : NH_4F : 0.204\ TPABr : 36\ H_2O$ was heated at 175°C for 7 days in a Teflon-lined autoclave after which it was cooled, filtered, washed with deionized water and dried at 90°C for 18 h. The calcined sample was prepared by heating the as-prepared zeolite at 550°C for 18 h in a layer 2 mm deep.

X-ray powder diffraction patterns were acquired on a Philips PW1050 diffractometer fitted with an Anton-Parr high-temperature attachment and a vertical goniometer using Cu Kα radiation ($\lambda = 1.5418$ Å) selected by a graphite monochromator in the diffracted beam. The sample was placed on the heating element (a strip of platinum metal) and heated to 900°C at a rate of 5°C min^{-1} in a static atmosphere. Diffraction patterns from 3 to 60° 2θ were recorded at 20, 300, 400, 500, 600, 900°C and again at 20°C after completion of the high-temperature treatment.

The chemical composition of the product was determined by atomic absorption using a Perkin-Elmer 2380 spectrophotometer. Before measurement the sample was dissolved in 40% HF and the solution suitably diluted with deionized water.

^{29}Si MAS NMR spectra were recorded at 79.5 MHz on a Bruker MSL-400 multinuclear NMR spectrometer. A typical radiofrequency pulse length was 5 μs with the recycle delay 30 s. Samples were spun at ca. 3 kHz in zirconia rotors using air as the driving gas.

EXAFS measurements were carried out at the SERC synchrotron radiation source at Daresbury using Station 7.1 with an electron energy of 1.998 GeV and ring current of 240 mA. Data were collected with a Si(111) double monochromator at the Fe K-edge (7.112 keV) at room temperature in transmission mode with 50% harmonic rejection. The results were analyzed using the Daresbury database programs EXCALIB (to convert motor step units into electronvolts, add spectra and remove erroneous data points), EXBACK (to subtract the spectral background) and EXCURV90[19] (to generate structural models, compare them with the experimental results and Fourier transform the data) on a CONVEX computer. Phase shifts calculated using EXCURV90 were modified slightly, so that Fe-O distances agreed to ±0.005 Å with published X-ray diffraction values for FePO$_4$.[20] These data were converted to radial distances using the CRAD subroutine of the Chemical Database Service. Pre-edge data were taken from EXBACK to the general plotting program PLOTEK where, following normalization of the spectra, the pre-edge peak areas were determined under identical magnification conditions.

Mössbauer spectra were measured using a standard constant acceleration spectrometer with a ca. 30 mCi ^{57}Co/Rh source. The spectra were recorded at room temperature in zero field and calibrated using a standard iron foil. All isomer shifts are given relative to α-Fe at 20°C. Calculations to determine the line fit and isomer shift (band positions) were performed using the MOSSJOB program (a standard least squares fitting routine) on the IBM 3081 mainframe computer.

RESULTS AND DISCUSSION

The powder X-ray diffraction pattern [Fig. 35-1(a)] shows that the as-prepared product is highly crystalline, orthorhombic, and free from impurities. *In situ* XRD patterns of thermally treated samples indicate that structural changes have occurred. Thus below 600°C, during the thermal decomposition of the template, there is a slight loss of crystallinity which we attribute to the removal of Fe from the zeolite framework and the consequent generation of structural defects. The intensity of the peaks at ca. 7.9 (101/011 reflections) and 9.0° (200/020 reflections) 2θ increases. These reflections are known to be sensitive to the presence and nature of the cations and the template,[21] and calcination has converted TPA$^+$-[Si,Fe]-ZSM-5 to H$^+$-[Si,Fe]-ZSM-5.

The crystallinity of the sample heated to 900°C is lower still, but there is no evidence of the formation of a crystalline oxide such as α-Fe$_2$O$_3$. The XRD pattern of thermally treated sample recorded at 20°C shows a ZSM-5 structure with lower crystallinity and orthorhombic structure with the characteristic 501/051 peak doublet at ca. 23° 2θ. The temperature at which the structure of ZSM-5 undergoes a transition from monoclinic to orthorhombic symmetry decreases with increased amounts of framework aluminium or other adsorbed species.[22] Thus the fact that the structure is orthorhombic at room temperature indicates that although *some* of the Fe atoms may have been ejected from the framework, the majority are unaffected. The chemical composition of the as-prepared [Si,Fe]-ZSM-5 is shown in Tab. 35-1. The mole ratio Si/Fe = 20.88, corresponds to the unit cell formula Si$_{91.61}$Fe$_{4.39}$O$_{192}$.

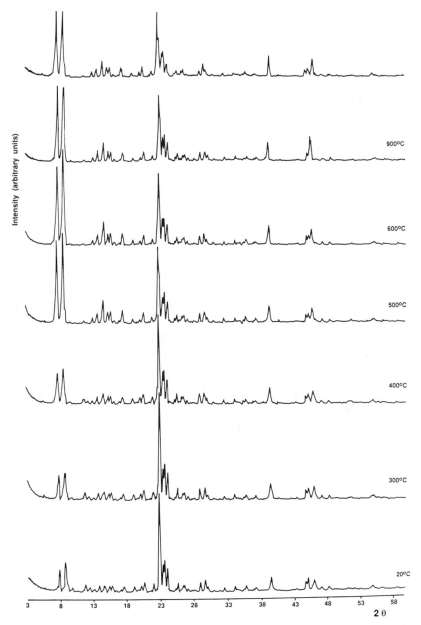

Fig. 35-1. Powder X-ray diffraction patterns of [Si,Fe]-ZSM-5 taken at 20, 300, 400, 500, 600, 900°C and at 20°C in a sample which underwent the high-temperature treatment.

Tab. 35-1.   Chemical composition of the sample (in weight %).

Species	Content
Si	41.8
Fe	3.98
Na	< 0.02
Al	< 0.02
H_2O	0.36
Template	9.75*

* corresponding to 0.96 occluded TPA^+ cations per channel intersection.

The ^{29}Si MAS NMR spectrum of the as-prepared [Si,Fe]-ZSM-5 [Fig. 35-2(a)] consists of signals at ca. -114 ppm and ca. -117 ppm, both corresponding to $Si[OSi]_4$ structural units, and a shoulder at ca. -110 ppm corresponding to $Si[OFe][OSi]_3$ units. The spectrum is similar to that of a conventional [Si,Al]-ZSM-5 with Si/Al = 25.[23]   Fig. 35-2(b) shows that the intensity of the shoulder at -110 ppm decreases upon calcination, indicating that some of the Fe has been removed from the framework. This agrees with the deterioration of crystallinity with increasing temperature observed by XRD. The ^{29}Si MAS NMR spectra of the as-prepared and calcined samples are not broadened by paramagnetic Fe^{3+} species.

While the Fe pre-K-edge region of an X-ray absorption spectrum (Fig. 35-3) cannot be modeled quantitatively, it can provide a qualitative indication of symmetry, absorber site geometry and electronic configuration.[24] Both samples of [Si,Fe]-ZSM-5 show a small pre-edge peak, due to forbidden transitions between 1s and 3d orbitals. This feature increases in intensity with the asymmetry of the electronic distribution,[25] and is more pronounced in 4-coordinated than in 6-coordinated Fe^{3+} species (Fig. 35-3 and Tab. 35-2).

It is clear that pre-edge absorption of as-prepared [Si,Fe]-ZSM-5 is similar to that of 4-coordinated Fe^{3+} in $FePO_4$, indicating that Fe in the zeolite is 4-coordinated, i.e. incorporated in the zeolitic framework. Calcination reduces the relative area of the pre-edge peak, indicating a change of coordination brought about by the expulsion of some Fe from the framework. If Fe in the extra-framework species is 6-coordinated (see below), then the pre-edge peak areas given in Tab. 35-2 are consistent with the calcined [Si,Fe]-ZSM-5 sample in which ca. 60% of Fe^{3+} is in the framework, the remainder occupying extra-framework sites.

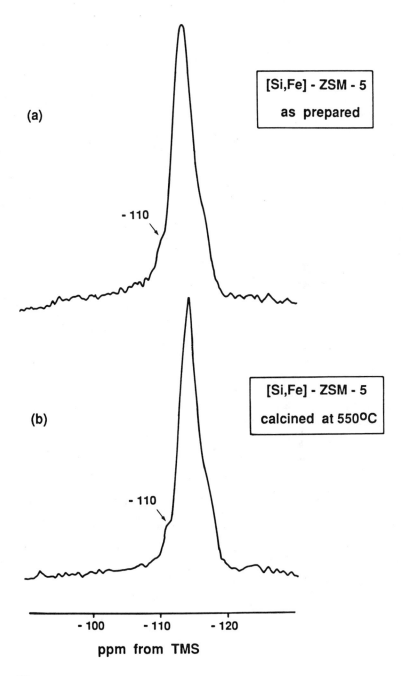

Fig. 35-2. ^{29}Si MAS NMR spectra of (a) as prepared (b) calcined [Si,Fe]-ZSM-5.

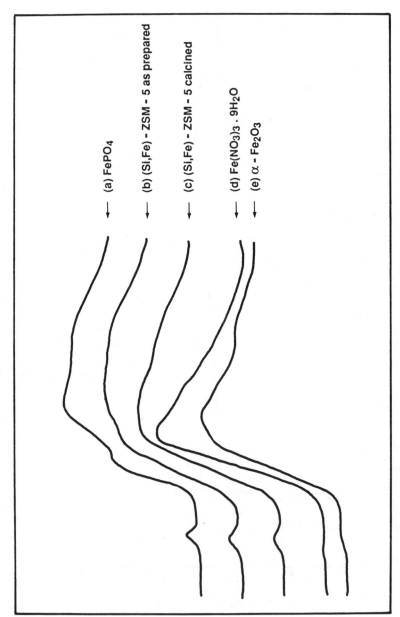

Fig. 35-3. Pre-edge region of the EXAFS spectrum for (a) $FePO_4$; (b) as-prepared [Si,Fe]-ZSM-5; (c) calcined [Si,Fe]-ZSM-5; (d) $Fe(NO_3)_3 \cdot 9H_2O$; and (e) α-Fe_2O_3.

Tab. 35-2.   Pre-edge   peak   areas   for   several Fe^{3+} compounds.

Sample	Coordination of Fe	Pre-edge area	Reference
$FePO_4$	4	0.26	25
[Si,Fe]-ZSM-5 as prepared	4	0.27	this work
[Si,Fe]-ZSM-5 calcined	4 & 6	0.19	this work
$Fe(NO_3)_3 \cdot 9H_2O$	6	0.048	this work
α-Fe_2O_3	6	0.051	25

Fig. 35-4 compares the EXAFS results and their Fourier transforms for the as-prepared and calcined samples. The major difference is that the height of the most intense peak in the Fourier transform for the calcined sample is reduced compared with the as-prepared sample. Subsequent spectral analysis was performed on k^3 weighted data sets using the EXCURV90[19] programs after standard calibration (EXCALIB) and background subtraction (EXBACK) procedures.

EXAFS fit for as-prepared [Si,Fe]-ZSM-5

Fig. 35-5 shows the experimental and calculated EXAFS spectrum for the as prepared [Si,Fe]-ZSM-5 and its Fourier transform using a 2-shell model, using the parameters listed in Tab. 35-3. The "goodness of fit" parameter R is defined as

$$R_{EXAFS} = \frac{\int (\chi_{exp} - \chi_{theor}) k^3 \, dk}{\int \chi_{exp} k^3 \, dk} \times 100\%$$

where χ is the amplitude of EXAFS absorption. The Debye-Waller term is defined in EXCURV90 as $2\sigma^2$, where σ is twice the mean square amplitude of vibration multiplied by a correlation term for an ordered monoatomic solid. The results are consistent with a structural model involving 4-coordinated framework Fe atom bonded to four O atoms with an Fe-O bond distance of 1.83 Å. This is larger than the Si-O bond distance in silicalite/ZSM-5 (1.56 - 1.63 Å),[26] as expected if one considers the relative Pauling ionic radia: 0.42 Å for Si^{4+} and 0.64 Å for Fe^{3+}. Of four Si atoms in the second-coordination shell two correspond to Fe-Si distance of 3.2 Å and two to 3.38 Å, compared with an average Si-Si distance in silicalite of 3.1 Å.[27] These distances are fully consistent with Fe substitution in the framework.

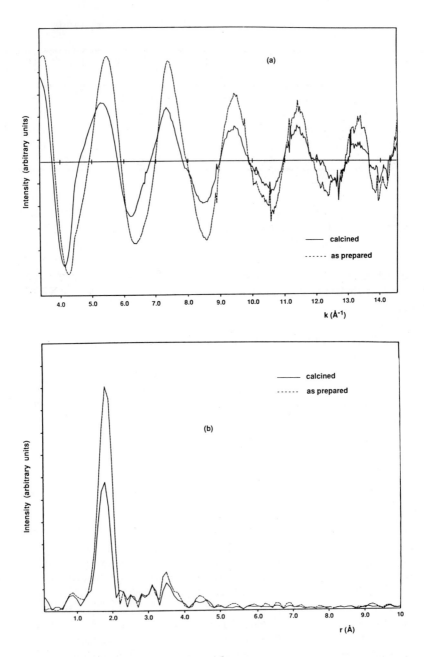

Figure. 35-4. (a) EXAFS data and (b) Fourier transforms for as-prepared (broken line) and calcined (solid line) [Si,Fe]-ZSM-5.

Tab. 35-3. Parameters used to fit EXAFS data for as-prepared [Si,Fe]-ZSM-5 using a 2-shell model.   R = 15.36.

Element	No. of atoms $\pm 20\%$[26]	Distance from Fe ($\text{Å} \pm 0.02$)	Debye-Waller term (Å^2)
O	4.5	1.83	0.006
Si	1.9	3.16	0.014
Si	2.0	3.38	0.008

EXAFS fit for calcined [Si,Fe]-ZSM-5

The *calcined* sample of [Si,Fe]-ZSM-5 gives a smaller pre-edge feature in the absorption spectrum (Fig. 35-3) and a reduced major peak in the EXAFS Fourier transform [Fig. 35-4(b)] compared with the as-prepared sample. If we attribute these effects to thermal expulsion of 40% of the 4-coordinated framework Fe^{3+} into octahedral extra-framework sites, we derive the model described in Tab. 35-4. The experimental and calculated EXAFS spectra and their Fourier transforms for the calcined sample are shown in Fig. 35-6.

The larger Debye-Waller term for the oxygen coordinated to the extra-framework iron (40% of 6-coordinate Fe with Fe-O distance of 1.96 Å) indicates a larger static distribution of Fe-O distances for this species, consistent with an amorphous iron oxide or oxyhydroxide. Many unsuccessful attempts were made to fit the spectra using the parameters given in Tab. 35-3 for the framework species. The extra-framework Fe was modeled as various iron oxides and oxyhydroxides.[28] A satisfactory EXAFS fit could not be found using models involving Fe_2O_3, Fe_3O_4 or α-FeO(OH). Better fits were obtained using a second shell of Fe atoms at 2.6 Å. Comparable Fe-Fe distances occur only in δ-FeO(OH),[28,29] but our attempt to fit experimental results on the basis of

Tab. 35-4. Parameters used to fit EXAFS data for calcined [Si,Fe]-ZSM-5.   R = 23.04.

Element	No. of atoms[26]	Distance from Fe ($\text{Å} \pm 0.02$)	Debye-Waller term (Å^2)
O	2.4	1.83	0.005
O	2.4	1.97	0.025
Si	1.2	3.18	0.008
Si	1.2	3.38	0.009

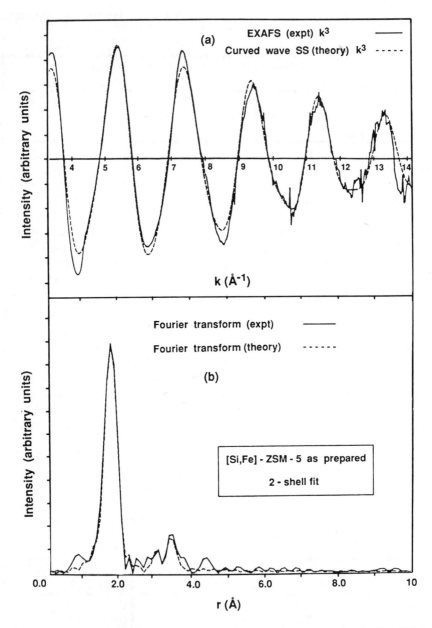

Fig. 35-5. EXAFS data (a) and Fourier transform (b) for as-prepared [Si,Fe]-ZSM-5 based on a 2-shell fit. Experimental results are given as the solid line, and theoretical simulation based on curved wave theory as the broken line.

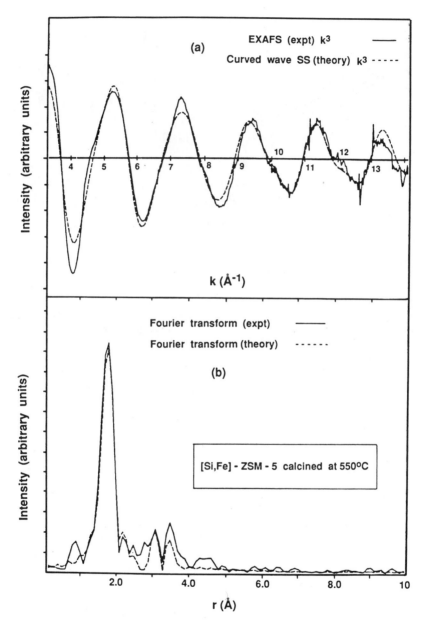

Fig. 35-6. EXAFS data (a) and Fourier transform (b) for calcined [Si,Fe]-ZSM-5 based on a 2-shell fit. Experimental results are given as the solid line, and theoretical simulation based on curved wave theory as the broken line.

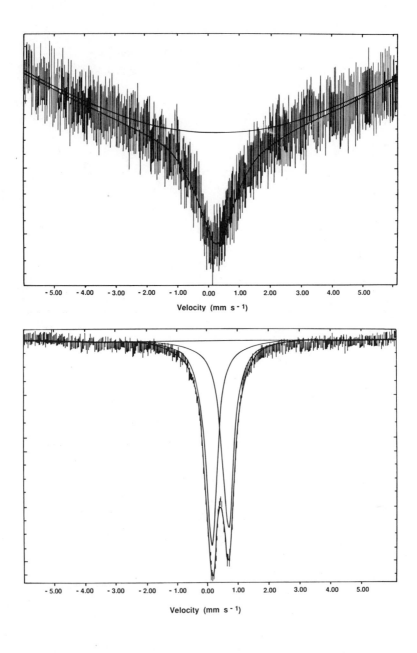

Fig. 35-7. Room temperature Mössbauer spectra of (a) as prepared and (b) calcined [Si,Fe]-ZSM-5.

40% δ-FeO(OH)[29] was inconclusive (R = 17.8). Also, multiple scattering would be expected for δ-FeO(OH), modifying the scattering distances for the outer shells. The presence of other amorphous Fe oxyhydroxides in addition to δ-FeO(OH) make fitting of the EXAFS spectrum to only two types of Fe species difficult. Therefore, while EXAFS clearly indicates the presence of both 4-coordinated and 6-coordinated Fe, we can only conclude that the extra-framework Fe is present as a mixture of oxides and oxyhydroxides and not as a pure oxide.

The room temperature Mössbauer spectrum of the as-prepared [Si,Fe]-ZSM-5 (see Fig. 35-7) shows one broad feature at an average isomer shift of 0.28 mm s^{-1}, in agreement with the estimate by Meagher et al.[7] that an isomer shift of ca. 0.25 mm s^{-1} should be observed for Fe incorporated in zeolite frameworks. Upon calcination at 550°C the signal splits into an inequivalent doublet with components at 0.15 and 0.68 mm s^{-1} (the average isomer shift is 0.41 mm s^{-1}), which corresponds to a combination of 6- and 4-coordinated Fe as discussed by Garten et al.[15] These results support the idea of complete initial incorporation of Fe in the framework, followed by expulsion of a large proportion of this after calcination, although the Mössbauer spectra yield no further structural information.

EXAFS thus provides direct evidence for the substitution of Fe in the zeolite framework during synthesis and its subsequent partial removal upon thermal treatment. The color change of the samples upon calcination (from off-white to pale brown) supports the conclusion that iron oxide/oxyhydroxide species form on the surface of the crystals. Attempts to remove the extra-framework iron by ion exchange with La^{3+} were unsuccessful, indicating that the extra-framework material is not Fe(H$_2$O)$_6$$^{3+}$ or a similar ionic species.

Acknowledgments. We are grateful to Professor C. R. A. Catlow and Professor Sir John Meurig Thomas, FRS, for access to EXAFS beamtime; to the SERC and Unilever Research, Port Sunlight, for support, and to Dr. W. Jones, Dr. Y. Mitsui, Dr. A. T. Steel, Dr. M. Cole, Professor C. M. B. Henderson, Dr. J. Charnoc and Dr. A. Dent for discussions. We acknowledge the use of the Chemical Database Service at Daresbury funded by the SERC.

References

1. Argauer, R. J.; Landolt, G. R. U.S. Patent NM 3 702 886, 1972 .

2. Kokotailo, G. T.; Lawton S. L.; Olson, D. H.; Meier, W. M. *Nature (London)* **1978**, *272*, 437.

3. Barrer, R. M. *Hydrothermal Chemistry of Zeolites*, Academic Press 1982.

4. Szostak, R.; Thomas, T. L.; *J. Catal.* **1986**, *100*, 555.

5.   Marosi, L.; Stabenow, J.; Schwarzmann, M. German Patent NM 2 831 611, 1980.

6.   Borade, R. B.; Halgeri, A. B.; Prasada Rao, T .S. R. in *New Developments in Zeolite Science and Technology*, Proc. 7th Int. Zeolite Conf. 1986, 851.

7.   Meagher, A.; Nair, V.; Szostak, R. *Zeolites* **1988**, *8*, 3.

8.   Calis, G.; Frenken, P.; de Boer, E.; Swolfs, A.; Hefni, M. A. *Zeolites* **1987**, *7*, 319.

9.   Carr, S. W.; Steel, A.; Townsend, R. P.; Thomas, J. M.; Dooryhee, E.; Greaves, G. N.; Catlow, C. R. A. in *Zeolites for the Nineties. Recent Research Reports*, 8th Int. Zeolite Conf. 1989, 219.

10.   Axon, S. A.; Huddersman, K.; Klinowski, J. *Chem. Phys. Lett.* **1990**, *172*, 398.

11.   Antonioli, G.; Vlaic, G.; Nardin, G.; Randaccio, L. *J. Chem. Soc., Dalton Trans.* **1990**, 943.

12.   Coddington, J. M.; Howe, R. F.; Yong, Y-S.; Asakura, K.; Iwasawa, Y. *J. Chem. Soc., Faraday Trans.* **1990**, *86*, 1015.

13.   Moller, K.; Bein, T.; Herron, N.; Mahler, W.; MacDougall, J.; Stucky, G. *Mol. Cryst. Liq. Cryst.* **1990**, *181*, 305.

14.   Bancroft, G. M. *Mössbauer Spectroscopy*, McGraw Hill 1973.

15.   Garten, R. L.; Degass, W. N.; Boudart, M. J. *J. Catal.* **1970**, *18*, 90.

16.   Teo, B. K. *Acc. Chem. Res.* **1980**, *13*, 412.

17.   Evans, J. *Chem. in Britain* **1986**, 803.

18.   Axon, S. A.; Klinowski, J. in *Recent Advances in Zeolite Science*, Proc. 1989 Meeting of the British Zeolite Association, Cambridge (Eds. J. Klinowski and P. J. Barrie) Elsevier, 1989, 113.

19.   Gurman, S. J.; Binstead, N.; Ross, I. *J. Phys. Chem.* **1984**, *17*, 143 ; SERC Daresbury Laboratory Update 1990.

20.   Goiffon, A.; Dumas, J.-C; Philippot, E. Rev. Chim. Minérale **1986**, *23*, 99. The FePO$_4$ EXAFS spectrum was kindly provided by Dr. A. T. Steel.

21.   von Ballmoos, R.; Higgins, J. B. *Zeolites* **1990**, *10*, 5.

22.   Hay, D. G.; Jaeger, H. *J. Chem. Soc., Chem. Comm.* **1984**, 1433.

23.   Boxhoorn, G.; van Santen, R. A.; van Erp, W. A.; Huis, R.; Clague, A. D. H. *J. Chem. Soc., Chem. Comm.* **1982**, 264.

24.   Teo, B. K.; Joy, D. *EXAFS Spectroscopy: Techniques and Applications*, Plenum Press, 1981.

25.   Calas, G.; Petiav, J. *Solid State Comm.* **1983**, *48*, 625.

26.   Sayers, D. E.; Bunker, B. A. in *X-Ray Absorption*, Koningsberger, D. C.; Prins, R. Eds., Ch. 6, Wiley, 1988.

27.   van Koningsveld, H.; van Bekkum, H.; Janssen, J. C. *Acta. Cryst.* **1987**, *B43*, 127.

28.   Allen, F. H ; Kennard, O.; Taylor, R. *Acc. Chem. Res.* **1983**, *16*, 146.

2 9 .   Patrat P. G.; de Bergevin F.; Pernet M.; Joubert J. C. *Acta. Cryst.* **1983,** *B39,* 165.

Index